ISAAC DEUTSCHER, who has been a specialist in Russian affairs for many years, is the author of *The Prophet Armed: Trotsky, 1879-1921; Stalin: A Political Biography; Russia in Transition; Soviet Trade Unions;* and *Russia: What Next?* His syndicated articles on current Soviet affairs appear regularly in the leading newspapers of fifteen countries and he contributes frequently to *The Reporter.* Polish by birth, he is now a British subject.

The Prophet Unarmed

TROTSKY : 1921–1929

Trotsky in exile, on a hunting trip

The
Prophet Unarmed

TROTSKY : 1921–1929

ISAAC DEUTSCHER

LONDON
OXFORD UNIVERSITY PRESS
NEW YORK   TORONTO
1959

© *Oxford University Press 1959*

Printed in the United States of America

PREFACE

CARLYLE once wrote that as Cromwell's biographer he had to drag out the Lord Protector from under a mountain of dead dogs, a huge load of calumny and oblivion. My job, as Trotsky's biographer, has been somewhat similar, with this difference, however, that when I set out to assail my mountain of dead dogs great events were about to strike at it with immense force. I had concluded *The Prophet Armed*, the first part of my study of Trotsky, while Stalin was still alive, and while his 'cult' appeared as indestructible as the stigma attached to Trotsky seemed indelible. Most reviewers of *The Prophet Armed* agreed with a British critic who wrote that 'that single book undoes three decades of Stalinist denigration'; but, of course, neither the book nor its documentation brought forth a single word of comment from the Soviet historians and critics, who usually devote an unconscionable amount of attention to every piece of 'Sovietology', no matter how trashy, that appears in the West. Then came Stalin's death, the Twentieth Congress, and Khrushchev's 'secret' speech. An earthquake shook the mountain of dead dogs, scattering half of it far and wide; and for a moment it looked as if the other half too was about to be blown away. Historically truthful references to Trotsky's part in the Russian Revolution began to appear in Soviet periodicals for the first time in three decades, although the paucity and timidity of the references suggested how close the connexion between history and politics still was in this case, and how delicate the problem.

When Stalin's idol was being smashed and the Stalinist falsification of history was being officially and emphatically denounced, the shade of Stalin's chief antagonist inevitably aroused fresh and lively, though bewildered, interest. In Moscow, Peking, Warsaw, and East Berlin people wondered anew what had been the significance and the moral of Trotsky's struggle against Stalin. Young historians, to whom the archives, hitherto kept under lock and key, had suddenly been thrown open, avidly looked for an answer in the unfamiliar records of Bolshevism. Khrushchev having declared that Stalin had destroyed his inner party critics by means of false and monstrous accusations, the historians naturally expected an explicit rehabilitation of the victims of the Great Purges. Here and there the rehabilitation was already taken for granted. In Poland, for instance, the writings of Trotsky and Bukharin, Rakovsky and Radek, were quoted, and even reprinted, as offering much needed illumination of the enigma of the Stalin era (and so were my own books and essays).

Soon thereafter, however, the assault on the 'mountain of dead dogs' was halted in its tracks. Towards the end of 1956, or early in 1957, during the reaction against the Hungarian turmoil, a halt was called in Moscow to the restitution of historical truth. The dilemmas and fluctuations of current policy became once again reflected in the writing of history, and were focused, as it were, in the treatment of Trotsky. Since then Stalin's discredited *Short Course of the History of the C.P.S.U.* has been replaced by a new official compendium of party history which attempts to reimpose, though in a modified and softened version, the anathema on Trotsky; and in Soviet periodicals the volume of writings designed surreptitiously to defame Trotsky has grown much larger than it ever was in the last decade or so of the Stalin era.

However, what was once a drama has now become sheer farce. The Stalinist anathema, absurd though it was, had its 'logic' and consistency: Stalin knew that he could not maintain it effectively without gross, unscrupulous, and systematic falsification of the past. Khrushchev tries to ban the truth about Trotsky without resorting to outright falsification—he contents himself with a 'moderate' dose of distortion; and by this alone he renders the anathema ludicrous. Thus, the authors of the new party-history extol the work of the Military Revolutionary Committee of 1917 and of the Commissariat of War of the civil war period, without mentioning in this connexion that Trotsky stood at the head of both these bodies; but almost in the same breath they do mention this fact when they have to find fault with the work of the same Committee or the same Commissariat. (It is as if one watched a child, who has not yet learned the meaning of hide-and-seek, pull at his mother's skirt and shout: '*Here* I am, look for me.') The Khrushchevite historians evidently assume that Soviet readers will not be intelligent enough to see that the praise and the blame are both directed alike at the same person. In his own, perverse way Stalin took a far higher view of the perspicacity of his subjects; and he preferred to give them no facts that might stimulate heretical guesswork, and to leave no scope for such guesswork. The new versions of party history also dwell one-sidedly on disagreements between Lenin and Trotsky; but by publishing Lenin's suppressed writings and throwing open the archives, the new party leaders have in fact done virtually everything that was needed for Trotsky's rehabilitation. Now all their attempts to banish him once again from the annals of the revolution are vain.

. . . .

Trotsky's ghost is evidently still haunting Stalin's successors. I trust that in these pages readers will find at least part of the explanation of this apparently bizarre fact. Despite all the great changes that have occurred in Soviet society since the 1920's, or rather because of these changes, some of the crucial issues of the controversy between Stalin and Trotsky are as alive today as ever. Trotsky denounced the 'bureaucratic degeneration' of the workers' state; and he confronted Stalin's 'monolithic' and 'infallibly' led party with the demand for freedom of expression, debate, and criticism, believing that on these alone could and should voluntary and genuine communist discipline be based. His voice was smothered in the Russia of the 1920's; but with the many-sided, industrial, educational, and social progress of the Soviet Union this his idea has come back to life, gaining possession of many communist minds. In their brief hour of truth, Khrushchev and Mikoyan, Mao and Gomulka, Kadar and Togliatti, not to speak of Tito and Nagy, had to pay their tribute to it. A substratum of 'Trotskyism' could be found in the contributions, however half-hearted and fragmentary, which each of them then made to 'de-Stalinization'. Indeed, in this hour of truth Trotsky appeared as the giant forebear of all of them, for none of them approached Stalinism with anything like the depth, the sweep, and the vigour of his critical thought. Since then, frightened by their own bravado, they have retraced their steps; and the Soviet régime and the Communist party, moving two steps forward and one step back, are still far from having overcome their 'bureaucratic deformation'.

The fact that as yet the issues posed by Trotsky have, at best, been only half resolved makes the story of his opposition to Stalinism more, not less, topical. Nor is Trotsky's antagonism to Stalinist bureaucracy the only aspect of his struggle which has a bearing on our times. A large part of this narrative centres on the conflict between his internationalism and the isolationist self-sufficiency of latter-day Bolshevism embodied in Stalin. This conflict reappeared and grew acute even before the close of the Stalin era; and since then the balance has begun to tilt towards internationalism. This is yet another unresolved issue lending fresh interest to the controversy of the 1920's.

Stalin's successors live in such grotesque horror of Trotsky's shade because they are afraid of coming to grips with the issues with which he, so much ahead of his time, did come to grips. Their behaviour may be explained in part by objective circumstances and in part by inertia, for Khrushchev and his associates, even in their rebellion against Stalinism, are still Stalin's epigones. But they act also from the narrowest motives of self-defence. The following incident, which occurred during a session of the Central Committee in June 1957, illustrates the nature of their predicament. At that session Khrushchev, speaking on the motion for the expulsion of Molotov, Kaganovich, and Malenkov, recalled the Great Purges, the subject invariably recurring in all the secret debates since Stalin's death. Pointing at Molotov and Kaganovich, he exclaimed: 'Your hands are stained with the blood of our party leaders and of innumerable innocent Bolsheviks!' 'So are yours!', Molotov and Kaganovich shouted back at him. 'Yes, so are mine', Khrushchev replied. 'I admit this. But during the Great Purges I only carried out your orders. I was not then a member of the Politbureau and I am not responsible for its decisions. You were.' When Mikoyan later reported the incident to the Comsomol in Moscow, he was asked why the accomplices of Stalin's crimes were not tried in court. 'We cannot try them', Mikoyan is said to have answered, 'because if we start putting such people in the dock, there is no knowing where we should be able to stop. We have all had some share in conducting the purges.' Thus, if only in order to safeguard their own immunity, Stalin's successors must still keep in the dock the ghosts of some of Stalin's victims. As to Trotsky, is it not safer indeed to leave him where he lies, under the half-shattered pyramid of slander, rather than transfer him to the Pantheon of the revolution?

.   .   .   .

I do not believe and have never believed that Trotsky's memory is in any need of rehabilitation by rulers or party leaders. (It is, I think, rather they who, if they can, ought to work for their exculpation!) Nothing, however, is farther from my intention than to indulge in any cult of Trotsky.

I do indeed consider Trotsky as one of the most outstanding revolutionary leaders of all times, outstanding as fighter, thinker, and martyr. But I am not seeking to present here the glorifying image of a man without blemish or blur. I have endeavoured to portray him as he was, in his real stature and strength but with all his weaknesses; I have tried to show the extraordinary power, fertility, and originality of his mind, but also his fallibility. In discussing the ideas which form his distinctive contribution to Marxism and modern thought, I have attempted to disentangle what in my view is, and for a long time is likely to remain, of objective and lasting value from that which reflected merely transient situations, subjective emotions, or errors of judgement. I have done my best to do justice to Trotsky's heroic character to which I find only very few equals in history. But I have also shown him in his many moments of irresolution and indecision: I describe the embattled Titan as he falters, and boggles, and yet goes out to meet his destiny. I see him as the representative figure of pre-Stalinist communism and the precursor of post-Stalinist communism. Yet I do not imagine that the future of communism lies in Trotskyism. I am inclined to think that the historic development is transcending both Stalinism and Trotskyism and is tending towards something broader than either of them. But each of them will probably be 'transcended' in a different manner. What the Soviet Union and communism take over from Stalinism is mainly its practical achievement; in other respects, as regards methods of government and political action, ideas, and 'moral climate', the legacy of the Stalin era is worse than empty; the sooner it is disposed of the better. But precisely in these respects Trotsky has still much to offer; and the political development can hardly transcend him otherwise than by absorbing all that is vital in his thought and applying it to realities which are far more advanced, varied, and complex than those he knew.

. . . .

In the preface to *The Prophet Armed*, I indicated that I intended to tell the whole story of Trotsky's life and work from 1921 onwards in a single volume entitled *The Prophet Unarmed*.[1] A reviewer, writing in *The Times Literary Supplement*, doubted whether the story could be told, on the appropriate scale, in one volume. His doubt has proved justified. *The Prophet Unarmed* ends with Trotsky's banishment from the Soviet Union in January 1929; another volume, *The Prophet Outcast*, is to cover the stormy twelve years of Trotsky's last exile and to give the final assessment of his role. These three volumes form part of a larger trilogy, of which one section, *Stalin, A Political Biography*, appeared in 1949, and another, a two-volumed *Life of Lenin*, is still in an early stage of preparation. (I also intend to supplement my biography of Stalin by a book *Stalin's Last Years*, if and when sufficient historical documentation becomes available.)

The three volumes of the present work are, of course, interconnected, as are also, more loosely, all the parts of the entire trilogy. But I have so planned them that each volume is as far as possible self-contained and can be read as an independent work. The narrative of this volume covers the years which were in many respects the formative period of the Soviet Union. It begins with 1921 and the aftermath of the civil war, with Trotsky still at the height of power; it ends in 1929, with Trotsky *en route* to Constantinople, and the Soviet Union entering the epoch of forced industrialization and collectivization. Between these years there unfolds the drama of the Bolshevik party which, after Lenin's death, found itself plunged into what was probably the fiercest and the most momentous political controversy of modern times, uncertain of its policies and groping for direction, caught in extraordinary

---

[1] It will be remembered that both these titles allude to Machiavelli's dictum that 'all armed prophets have conquered, and the unarmed ones have been destroyed'. (See the text from *The Prince* quoted in *The Prophet Armed*, p. xii.)

social and political tensions and in the logic of the single-party system, and succumbing to Stalin's autocracy. Throughout, Trotsky is at the centre of the struggle as Stalin's chief adversary, the only alternative candidate to the Bolshevik leadership, the 'premature' advocate of industrialization and planned economy, the critic of Socialism in One Country, and the champion of 'proletarian democracy'.

Much of the documentation on which this narrative is based has hitherto been unknown. I have drawn heavily on Trotsky's Archives, which offer rich insights into the proceedings of the Politbureau and the Central Committee and into the work of all the factions of the Bolshevik party; on the voluminous and revealing correspondence between Trotsky, Radek, Rakovsky, Preobrazhensky, Sosnovsky, and many other eminent Bolsheviks; on the records of party congresses and conferences, on files of contemporary Russian and non-Russian newspapers and periodicals; and on published and unpublished eye-witness accounts. I have benefited from personal contact with Natalya Sedova, Trotsky's widow, Heinrich Brandler, Alfred Rosmer, Max Eastman, and other participants and survivors of the struggle who have been good enough to answer my queries and sometimes to submit to prolonged and repeated questioning. In my attempt to reproduce the background and the 'climate' of the time, my own experience may have been of some value. From the middle 1920's I was active in the Polish Communist party which stood closer to Bolshevism than did any other party; soon thereafter I was leading spokesman of an inner-party opposition strongly influenced by Trotsky's ideas; and in 1932 I had the somewhat curious distinction of being the first member ever to be expelled from the Polish party for his anti-Stalinism.

Access to untapped sources has, I think, enabled me to give either wholly or partly new versions of many crucial events and episodes. The relations between Lenin and Trotsky in Lenin's last years; the vicissitudes of the subsequent struggles; the relations between Trotsky, Bukharin, Zinoviev, Kamenev, Radek, and other leaders; the formation and the defeat of the various anti-Stalinist oppositions; the events of Trotsky's first year of exile near the Soviet-Chinese frontier, especially the divisions which had already appeared in the Trotskyist Opposition, foreshadowing its collapse many years before the Moscow trials—nearly all of these are narrated or interpreted in the light of some hitherto unknown facts. I have also, as in the previous volume, paid special attention to Trotsky the man of letters and devoted many pages to his views on science, literature, and the arts, in particular to his work as Russia's leading literary critic in the early 1920's. That work, remarkable for the largeness of his views and his clear-sighted rejection of any party tutelage over science and art, has also a special relevance to the present situation: such progress in these fields as was achieved in the Soviet Union during the post-Stalinist 'thaw' went in the direction of Trotsky's ideas, although it will probably still take a long time before views as undogmatic and bold as his make their appearance again in the Soviet Union.

Much as I have been concerned with the restoration of the various features and details of the historic drama, I have never been able to dismiss from my thoughts the tragic theme that runs through it from beginning to end and affects nearly all the characters involved. Here is modern tragedy in the sense in which Trotsky himself has defined it (see Chapter III, p. 193): 'As long as man is not yet master of his social organization, that organization towers above him like Fate itself. . . . The stuff of contemporary tragedy is found in the clash between the individual and a collective, or between hostile collectives represented by individuals.' Trotsky found it 'difficult to foresee whether the dramatist of the revolution will create "high" tragedy'. The Soviet dramatist has certainly not yet created it; but what modern Sophocles or Aeschylus could possibly produce tragedy as high as Trotsky's own life? Is it too much to hope that this is nevertheless an 'optimistic tragedy', one in which not all the suffering and sacrifice have been in vain?

· · · · · · · · · ·

I am greatly indebted to Mr. Donald Tyerman who has read in manuscript this volume as well as all my previous books and has been a constant source of encouragement to me; and my thanks are due to Mr. Dan Davin and Mr. John Bell for most valuable stylistical criticisms and suggestions. My wife has as ever been my only research assistant and also my first, the most severe and the most indulgent, critic. I. D.

CONTENTS

LIST OF PLATES

CHAPTER I

The Power and the Dream

THE Bolsheviks made their October Revolution of 1917 in the conviction that what they had begun was mankind's 'leap from the realm of necessity to the realm of freedom'. They saw the bourgeois order dissolving and class society crumbling all over the world, not merely in Russia. They believed that everywhere the peoples were at last in revolt against being the playthings of socially unorganized productive forces, and against the anarchy of their own existence. They imagined that the world was fully ready to free itself from the necessity to slave and sweat for the means of its subsistence—and ready also to put an end to man's domination by man. They greeted the dawn of the new age in which the human being, all his energies and capacities released, would achieve self-fulfilment. They were proud to have opened for humanity 'the passage from pre-history to history'.

This brilliant vision inspired the minds and hearts not only of the leaders, ideologues, and dreamers of Bolshevism. It sustained the hope and ardour of the mass of their followers as well. They fought in the civil war with no mercy for their enemies and no pity for themselves because they believed that by doing so they were ensuring for Russia and the world the chance of accomplishing the great leap from necessity to freedom.

When victory was theirs at last they found that revolutionary Russia had overreached herself and was hurled down to the bottom of a horrible pit. No other nation had followed her revolutionary example. Surrounded by a hostile or, at best, indifferent world Russia stood alone, bled white, starving, shivering with cold, consumed by disease, and overcome with gloom. In the stench of blood and death her people scrambled wildly for a breath of air, a faint gleam of light, a crust of bread. 'Is this', they asked, 'the realm of freedom? Is this where the great leap has taken us?'

What answer could the leaders give? They replied that the

great and celebrated revolutions of earlier ages had suffered similar cruel setbacks but had nevertheless justified themselves and their work in the eyes of posterity; and that the Russian Revolution too would emerge triumphant. Nobody argued thus with greater power of conviction than did the chief character of this book. Before the hungry crowds of Petrograd and Moscow Trotsky recalled the privations and the distress which revolutionary France endured many years after the destruction of the Bastille; and he told them how the First Consul of the Republic every morning visited in person the *Halles* of Paris, anxiously watched the few peasant carts bringing food from the country, and went away every morning knowing that the people of Paris would continue to starve.[1] The analogy was all too real; but consoling historical parallels, no matter how true and relevant, could not fill Russia's empty stomach.

Nobody was able to gauge the depth to which the nation had slumped. Down below hands and feet fumblingly searched for solid points of support, for something to lean on and something to grasp at—in order to climb up. Once revolutionary Russia had climbed up she would surely resume the leap from necessity to freedom. But how was the ascent to be accomplished? How was the pandemonium at the bottom to be calmed? How were the desperate multitudes to be disciplined and led for the ascent? How could the Soviet republic overcome its appalling misery and chaos and then go on to fulfil the promise of socialism?

At first the Bolshevik leaders did not try to belittle or embellish the predicament or deceive their followers. They attempted to uphold their courage and hope with words of truth. But the truth, unvarnished, was too harsh to mitigate misery and allay despair. And so it began to make room for the soothing lie which at first sought merely to conceal the chasm between dream and reality but soon insisted that the realm of freedom had already been reached—and that it lay there at the bottom of the pit. 'If people refused to believe, they had to be made to believe by force.' The lie grew by degrees until it became elaborate, complex, and vast—as vast as the chasm it was designed to cover up. It found among the Bolshevik leaders its mouthpieces and dedicated supporters, who felt that without the lie and the force which supported it the nation could not be

[1] Trotsky, *Sochinenya*, vol. xii, pp. 318–29.

dragged out of the mire. The salutary lie, however, could bear no confrontation with the original message of the revolution. Nor as the lie grew could its expounders stand face to face or side by side with the genuine leaders of the October Revolution to whom the revolution's message was and remained inviolable. The latter did not at once raise their voices in protest. They did not even at once recognize the falsehood for what it was, for it insinuated itself slowly and imperceptibly. The leaders of the revolution could not help being entangled in it at the outset; but then, one after another, hesitantly and falteringly, they rose to expose and denounce the lie and to invoke against it the revolution's broken promise. Their voices, however, once so powerful and inspiring, sounded hollow at the bottom of the pit and brought forth no response from hungry, weary, and cowed multitudes. Of all these voices none vibrated with such deep and angry conviction as did that of Trotsky. He now began to rise to his height as the revolution's prophet unarmed, who, instead of imposing his faith by force, could rely only on the force of his faith.

.   .   .   .

The year 1921 at length brought peace to Bolshevik Russia. The last shots died down on the battlefields of the civil war. The White Armies had dissolved and vanished. The armies of intervention had withdrawn. Peace was concluded with Poland. The European frontiers of the Soviet Federation were drawn and fixed.

Amid the silence which had fallen on the battlefields Bolshevik Russia listened intently to sounds from the outside world and was becoming poignantly aware of her isolation. Since the summer of 1920, when the Red Army was defeated at the gates of Warsaw, the revolutionary fever in Europe had subsided. The old order there found some balance, unstable yet real enough to allow the conservative forces to recover from disarray and panic. Communists could not hope for imminent revolutionary developments; and attempts to provoke such developments could result only in costly failures. This was demonstrated in March 1921, when a desperate and ill-prepared communist rising broke out in central Germany. The rising had been encouraged and in part instigated by Zinoviev, the President of the Communist International, and Bela Kun, the luckless leader of the Hungarian Revolution of 1919, who believed that the rising might 'electrify' and spur to action the apathetic mass of the German working class.[1] The mass failed to respond, however; and the German government suppressed the rising without much difficulty. The fiasco threw German communism into confusion; and amid bitter recriminations the leader of the German Communist party, Paul Levy, broke with the International. The March rising thus weakened even further the forces of communism in Europe and deepened the sense of isolation in Bolshevik Russia.

The nation ruled by Lenin's party was in a state of near dissolution. The material foundations of its existence were shattered. It will be enough to recall that by the end of the civil war Russia's national income amounted to only one-third of her income in 1913, that industry produced less than one-fifth of the goods produced before the war, that the coal-mines turned out less than one-tenth and the iron foundries only one-fortieth of their normal output, that the railways were destroyed, that all stocks and reserves on which any economy depends for its work were utterly exhausted, that the exchange of goods between town and country had come to a standstill, that Russia's cities and towns had become so depopulated that in 1921 Moscow had only one-half and Petrograd one-third of its former inhabitants, and that the people of the two capitals had for many months lived on a food ration of two ounces of bread and a few frozen potatoes and had heated their dwellings with the wood of their furniture—and we shall obtain some idea of the condition in which the nation found itself in the fourth year of the revolution.[2]

The Bolsheviks were in no mood to celebrate victory. The Kronstadt rising had finally compelled them to give up war communism and to promulgate N.E.P.—the New Economic Policy. Their immediate purpose was to induce peasants to sell food and private merchants to bring the food from country to town, from producer to consumer. This was the beginning of

[1] Trotsky, *Pyat Let Kominterna*, pp. 284–7; Radek, *Pyat Let Kominterna*, vol. ii, pp. 464–5; *Tretii Vsemirnyi Kongress Kominterna*, pp. 58 ff., 308 ff; Lenin, *Sochinenya*, vol. xxxii, pp. 444–50 *passim*.

[2] Kritsman, *Geroicheskii Period Velikoi Russkoi Revolutsii*, pp. 150 ff.; *3 Syezd Profsoyuzov*, pp. 79–86 and Miliutin's report in *4 Syezd Profsoyuzov*, pp. 72–77.

a long series of concessions to private farming and trade, the beginning of that 'forced retreat' which Lenin avowed his government was compelled to beat before the anarchic elements of small property which were predominant in the country.

Presently calamity struck the nation. One of the worst famines in history visited the populous farming land on the Volga. Already in the spring of 1921, just after the Kronstadt rising, Moscow had been alarmed by reports about droughts, sand blizzards, and an invasion of locusts in the southern and south-eastern provinces. The government swallowed its pride and appealed for help to bourgeois charitable organizations abroad. In July it was feared that 10 million peasants would be hit by the famine. By the end of the year the number of sufferers had risen to 36 million.[1] Uncounted multitudes fled before the sand blizzards and the locust and wandered in aimless despair over the vast plains. Cannibalism reappeared, a ghastly mockery of the high socialist ideals and aspirations emanating from the capital cities.

Seven years of world war, revolution, civil war, intervention, and war communism had wrought such changes in society that customary political notions, ideas, and slogans became almost meaningless. Russia's social structure had been not merely overturned; it was smashed and destroyed. The social classes which had so implacably and furiously wrestled with one another in the civil war were all, with the partial exception of the peasantry, either exhausted and prostrate or pulverized. The landed gentry had perished in their burning mansions and on the battlefields of the civil war; survivors escaped abroad with remnants of the White Armies which scattered to the winds. Of the bourgeoisie, never very numerous or politically confident, many had also perished or emigrated. Those who saved their skins, stayed in Russia, and attempted to adjust themselves to the new régime, were merely the wreckage of their class. The old intelligentsia, and to a lesser degree the bureaucracy, shared the fate of the bourgeoisie proper: some ate the émigré's bread in the West; others served Russia's new masters as 'specialists'. With the revival of private trade, a new upstart middle class made its appearance. Its members, contemptuously labelled N.E.P.-men, were bent on exploiting quickly the opportunities N.E.P.

[1] See Kalinin's report in *9 Vserossiiskii Syezd Sovietov*, pp. 23–26.

offered them, amassed mushroom-fortunes, and enjoyed their day with the feeling that one deluge was behind them and another ahead of them. Despised even by the survivors of the old bourgeoisie, this new middle class did not aspire to develop a political mind of its own. *Sukharevka*, Moscow's sprawling and squalid black market, was the symbol of its social existence and morality.

It was a grim and paradoxical outcome of the struggle that the industrial working class, which was supposed now to exercise its dictatorship, was also pulverized. The most courageous and politically minded workers had either laid down their lives in the civil war or occupied responsible posts in the new administration, the army, the police, the industrial managements, and a host of newly created institutions and public bodies. Proudly conscious of their origin, these proletarians turned Commissars did not in fact belong to the working class any longer. With the passage of time many of them became estranged from the workers and assimilated with the bureaucratic environment. The bulk of the proletariat too became *déclassé*. Masses of workers fled from town to country during the hungry years; and being mostly town dwellers in the first generation and not having lost roots in the country, they were easily reabsorbed by the peasantry. In the early years of N.E.P. there started a migration in the opposite direction, an exodus from country to town. Some old workers returned to the cities; but most of the new-comers were raw and illiterate peasants, without any political, let alone cultural, tradition. However, in 1921 and 1922 the migration from country to town was only a trickle.

The dispersal of the old working class created a vacuum in urban Russia. The old, self-reliant, and class-conscious labour movement with its many institutions and organizations, trade unions, co-operatives, and educational clubs, which used to resound with loud and passionate debate and seethe with political activity—that movement was now an empty shell. Here and there small groups of veterans of the class struggle met and argued about the prospects of the revolution. They had once formed the real 'vanguard' of the working class. Now they were a mere handful; and they could not see behind them the main force of their class which had once listened to them, taken its cue from them, and followed them into the thick of social struggle.[1]

The proletarian dictatorship was triumphant but the proletariat had nearly vanished. It had never been more than a small minority of the nation; and it had played a decisive part in three revolutions not because of its numbers but because of the extraordinary strength of its political mind, initiative, and organization. At the best of times Russia's large-scale industry employed not much more than three million workers. After the end of the civil war only about half that number remained in employment. Even of these many were in fact idle because the plant was idle. The government kept them on industrial payrolls as a matter of social policy, in order to save a nucleus of the working class for the future. These workers were, in fact, paupers. If a worker received his wage in money, the wage was worthless because of the catastrophic depreciation of the rouble. He made his living, such as it was, by doing odd jobs, trading on black markets, and scouring nearby villages for food. If he received his wage in kind, especially in the produce of his factory, he rushed from the bench to the black market to barter away a pair of shoes or a piece of cloth for bread and potatoes. Left with nothing to barter, he would return to the factory to steal a tool, a few nails, or a sack of coal, and he went back to the black market. Theft in factories was so common that it was estimated that half the workers normally stole the things they themselves produced.[2] It may be imagined what effect the hunger, the cold, the terrifying idleness at the factory bench and the hurly-burly of the black markets, the cheating and the stealing—the almost zoological struggle for survival—had on the morale of the people who were supposed to be the ruling class of the new state.

As a social class the peasantry alone emerged unbroken. World war, civil war, and famine had, of course, taken their toll; but they had not cracked the mainsprings of the peasantry's life. They had not reduced its resilience and powers of regeneration. Not even the worst calamity could disperse the heavy bulk

---

[1] See *4 Syezd Profsoyuzov*, Bukharin's, Lozovsky's, and Miliutin's reports.

[2] Lozovsky claimed that 50 per cent. of the produce was stolen in some factories; and it was estimated that wages covered only one-fifth of a worker's cost of living. Ibid., p. 119.

of the peasantry which, indestructible almost like nature itself, needed to work in contact only with nature in order to keep alive, while the industrial workers dispersed when the artificial industrial machinery on which their existence depended had collapsed. The peasantry had preserved its character and its place in society. It had enhanced its position at the expense of the landed gentry. It could now afford to count the gains as well as the losses the revolution had brought it. The requisitions having ceased, the peasants hoped to gather in at last the full harvest from their enlarged possessions. True, they lived in utter poverty. But this and the backwardness which went with it were part and parcel of their social heritage. Freed from seignoral overlordship, the peasants preferred poverty on their own small-holdings to the incomprehensible vistas of abundance under communism which the urban agitators unfolded before them. The muzhiks were no longer greatly disturbed by the agitators' talk. They noticed that of late these had become chary of offending them and even sought to befriend them and to flatter them. For the time being, the muzhik was indeed the Benjamin of the Bolshevik government which was anxious to re-establish the 'link' between town and country and the 'alliance between workers and peasants'. Since the working class could not make its weight felt, the peasantry's weight was all the heavier. Every month, every week, brought the farmer a thousand fresh proofs of his new importance; and his self-confidence was heightened accordingly.

Yet this social class which alone had preserved its character and place in society was by its very nature politically impotent. Karl Marx once described in a striking image the 'idiocy of rural life' which in the last century prevented the French peasantry from 'asserting their class interests in their own name'; and his image fits well the Russian peasantry of the 1920's:

Throughout the country they live in almost identical conditions, but enter very little into relationships with one another. Their mode of production isolates them, instead of bringing them into mutual contact. The isolation is intensified by the inadequacy of the means of communication . . . and poverty. Their farms are so small that there is practically no scope for division of labour. . . . Among the peasantry therefore there can be no diversity of development, no differentiation of talents, no wealth of social relationship. Each family is almost self-sufficient, producing on its own plot of land the greater part of its requirements, and thus providing itself with the necessaries of life through an interchange with nature rather than by means of intercourse with society. Here is a small plot of land, with the peasant-farmer and his family; there is another plot of land, another peasant with wife and children. A score or two of these atoms make up a village, and a few scores of villages make up a *département*. In this way, the great mass of the French nation is formed by the simple addition of like entities, much as a sack of potatoes consists of a lot of potatoes huddled into a sack.[1]

The huge sack of potatoes that was rural Russia also proved quite incapable of asserting itself 'in its own name'. Once the Populist, or Social Revolutionary intelligentsia represented it, and spoke on its behalf. But the Social Revolutionary party, discredited by its own refusal to countenance the agrarian revolution and then driven underground and destroyed by the Bolsheviks, had played out its role. The sack of potatoes lay vast, formidable, and mute. Nobody could take his eyes off it; nobody could ignore it, or trample on it with impunity: it had already hit urban Russia on the head; and the Bolshevik rulers had to bow to it. But the sack of potatoes could not give backbone, form, will, and voice to a shapeless and disintegrated society.

.   .   .   .

Thus a few years after the revolution the nation was incapable of managing its own affairs and of asserting itself through its own authentic representatives. The old ruling classes were crushed; and the new ruling class, the proletariat, was only a shadow of its former self. No party could claim to represent the dispersed working class; and the workers could not control the party which claimed to speak for them and to rule the country on their behalf.

Whom then did the Bolshevik party represent? It represented only itself, that is, its past association with the working class, its present aspiration to act as the guardian of the proletarian class interest, and its intention to reassemble in the course of economic reconstruction a new working class which should be able in due time to take the country's destinies into its hands. In the meantime, the Bolshevik party maintained itself in power

[1] Marx, *18 Brumaire of Louis Bonaparte.*

by usurpation. Not only its enemies saw it as a usurper—the party appeared as a usurper even in the light of its own standards and its own conception of the revolutionary state.

The enemies of Bolshevism, we remember, had from the outset denounced the October Revolution and then the dispersal of the Constituent Assembly in 1918 as acts of usurpation. The Bolsheviks did not take this accusation to heart: they replied that the government from which they seized power in October had not been based on any elective representative body; and that the revolution vested power in a government backed by the overwhelming majority of the elected and representative Councils of Workers' and Soldiers' Deputies. The Soviets had been a class representation and by definition an organ of proletarian dictatorship. They had not been elected on the basis of universal suffrage. The gentry and the bourgeoisie had been disfranchised; and the peasantry was represented only in such a proportion as was compatible with the predominance of the urban workers. The workers cast their votes not as individuals in traditional constituencies but in factories and workshops as members of those productive units of which their class consisted. It was only this class representation that the Bolsheviks had since 1917 considered as valid and legitimate.[1]

Yet it was precisely in the terms of the Bolshevik conception of the workers' state that Lenin's government had gradually ceased to be representative. Nominally, it was still based on the Soviets. But the Soviets of 1921–2, unlike those of 1917, were not and could not be representative—they could not possibly represent a virtually non-existent working class. They were the creatures of the Bolshevik party; and so when Lenin's government claimed to derive its prerogatives from the Soviets, it was in fact deriving them from itself.

The Bolshevik party had the usurper's role thrust upon it. It had become impossible for it to live up to its principle once the working class had disintegrated. What could or should the party have done under these circumstances? Should it have thrown up its hands and surrendered power? A revolutionary government which has waged a cruel and devastating civil war does not abdicate on the day after its victory and does not surrender to its defeated enemies and to their revenge even if it

[1] Lenin, *Sochinenya*, vol. xxvi, pp. 396–400; Trotsky, *Kommunizm i Terrorizm*.

discovers that it cannot rule in accordance with its own ideas and that it no longer enjoys the support it commanded when it entered the civil war. The Bolsheviks lost that support not because of any clear-cut change in the minds of their erstwhile followers, but because of the latters' dispersal. They knew that their mandate to rule the republic had not been properly renewed by the working class—not to speak of the peasantry. But they also knew that they were surrounded by a vacuum; that the vacuum could be filled only slowly over the years, and that for the time being nobody could either prolong or invalidate their mandate. A social catastrophe, a *force majeure*, had turned them into usurpers; and so they refused to consider themselves as such.

The disappearance in so short a time of a vigorous and militant social class from the political stage and the atrophy of society consequent upon civil war formed a strange but not a unique historic phenomenon. In other great revolutions, too, society broke down exhausted, and revolutionary government was similarly transformed. The English Puritan Revolution and the French Great Revolution had each first upheld a new principle of representative government against the *ancien régime*. The Puritans asserted the rights of Parliament against the Crown. The leaders of the French Third Estate did likewise when they constituted themselves as the National Assembly. Upheaval and civil strife followed, in consequence of which the forces of the *ancien régime* were no longer able to dominate society while the classes which had supported the revolution were too strongly divided against themselves and too exhausted to exercise power. No representative government was therefore possible. The army was the only body with enough unity of will, organization, and discipline to master the chaos. It proclaimed itself the guardian of society; and it established the rule of the sword, a nakedly usurpatory form of government. In England the two broad phases of the revolution were embodied in the same person: Cromwell first led the Commons against the Crown and then as Lord Protector usurped the prerogatives of both Crown and Commons. In France there was a definite break between the two phases; and in each different men came to the fore: the usurper Bonaparte played no significant part in the early acts of the revolution.

In Russia the Bolshevik party provided that closely knit and disciplined body of men, inspired by a single will, which was capable of ruling and unifying the disintegrated nation. No such party had existed in previous revolutions. The main strength of the Puritans lay in Cromwell's army; and so they came to be dominated by the army. The Jacobin party came into being only in the course of the upheaval. It was part of the fluctuating revolutionary tide. It broke up and vanished at the ebb of the tide. The Bolshevik party, on the contrary, formed a solid and centralized organization long before 1917. This enabled it to assume leadership in the revolution and, after the ebb of the tide, to play for many decades the part the army had played in revolutionary England and France, to secure stable government and to work towards the integration and remodelling of the national life.

By its cast of mind and political tradition the Bolshevik party was extremely well prepared and yet peculiarly ill adapted for the usurper's part. Lenin had trained his disciples as the 'vanguard' and the *élite* of the Labour movement. The Bolsheviks had never contented themselves with giving expression to the actual moods or aspirations of the working class. They regarded it as their mission to shape those moods and to prompt and develop those aspirations. They looked upon themselves as political tutors of the working class and were convinced that as consistent Marxists they knew better than the oppressed and unenlightened working class could know what was its real historic interest and what should be done to promote it. It was because of this, we remember, that the young Trotsky had charged them with the inclination to 'substitute' their own party for the working class and to disregard the workers' genuine wishes and desires.[1] The charge, when Trotsky first levelled it, in 1904, ran far ahead of the facts. In 1917 as in 1905 the Bolsheviks made their own intervention in the revolution wholly dependent on the degree of proletarian mass support they could muster. Lenin and his staff scrutinized with cold and sober eyes even the most minute fluctuations in the workers' political temper; and to these they carefully related their own policies. It never then occurred to them that they could seize or hold power without the approval of the majority of the

[1] See *The Prophet Armed*, pp. 89–97 and *passim*.

workers or of the workers and the peasants. Up to the revolution, during it, and for some time afterwards, they were always willing to submit their own policies to the 'verdict of proletarian democracy', i.e. to the vote of the working class.

Towards the end of the civil war, however, the 'verdict of proletarian democracy' had become a meaningless phrase. How could that verdict express itself when the working class was scattered and *déclassé*? Through elections to the Soviets? Through 'normal' procedures of Soviet democracy? The Bolsheviks thought that it would be the height of folly on their part to be guided in their actions by the vote of a desperate remnant of the working class and by the moods of accidental majorities which might form themselves within the shadowy Soviets. At last they —and Trotsky with them—did in fact substitute their own party for the working class. They identified their own will and ideas with what they believed would have been the will and the ideas of a full-blooded working class, if such a working class had existed. Their habit of regarding themselves as *the* interpreters of the proletarian-class interest made that substitution all the easier. As the old vanguard, the party found it natural for itself to act as the *locum tenens* for the working class during that strange and, it hoped, short interval when that class was in a state of dissolution. Thus the Bolsheviks drew a moral justification for their usurpatory role from their own tradition as well as from the actual state of society.

The Bolshevik tradition, however, was a subtle combination of diverse elements. The party's moral self-reliance, its superiority, its sense of revolutionary mission, its inner discipline, and its deeply ingrained conviction that authority was indispensable to proletarian revolution—all these qualities had formed the authoritarian strands in Bolshevism. These, however, had been held in check by the party's intimate closeness to the real, and not merely to the theoretical, working class, by its genuine devotion to it, by its burning belief that the weal of the exploited and the oppressed was the beginning and the end of the revolution and that the worker should eventually be the real master in the new state, because in the end History would through his mouth pronounce a severe and just verdict on all parties, including the Bolsheviks, and all their deeds. The idea of proletarian democracy was inseparable from this attitude.

When the Bolshevik evoked it he expressed his contempt for the formal and deceptive democracy of the bourgeoisie, his readiness to ride roughshod, if need be, over all the non-proletarian classes, but also his feeling that he was bound in duty to respect the will of the working class even when momentarily he dissented from it. In the early stages of the revolution the proletarian-democratic strand was pre-eminent in the Bolshevik character. Now the bent towards authoritarian leadership was on top of it. Acting without the normal working class in the background, the Bolshevik from long habit still invoked the will of that class in order to justify whatever he did. But he invoked it only as a theoretical surmise and an ideal standard of behaviour, in short, something of a myth. He began to see in his party the repository not only of the ideal of socialism in the abstract, but also of the desires of the working class in the concrete. When a Bolshevik, from the Politbureau member to the humblest man in a cell, declaimed that 'the proletariat insists' or 'demands' or 'would never agree' to this or that, he meant that his party or its leaders 'insisted', 'demanded', and 'would never agree'. Without this half-conscious mystification the Bolshevik mind could not work. The party could not admit even to itself that it had no longer any basis in proletarian democracy. True, at intervals of cruel lucidity the Bolshevik leaders themselves spoke frankly about their predicament. But they hoped that time, economic recovery, and the reconstitution of the working class would solve it; and they went on to speak and to act as if the predicament had never arisen and as if they still acted on a clear and valid mandate from the working class.[1]

. . . .

[1] At a congress of Soviets in December 1921 Lenin arguing against those who all too often referred to themselves as 'representatives of the proletariat' said: 'Excuse me, but what do you describe as proletariat? That class of labourers which is employed in large-scale industry. But where is [your] large-scale industry? What sort of a proletariat is this? Where is your industry? Why is it idle?' (*Sochinenya*, vol. xxxiii, p. 148.) In March 1922, at the eleventh party congress, Lenin again argued: 'Since the war it is not at all working-class people but malingerers that have gone to the factories. And are our social and economic conditions at present such that genuine proletarians go to the factories? No. They should go, according to Marx. But Marx wrote not about Russia—he wrote about capitalism in general, capitalism as it has developed since the fifteenth century. All this has been correct for 600 years, but is incorrect in present-day Russia.' (Op. cit., p. 268.) Shlyapnikov, speaking on behalf of the Workers' Opposition, thus replied

The Bolsheviks had by now finally suppressed all other parties and established their own political monopoly. They saw that only at the gravest peril to themselves and the revolution could they allow their adversaries to express themselves freely and to appeal to the Soviet electorate. An organized opposition could turn the chaos and discontent to its advantage all the more easily because the Bolsheviks were unable to mobilize the energies of the working class. They refused to expose themselves and the revolution to this peril. As the party substituted itself for the proletariat it also substituted its own dictatorship for that of the proletariat. 'Proletarian dictatorship' was no longer the rule of the working class which, organized in Soviets, had delegated power to the Bolsheviks but was constitutionally entitled to depose them or 'revoke' them from office. Proletarian dictatorship had now become synonymous with the exclusive rule of the Bolshevik party. The proletariat could 'revoke' or depose the Bolsheviks as little as it could 'revoke' or depose itself.

In suppressing all parties, the Bolsheviks wrought so radical a change in their political environment that they themselves could not remain unaffected. They had grown up under the Tsarist régime within a half-open and half-clandestine multiparty system, in an atmosphere of intense controversy and political competition. Although as a combatant body of revolutionaries they had had their own doctrine and discipline which even then set them apart from all other parties, they nevertheless breathed the air of their environment; and the multiparty system determined the inner life of their own party. Constantly engaged in controversy with their adversaries, the Bolsheviks cultivated controversy in their own ranks as well. Before a party member took the platform in order to oppose a Cadet or a Menshevik, he thrashed out within his own party cell or committee the issues which occupied him, the adversary's case, the reply to it, and the party's attitude and tactical moves. If he thought that the party was wrong on any point or its leadership inadequate, he said so without fear or favour, and tried to convert his comrades to his view. As long as the party fought for the

to Lenin: 'Vladimir Ilyich said yesterday that the proletariat as a class, in the Marxian sense, did not exist [in Russia]. Permit me to congratulate you on being the vanguard of a non-existing class.' *11 Syezd RKP (b)*, p. 109. The taunt expressed a bitter truth. See also Zinoviev's speech, ibid., pp. 408-9.

workers' democratic rights, it could not refuse those rights to its own members within its own organization.[1]

Destroying the multiparty system the Bolsheviks had no inkling of the consequences to themselves. They imagined that outside that system they would still remain what they had always been: a disciplined but free association of militant Marxists. They took it for granted that the collective mind of the party would continue to be shaped by the customary exchange of opinion, the give and take of theoretical and political argument. They did not realize that they could not ban all controversy outside their ranks and keep it alive within their ranks: they could not abolish democratic rights for society at large and preserve those rights for themselves alone.

The single-party system was a contradiction in terms: the single party itself could not remain a party in the accepted sense. Its inner life was bound to shrink and wither. Of 'democratic centralism', the master principle of Bolshevik organization, only centralism survived. The party maintained its discipline, not its democratic freedom. It could not be otherwise. If the Bolsheviks were now to engage freely in controversy, if their leaders were to thrash out their differences in public, and if the rank and file were to criticize the leaders and their policy, they would set an example to non-Bolsheviks who could not then be expected to refrain from argument and criticism. If members of the ruling party were to be permitted to form factions and groups in order to advance specific views within the party, how could people outside the party be forbidden to form their own associations and formulate their own political programmes? No body politic can be nine-tenths mute and one-tenth vocal. Having imposed silence on non-Bolshevik Russia, Lenin's party had in the end to impose silence on itself as well.

The party could not easily reconcile itself to this. Revolutionaries accustomed to take no authority for granted, to question accepted truth, and to examine critically their own party, could not suddenly bow to authority with unquestioning obedience.

---

[1] How unreconciled to their own single-party system the Bolsheviks were even in the fifth year of the revolution can be seen *inter alia* from this passage of Zinoviev's speech at the eleventh congress: '. . . we are the only party which exists legally . . . we have, so to speak, a monopoly. . . . *This jars on the ears of our party patriotism* . . . we have denied our adversaries political freedom . . . but we could not act otherwise. . . .' Ibid., pp. 412–13. (My italics, I.D.)

Even while they obeyed they continued to question. After the tenth congress had, in 1921, declared the ban on inner party factions, Bolshevik assemblies still resounded with controversy. Like-minded members still formed themselves into leagues, produced their 'platforms' and 'theses', and made scathing attacks on the leaders. In so doing, they threatened to undermine the basis of the single-party system. Having suppressed all its enemies and adversaries, the Bolshevik party could not continue to exist otherwise than by a process of permanent self-suppression.

The very circumstances of its own growth and success drove the party to adopt this course. Early in 1917 it had no more than 23,000 members in the whole of Russia. During the revolution the membership trebled and quadrupled. At the height of the civil war, in 1919, a quarter of a million people had joined its ranks. This growth reflected the party's genuine pull on the working class. Between 1919 and 1922 the membership trebled once again, rising from 250,000 to 700,000. Most of this growth, however, was already spurious. By now the rush to the victors' bandwagon was in full progress. The party had to fill innumerable posts in the government, in industry, in trade unions, and so on; and it was an advantage to fill them with people who accepted party discipline. In this mass of new-comers the authentic Bolsheviks were reduced to a small minority.[1] They felt that they were swamped by alien elements; and they were alarmed and anxious to winnow the chaff from the wheat.

But how was this to be done? It was difficult enough to tell those who joined the party for the sake of a disinterested conviction from the turncoats and pot-hunters. It was more difficult still to determine whether even those who sought membership from no disreputable motive grasped the party's aims and aspirations and were ready to fight for them. As long as several parties expounded their programmes and recruited followers, their perpetual contest secured the proper selection of the human material and its distribution between the parties. The new-comer to politics had then every opportunity to compare the competing programmes, methods of action, and slogans. If he joined the Bolsheviks he did so by an act of conscious choice. But those who entered politics in the years 1921–2 could make

---

[1] According to Zinoviev, Bolsheviks who had fought in the underground before February 1917 formed only 2 per cent. of the membership in 1922. Ibid., p. 420.

no such choice. They knew the Bolshevik party only. In other circumstances their inclinations might have led them to join the Mensheviks, the Social Revolutionaries, or any other group. Now their urge for political action led them to the only party in existence, the only one which offered an outlet to their energy and ardour. Many of the new entrants were, as Zinoviev called them, 'unconscious Mensheviks' or 'unconscious Social Revolutionaries',[1] who sincerely thought of themselves as 'good Bolsheviks'. The influx of such elements threatened to falsify the party's character and to dilute its tradition. Indeed, at the eleventh party congress, in 1922, Zinoviev claimed that there were already within the Bolshevik organization two or more potential parties formed by those who honestly mistook themselves for Bolsheviks. Thus by the mere fact that it was the single party, the party was losing its single mind; and inchoate substitutes for the parties it had banned began to appear in its own midst. The social background with all its repressed diversity of interest and political mentality reasserted itself and pressed on the only existing political organization and infiltrated it from all sides.

The leaders resolved to defend the party against this infiltration. They began a purge. The demand for a purge had come from the Workers' Opposition at the tenth congress; and the first purge was carried out in 1921. Police and the courts had nothing to do with the procedure. At public meetings the Control Commissions, i.e. party tribunals, examined the record and the morals of every party member, high or low. Every man and woman in the audience could come forward and testify for or against the investigated individual, whom the Control Commission then declared either worthy or unworthy of continued membership. The unworthy bore no punishment; but the loss of membership in the ruling party was likely to deprive him of chances of promotion or of a responsible post.

Within a short time 200,000 members, about one-third of the total membership, were thus expelled. The Control Commission classified those expelled into several categories: vulgar careerists; former members of anti-Bolshevik parties, especially former Mensheviks, who joined after the end of the civil war; Bolsheviks corrupted by power and privilege; and, finally, the politically

[1] Ibid., pp. 413–14.

immature who lacked an elementary grasp of the party's principles.[1] It seems that people whose only fault was that they had criticized the party's policy or its leaders were not expelled. But it soon became clear that the purge, needed though it was, was a double-edged weapon. It provided the unscrupulous with opportunities for intimidation and with pretexts for settling private accounts. The rank and file applauded the expulsion of turncoats and corrupt commissars, but were bewildered by the scope of the purge. It was known that purges would be repeated periodically; and so people wondered what, if a third of the membership could be expelled in a single year, would happen next year or the year after. The humble and the cautious began to think twice before they ventured to make a risky remark or take a step which at the next purge might bring upon them the reproach of political immaturity or backwardness. Initiated as a means of cleansing the party and safeguarding its character, the purge was destined to serve the party as the most deadly instrument of self-suppression.

We have seen that, when the working class had vanished as an effective social force, the party in all its formidable reality substituted itself for the class. But now the party too appeared to turn into an entity as elusive and phantom-like as that for which it had substituted itself. Was there any real substance, and could there be any autonomous life, in a party which in a single year declared a full third of its members unworthy and expelled them? The 200,000 purged men and women had presumably taken part in all normal procedures of party life up till now, voted for resolutions, elected delegates to congresses, and had thus had a large formal share in determining the party's policy. Yet their expulsion brought about no perceptible change or modification of policy. Not even a trace could be found in the party's outlook of the great surgical operation by which one third of its body had been slashed off. This fact alone proved that for some time past the mass of members had had no influence whatsoever on the conduct of affairs. Bolshevik policy was determined by a small section of the party which substituted itself for the whole.

Who constituted that section? Lenin himself answered the question in no uncertain terms. In March 1922 he wrote to Molotov who was then secretary to the Central Committee: 'If one

[1] *Izvestya Ts. K.* of 15 November 1921 (Nr. 34). Popov, N. *Outline History of the CPSU (b),* vol. ii, p. 150.

does not wish to shut one's eyes to reality, one ought to admit that at present the proletarian character of the party's policy is determined not by the class composition of the membership but by the enormous and undivided authority of that very thin stratum of members who might be described as the party's old guard.'[1] In that Guard Lenin now saw the only repository of the ideal of socialism, the party's trustee, and ultimately the *locum tenens* of the working class. The Guard consisted in all of a few thousand authentic veterans of revolution. The bulk of the party was, in Lenin's present view, a mushroom growth exposed to all the corrupting influences of a deranged and anarchic society. Even the best of the young members needed patient training and political education before they could become 'real Bolsheviks'. Thus, the identification of proletariat and party turned out to be an even more narrow identification of the proletariat with the Old Guard.

Yet even that Guard could not easily maintain itself on the dizzy height to which it had risen; it too might not be able to withstand the debasing influences of time, weariness, corruption by power, and the pressures of the social environment. Already there were cracks in the unity of the Old Guard. In his letter to Molotov Lenin remarked: 'Even a slight dissension in this stratum may be enough to weaken . . . its authority to such an extent that [the Old Guard] should forfeit its power of decision' and become unable to control events. At all costs, therefore, it was necessary to maintain the solidarity of the Old Guard, to keep alive in it the sense of its high mission, and to secure its political supremacy. Periodic purges of the party were not enough. Severe restrictions were to be placed on the admission of members; and new entrants were to be subjected to the most exacting tests. Finally, inside the party, Lenin suggested, it was necessary to establish a special hierarchy based on merit and length of revolutionary service. Certain important offices could be held only by people who had joined the party at least early in the civil war. Other posts involving still higher responsibility were to be available to those only who had served the party since the beginning of the revolution, while top positions were to be reserved for veterans of the clandestine struggle against Tsardom.[2]

---

[1] Lenin, *Sochinenya*, vol. xxxiii, pp. 228–30.

[2] See the resolutions of the eleventh party conference and the eleventh congress in *KPSS v Rezolutsyakh*, vol. i, pp. 595–6, 612, 628–30.

There was as yet no flavour of vulgar patronage about these rules. The Old Guard still lived by its austere code of revolutionary morality. Under the *partmaximum* a party member, even one who held the highest office, was not allowed to earn more than the wages of a skilled factory worker. True, some dignitaries were already availing themselves of loopholes and supplemented meagre earnings by all kinds of benefits. But such evasions were still the exception. The new regulations about the distribution of offices were designed not to bribe the Old Guard but to make sure that party and state should remain in its hands unfailing instruments for the building of socialism.

The Old Guard was a formidable body of men. They were bound together by the memory of heroic struggles fought in common, by an unshakeable belief in socialism, and by the conviction that, amid universal dissolution and apathy, the chances of socialism depended on them and almost on them alone. They acted with authority but often also with arrogance. They were selfless yet ambitious. They were animated by the loftiest sentiments and were capable of unscrupulous ruthlessness. They identified themselves with the revolution's historic destiny but they also identified that destiny with themselves. In their intense devotion to socialism they came to regard the struggle for it as their exclusive affair and almost private business; and they were inclined to justify their behaviour and even their private ambitions in the theoretical terms of socialism.

Amid the tribulations of these years the moral strength of the Old Guard was an invaluable asset to Bolshevism. The revival of private trade and the partial rehabilitation of property spread despondency in party ranks. Many a communist wondered uneasily where the 'retreat' Lenin had ordered would take the revolution: Lenin seemed ready to go to any length to encourage the merchant and the private farmer. Since the peasant refused to sell food for worthless bank-notes, money, despised under War Communism as a relic of the old society, was 'rehabilitated' and then stabilized. Nothing was to be had without it. The government cut down the subsidies it had paid to state-owned concerns; and workers who had clung to the factory bench through the worst times lost employment. The state banks used their scanty resources to encourage private enterprise with credits. The Central Committee assured the party that, nevertheless, by holding the 'commanding heights' of large-scale industry, the state would in any event be able to control the national economy. But these 'commanding heights' had a sad and unpromising look: state-owned industry was at a standstill while private trade began to flourish. Then Lenin invited the old concessionaires and foreign investors to return to Russia and do business; and only because the investors did not respond did an important element of capitalism fail to reappear. But what would happen, Bolsheviks wondered, if the concessionaires were to respond after all? In the meantime the N.E.P.-man grew self-confident, feasted in famished towns, and mocked at the revolution. In the country, the kulak tried to get the farm labourer under his thumb once again; and here and there he and his dependants began to dominate the rural Soviet, while his son became ringleader in the local branch of the Communist Youth. At universities teachers and students staged anti-communist demonstrations and strikes, and communists were man-handled for singing the *Internationale*, the revolution's anthem. Where was the retreat going to end? The Workers' Opposition threw the question at Lenin during the sessions of the Central Committee and at public assemblies. Repeatedly he promised to halt the retreat; and repeatedly events compelled him to retreat even farther. The idealists were shocked. Cries of 'betrayal' came from the ranks. Often a worker, a veteran of the Red Guards, appeared before his party committee, tore in disgust his member card and threw it in the face of the party secretary. So much was this a sign of the times that the description of such scenes can be found in many a contemporary novel and party chiefs spoke about them with undisguised anxiety.[1]

Amid all this dejection it seemed that the revolution could rely only on the Old Guard, on its steadfast faith and iron will. But could it?

.   .   .   .

At the end of the civil war Trotsky descended from the military train which had served him as field headquarters and in

---

[1] Manuilsky, for instance, protested at the eleventh congress against the fact that the veteran of the civil war who tore his party card was surrounded by a halo of heroism, whereas he should be treated as a traitor. He compared the prevalent mood to the depression which followed the defeats of revolution in 1849 and 1907. *11 Syezd RKP (b)*, pp. 461-3.

which he had, for three fateful years, rushed from danger spot to danger spot along a frontline of 5,000 miles, interrupting his journeys only for brief consultations and public appearances in Moscow. The military train was placed in a museum; its crew of drivers, mechanics, machine-gunners, and secretaries was disbanded; and Trotsky took his first holiday since the revolution. He spent it in the country not far from Moscow—hunting, fishing, writing, and preparing for a new chapter in his life. When he returned to Moscow, as whose voice he had spoken all these years, he was almost a stranger there. He had his first glimpse of the old capital at the turn of the century when he was brought to Butyrki prison to await deportation to Siberia; and so it was from behind the bars of a prison van that he first viewed the city of his future triumphs and defeats. He did not return to Moscow until twenty years later, in March 1918, during the Brest Litovsk crisis, after the Bolshevik government had evacuated Petrograd and established itself in the Kremlin. Presently he left for the fronts. Whenever he returned he felt as if out of place in the sprawling 'village of the Tsars', the Third Rome of the Slavophiles, with its Byzantine churches and Asiatic bazaars and its listless oriental fatalism. His revolutionary associations both in 1905 and in 1917 had been with Petrograd, Moscow's rival and Russia's window to Europe; and he always felt more at ease with the engineers, shipbuilders, and electrical workers of Petrograd than with the workers of Moscow who, employed mostly in textile mills, still looked and behaved more like muzhiks than like city dwellers.

He felt even more out of place within the walls and towers of the Kremlin, in the narrow tortuous streets of the old fortress, in the shade of its battlements reverberating with ancient bells, amid its cathedrals, arsenals, barracks, prison towers, and belfries, in the gilded halls of its palaces, surrounded by innumerable miraculous ikons the Tsars had assembled from all their conquered lands. With his wife and children he occupied four small rooms in the Kavalersky building, the former quarters of Court officials. Across the corridor lived Lenin and Krupskaya; the two families shared the dining-room and bath-room—in the corridor or in the bath-room Lenin might often be caught playing with the Trotsky children. Now and then an old friend, Rakovsky, Manuilsky, or someone else, arriving from the provinces on government business, stayed with the family. The Trotskys' domestic life was still as modest as it was when as exiles they lived in a garret in Paris or in a tenement block in Vienna. It was perhaps poorer, for food was scarce even in the Kremlin.[1] The children—Lyova was fifteen and Seryozha thirteen in 1921 —enjoyed little parental care: they saw even their mother for brief moments only; she spent her days at the Commissariat of Education and headed its Arts Department.

The magnificent setting of the Kremlin contrasted strangely with the way of life of its new inhabitants. Trotsky describes the family's amused embarrassment when they were first attended by an old Court butler who served meals on plates bearing the Tsar's coat of arms and carefully turned and manipulated the plates in front of the grown-ups and the children so that the Tsar's eagles should never, God forbid, be placed upside down.[2] From every corner 'the heavy barbarism of Moscow' stared at the Bolshevik leaders; and when the chimes of the old bells intruded in their conversation, Trotsky and Lenin 'looked at each other as if we had both caught ourselves thinking the same thing; we were being overheard by the past . . .'. They were not merely being overheard—the past was fighting back against them. In any case, Trotsky, as he confesses, never merged with the Kremlin background. He kept his distance from it; and only his sense of historic irony was tickled by the revolution's intrusion into Muscovy's holy of holies.

He had a gnawing feeling that the end of the civil war was an anticlimax in his fortunes. He repressed this feeling by an effort of conscious optimism, the optimism which should never abandon the revolutionary; and he looked forward to new triumphs for his cause and for himself. But scattered in his speeches and writings there were already nostalgic notes about the heroic era of revolution and civil war now closed. It was not that he idealized that era during which, as he put it, the muzhik's club served the revolution as its 'finest tool', that primordial club with which the peasants had once driven Napoleon and with which they had now driven the landlord

---

[1] Arthur Ransome relates that when, in 1919, he gave Bukharin a little saccharine for tea, this was quite a treat; a meal at Zinoviev's headquarters consisted of 'soup with shreds of horseflesh . . . a little *kasha* . . . tea and a lump of sugar'. *Six Weeks in Russia*, pp. 13, 56.

[2] *Moya Zhizn*, vol. ii, p. 77.

PLATE I

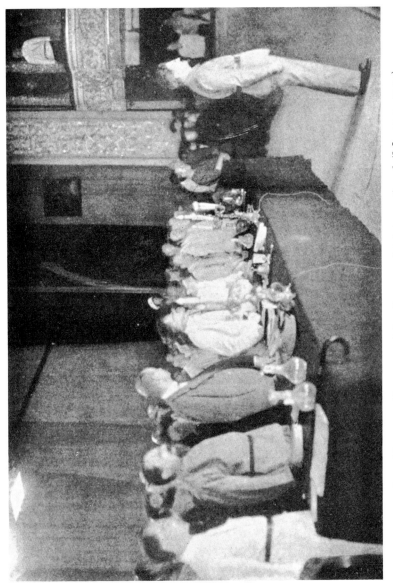

Trotsky addresses a Congress of the Communist International (Moscow, 1920)

PLATE II

(*a*) Trotsky as Commissar of War

(*b*) Inspecting troops in Moscow in 1921, after the end of the civil war

from Russia. Nor did he overlook the heavy legacy of that era—the destructive furies let loose by civil war which were revenging themselves on the Soviet Republic as it turned towards its constructive tasks. But for all their miseries, squalor, and cruelty, the years of destruction had also been years of creation; and he harked back to their mighty sweep, courage, and soaring hope; and he sensed the gap they left behind.[1]

His brain and energy were now only half occupied. The Commissariat of War was no longer the hub of government. The army was demobilized. By the beginning of 1922 it had been reduced to one-third of its establishment. It was also losing its revolutionary idealism and fervour. The veterans of the civil war had left; and the freshly mobilized age-groups in the barracks seemed as listless and apathetic as had been the peasant sons who came to the same barracks in the days of the Tsar. Circumstances compelled the Commissar of War to shelve his cherished plans for transforming the army into a modern, democratic, and socialist militia, and imposed on him the humdrum routine of administration and training. He spent his time delousing the army, teaching it to grease its boots, and clean its rifles, and entreating the best commanders and commissars to stay on in their jobs. He urged the Central Committee to arrest the mass exodus of Communists from the army; and the Central Committee tried formal prohibitions and bans. But these were ineffective. At national conferences Trotsky again and again implored political commissars to resist the 'infectious pacifist mood' and he lamented the Red Army's sagging morale. He struggled to keep the army uncontaminated by the 'spirit of Sukharevka' and to use it as an instrument of a Marxist *Kulturkampf* against the filth, backwardness, and superstition of Mother Russia, and above all, to keep alive in it revolutionary tradition and internationalist awareness.[2]

This was the time when the young commanders of the civil war, among them the future marshals of the Second World War, obtained serious training, and the Red Army received its rules and regulations. Of these Trotsky was the inspirer and part

---

[1] See, for instance, Trotsky's address to the commanding officers and commissars of the Moscow garrison of 25 October and his speech at the end of army manœuvres in September 1921, *Kak Vooruzhalas Revolutsia*, vol. iii, book 1.

[2] See Annual Report of C.C. in Appendix to *11 Syezd RKP (b)*, pp. 637–64; *Pyat Let Sovietskoi Vlasti*; and *Kak Vooruzhalas Revolutsia*, vol. iii, *passim*.

author. It is curious, for instance, to note the affinity between Trotsky's 'Infantry Regulations' and the Cromwellian Soldier's Catechism. 'You are an equal among comrades', the Infantry Regulations instructed the Red Army man, 'Your superiors are your more experienced and better educated brothers. In combat, during training, in the barracks, or at work you must obey them. Once you have left the barracks you are absolutely free. . . .' 'If you are asked in what way you fight, you answer: "I fight with the rifle, the bayonet, and the machine-gun. But I also fight with the word of truth. I address that to the enemy's soldiers who are themselves workers and peasants so that they should know that in truth I am their brother, not their enemy."'

His love of words, the simple words as well as the rich, and his sense of form and colour went into the making of a new pageantry with which he sought to appeal to the recruit's imagination and to develop in the army the feeling that it was not merely regimented cannon-fodder. On May Day and on the anniversaries of the revolution, flanked by the commanders of the Moscow garrison, he rode out on horseback, through the Kremlin's Spasky gates to the Red Square to review the massed columns of the garrison. To his greeting 'Salute, Comrades!' the troops replied: 'We Serve the Revolution!'; and the echo thundered against the spires of the Vassily Cathedral and over the graves of the revolution's martyrs along the Kremlin wall. There was as yet no mechanical pomp or ceremony. After the review the Commissar of War joined the other members of the Central Committee who from a wooden ramshackle stand or from a crowded army lorry took the parade of soldiers and workers.[1]

Trotsky's appearance and speech still thrilled the crowds. But he no longer seemed to find the intimate contact with his audiences which he found unerringly during the civil war, the contact which Lenin invariably established by his unobtrusive appearance and simple expression. Trotsky on the platform

---

[1] Morizet, *Chez Lénine et Trotski*, pp. 108–11. Serge and Rosmer give in their writings vivid and friendly descriptions of Trotsky in these years. Of the many eye-witness accounts and character-sketches, friendly and hostile, only a few will be mentioned here: L.-O. Frossard, *Sous le Signe de Jaurès* and *De Jaurès à Lénine*; B. Bajanov, *Avec Staline dans le Kremlin*; R. Fischer, *Stalin and German Communism*; F. Brupbacher, *60 Jahre Ketzer*; Clare Sheridan, *Russian Portraits*, the early writings of Radek, Bukharin, Sadoul, Eastman, Holitscher, L. Fischer.

appeared more than life-size; and his speech resounded with all its old heroic tones. Yet the country was tired of heroism, of great vistas, high hopes, and sweeping gestures; and Trotsky still suffered from the slump in his popularity caused by his recent attempts to militarize labour. His oratorical genius still cast its spell on any assembly. But the spell was already shot through with doubt and even suspicion. His greatness and revolutionary merits were not doubted; but was he not too spectacular, too flamboyant, and perhaps too ambitious?

His theatrical manner and heroic style had not struck people as odd in earlier years when they accorded with the drama of the time. Now they carried with them a suggestion of histrionics. Yet he behaved as he did because he could not behave otherwise. He did not posture to appear more than life-size—he could not help appearing it. He spoke an intense and dramatic language not from affectation or craving for stage effect, but because this was his most natural language, best suited to express his dramatic thought and intense emotion. One might apply to him the words in which Hazlitt described a man as different from him as Burke. He 'gave a hold to his antagonists by mixing up sentiment and imagery with his reasoning', and 'being unused to such a sight in the region of politics' people 'were deceived, and could not discern the fruit from the flowers . . .'. 'The generality of the world' was as always 'concerned in discouraging any example of unnecessary brilliancy.' But 'his gold was not the less valuable for being wrought into elegant shapes'; and 'the strength of a man's understanding is not always to be estimated in exact proportion to his want of imagination. His understanding was not the less real because it was not the only faculty he possessed.'

Like Burke, Trotsky was 'communicative, diffuse, magnificent'. He too conversed in private as he spoke in public, and addressed his family and friends in the same images, with the same wit, and even in the same rhythmical cadences which he used on the platform and in his writings. If he was an actor, then he was one to whom it made no difference whether he found himself on the proscenium, in the green-room, or at his home—one to whom theatre and life were one. He was indeed the heroic character in historic action; and because of this he must appear unreal and unnatural to a prosy or jaundiced generation; and because of this he seemed out of place—a stranger—in the unheroic atmosphere of the early N.E.P. There is no need, however, to overdraw the romantic aspect of Trotsky's character. He remained as strong in his realism as ever. In any case, he was not the veteran 'superfluously lagging on the stage'. He threw himself with zest upon the new economic and social issues posed by N.E.P.; and he did not by any means view N.E.P. through the prism of revolutionary fundamentalism. Absorbed in problems of finance, industry, trade, and agriculture, he placed before the Politbureau and the Central Committee specific proposals of policy about which more will be said later. He used all his inspiring eloquence to defend the uninspiring 'retreat'; and he appeared as the expounder of N.E.P. before the Communist International at its third and fourth congresses in 1921 and 1922.[1] He gave more of his time and energy than before to the International, on the Executive of which he resisted Zinoviev's and Bukharin's inclination to encourage untimely and reckless risings abroad, such as the German *Märzaktion*. He presided over the French Commission of the Comintern and intervened in the conduct of affairs of every major section of the International.

However, the Commissariat of War, domestic economic preoccupations, and the Comintern, did not absorb his whole energy. He was busy with a host of other assignments each of which would have made a full-time job for any man of less vitality and ability. He led, for instance, the Society of the Godless before Yaroslavsky took over its direction. He led it in a spirit of philosophical enlightenment which was least likely to produce those excesses, offensive to the sentiment of the believers, which marred the Society's work under Yaroslavsky. (He even headed a secret Commission for the confiscation and collection of ecclesiastical treasures which were to be used as payment for food imported from abroad to alleviate the famine on the Volga.[2]) He was at this time Russia's chief intellectual inspirer and leading literary critic. He frequently addressed audiences of scientists, doctors, librarians, journalists, and men of other professions, explaining to them where Marxism stood in relation to the issues which occupied them. At the same time he

---

[1] *Chetvertyi Vsemirnyi Kongress Kominterna*, pp. 74–111; and Trotsky, *Pyat Let Kominterna*, pp. 233–40, 460–510.   [2] *The Trotsky Archives.*

resisted within the party the tendency which was already becoming apparent to impose a deadly uniformity upon the country's cultural life.[1] In many articles and speeches he insisted in a more popular vein on the need to civilize the uncouth Russian way of life, to cultivate manners, to raise hygiene, to improve the spoken and written language which had been debased since the revolution, to widen and to humanize the interests of party members, and so on, and so on. With Lenin already somewhat withdrawn from the public eye, he was the party's chief and most authoritative spokesman in these, the last years of the Lenin era.

Nor did his romantic temperament revolt as yet against the harsh realism with which the party, or rather the Old Guard, established and consolidated its political monopoly. After as before the promulgation of N.E.P., he was indeed one of the sternest disciplinarians, although his call for discipline was based on persuasive argument and appeal to reason. He still extolled the party's 'historic birthright';[2] and he argued that the procedures of proletarian democracy could not be observed in conditions of social unsettlement and chaos, that the fate of the revolution should not be made dependent on the unstable moods of a shrunken and demoralized working class, and that it was the Bolsheviks' duty towards socialism to maintain their 'iron dictatorship' by every means at their disposal. He had once intimated that the party's political monopoly was an emergency measure to be revoked as soon as the emergency was over; but this was not what he said now. More than a year after the Kronstadt rising, writing in *Pravda* on the signs of economic recovery and on the 'upward movement' noticeable in all fields, he posed the question whether the time had not come to put an end to the single-party system and to lift the ban at least on the Mensheviks. His answer was a categorical No.[3] He now justified the monopoly not so much by the republic's internal difficulties as by the fact that the republic was a 'besieged fortress' within which no opposition, not even a feeble one, could be tolerated. He pleaded for the enforcement of the single-party system during the whole period of Russia's international isolation, which he did not, however, expect to last as long as it was to last. Recalling

---

[1] See Chapter III.   [2] See *The Prophet Armed*, pp. 508–9.
[3] *Pravda*, 10 May 1922; and *Pyat Let Kominterna*, pp. 373–4.

that he himself had once ridiculed attempts made by governments to suppress political opposition and had demonstrated their ultimate inefficacy, he excused his apparent change of attitude with the following argument which would be flung back at him one day: 'Repressive measures', he wrote, 'fail to achieve their aim when an anachronistic government and régime applies them against new and progressive historic forces. But in the hands of a historically progressive government they may serve as very real means for a rapid cleansing of the arena from forces which have outlived their day.'

He reasserted this view in June 1922, during the famous trial of the Social Revolutionaries. He produced a brilliant and ferocious exposure of the defendants, holding them to be politically responsible for Dora Kaplan's attempt on Lenin's life and for other terroristic acts. The trial took place at the time of the 'conference of three Internationals' in Berlin. At that conference, which aimed at establishing a 'united front' between Communist and Socialist parties in the west, Bukharin and Radek represented the Bolsheviks. Western Social Democratic leaders protested against the trial; and to smooth negotiations Bukharin and Radek promised that the defendants would not be sentenced to death. Lenin was indignant at Bukharin's and Radek's 'yielding to blackmail' and at their allowing European reformists to interfere with domestic Soviet affairs. Trotsky was not less indignant. But in order to avoid a breach of the undertaking he proposed a compromise by which death sentence was pronounced but then suspended on the express condition that the Social Revolutionary party refrained from committing and encouraging further terroristic attempts.[1]

Trotsky's disciplinarian attitude showed itself inside the party as well. On behalf of the Central Committee he indicted the Workers' Opposition before the party and Communist International. Since the tenth congress, at which its activity and views had been condemned, the Workers' Opposition had continued to attack the party leadership with increasing bitterness. Shlyapnikov and Kollontai charged the government with promoting the interests of the new bourgeoisie and of the kulaks,

---

[1] *Pravda*, 16, 18 May and 18 June 1922; Lenin, *Sochinenya*, vol. xxxiii, pp. 294–8; *The Second and Third International and the Vienna Union*; Trotsky, *Moya Zhizn*, vol. ii, pp. 211–12.

with trampling upon the workers' rights, and with the gross betrayal of the revolution. Defeated in the party and threatened by Lenin with expulsion, they appealed against Lenin to the Communist International. At the Executive of the International Trotsky presented the case against them and obtained the dismissal of their appeal.[1] Then, at the eleventh congress of the Russian party, in the spring of 1922, which was again called upon to pronounce itself on the matter, Trotsky once again acted as counsel for the prosecution.[2] He spoke without ill will or rancour and even with a certain warmth of sympathy for the Opposition; but he nevertheless firmly upheld the indictment. The Workers' Opposition, he said, acted within its rights when it took the unprecedented step of appealing against the Russian party to the International. What he held against Shlyapnikov and Kollontai was that they had introduced an intolerably violent tone into the dispute and that they spoke of themselves and the party in terms of 'we' and 'they', as if Shlyapnikov and Kollontai 'had already another party in reserve'. Such an attitude, he said, led to schism and provided grist to the mills of the enemies of the revolution. He defended the government, its rural policy, its concessions to private property, and also its view, equally strongly attacked, that ahead lay 'a long period of peaceful coexistence and of business-like cooperation with bourgeois countries'.[3]

The Workers' Opposition was not alone in voicing disillusionment. At the eleventh congress, the last attended by Lenin, Trotsky saw himself and Lenin attacked by old and intimate friends: Antonov-Ovseenko, who spoke about the party's surrender to the kulak and foreign capitalism;[4] Ryazanov, who thundered against the prevalent political demoralization and the arbitrary manner in which the Politbureau ruled the party;[5] Lozovsky and Skrypnik, the Ukrainian commissar, who protested against the over-centralistic method of government, which, he said, was all too reminiscent of the 'one and indivisible' Russia of old;[6] Bubnov, still the Decemist, who spoke about the danger of the party's 'petty bourgeois degeneration';[7] and Preobrazhensky, one of the leading economic theorists and former

---

[1] *11 Syezd RKP (b)*, appendix.   [2] Ibid., pp. 138–57.
[3] Ibid., p. 144.   [4] Ibid., pp. 80–83.   [5] Ibid., pp. 83–87.
[6] Ibid., pp. 77–79.   [7] Ibid., pp. 458–60.

secretary of the Central Committee.[1] One day most of the critics would be eminent members of the 'Trotskyist' Opposition; and one day Trotsky himself would appeal, as Shlyapnikov and Kollontai had done, against the Russian Central Committee to the International. But for the time being, heartily applauded by Lenin, he confronted the Opposition as mouthpiece of the Bolshevik Old Guard, demanding discipline, discipline, and once again discipline.

And yet he remained a stranger in the Old Guard as well—in it but not of it. Even at this Congress of 1922 Mikoyan, then still a young Armenian delegate, stated this from the platform without being contradicted. In the course of the debate Lenin, Zinoviev, and Trotsky had expressed uneasiness over the merger of party and state, and had spoken about the need to separate in some measure their respective functions. Mikoyan then remarked that he was not surprised to hear this view from Trotsky who was 'a man of the state but not of the party'; but how could Lenin and Zinoviev propound such ideas?[2] Mikoyan did not speak from his own inspiration. He summed up what many members of the Old Guard thought but did not yet utter in public: in their eyes Trotsky was the man of the state but not of the party.

Now, when the Old Guard found itself elevated to an undreamt-of height, above the people, the working class, and the party, it began to cultivate its own past, and also the legends about it, with a pietism which is never quite absent from any group of veterans with memories of great battles fought and great victories won in common. The nation had known little or nothing about the men who, having risen from the obscurity of an underground movement, stood at its head. It was time to tell the people who those men were and what they had done. The party historians dug up the archives and set out to reconstruct their epic story. The tale they told was one of almost superhuman heroism, wisdom, and devotion to the cause. They did not by any means concoct the tale in cold blood. Much of it was true; and they sincerely believed even in that which was not quite true. As the members of the Old Guard viewed themselves in the dim mirror of the past, they inevitably saw that mirror brightened up and their own reflections in it enlarged by the

[1] Ibid., pp. 89–90.   [2] Ibid., pp. 453–7.

PLATE III

Trotsky leaving his 'military train'

PLATE IV

Trotsky and General Muralov, Army Inspector and Commander of Moscow

retrospective glare of the victorious revolution. But as they looked into that mirror they invariably saw in it Trotsky as their antagonist, the Menshevik, the ally of the Mensheviks, the leader of the August bloc, and the bitter polemicist who had been dangerous to them even when he stood alone. They re-read all the excoriating epithets he and Lenin had once exchanged in open controversy; and the archives, which contained unknown manuscripts and letters, yielded many other rough remarks the two men had made about each other. Every document bearing on the party's past, no matter how trivial, was treasured and published with reverence. The question arose whether Trotsky's old anti-Bolshevik tirades should be withheld from publication. Olminsky, the comptroller of the party archives, put the question to him, when Trotsky's letter to Chkheidze, written in 1912, and describing Lenin as 'intriguer', 'disorganizer', and 'exploiter of Russian backwardness', was discovered in the files of the Tsarist gendarmerie.[1] Trotsky objected to publication: it would be foolish, he said, to draw attention to disagreements which had long since been lived down; besides, he did not think that he had been wrong in all that he had ever said against the Bolsheviks, but he was not inclined to go into involved historical explanations. The offensive document did not appear in print; but its contents were too piquant for copies not to be circulated among old and trusted party men. So this, they commented, is how Trotsky denigrated Lenin in a letter. And to whom?— To Chkheidze, the old traitor; and he still says that he was not altogether wrong! True, Trotsky had since made ample amends, if these were at all needed; in 1920, when Lenin was fifty years old, Trotsky paid his tribute to Lenin and wrote a character sketch of him, which was as incisive in its psychological truth as it was full of admiration.[2] All the same, the odd episodes from the past reminded those who had never felt anything but adoration for the party's founder how relatively recent was Trotsky's conversion to Bolshevism.

Not only memories of old feuds prevented the Old Guard from acknowledging Trotsky as its man. His strong personality had not become submerged in the Old Guard or taken on its protective colouring. He towered above the old 'Leninists' by

---

[1] *The Trotsky Archives*. Trotsky's letter to Olminsky is dated 6 December 1921.
[2] *Pravda*, 23 April 1920.

sheer strength of mind and vigour of will. He usually arrived at his conclusions, even when they coincided with those of others, from his own premises, in his own way, and without reference to the axioms consecrated by party tradition. He stated his opinion with an ease and freedom which contrasted strikingly with the laboured style of orthodox formulas in which most of Lenin's disciples expressed themselves. He spoke with authority, not as one of the scribes. The very width and variety of his intellectual interests aroused a sneaking suspicion in men who, from necessity, self-denial, or inclination, had accustomed themselves to concentrate narrowly on politics and organization and who prided themselves on their narrowness as on their virtue.

Thus, almost everything in him, his fertile mind, his oratorical boldness, his literary originality, his administrative ability and drive, his precise methods of work, the exacting demands he made on associates and subordinates, his aloofness, the absence of triviality in him, and even his incapacity for small talk—all this induced in the members of the Old Guard a sense of inferiority. He never bothered to stoop down to them and he was not even aware that he might do so. Not only did he not suffer fools gladly—he always made them feel that they were fools. The men of the Old Guard were much more at ease with Lenin whose leadership they had always accepted and who usually spared their susceptibilities. When Lenin, for instance, attacked a political attitude which he knew that some of his followers shared, he was careful not to attribute that attitude to those whom he hoped to wean from it; and so he always allowed them to retreat without losing face. When he was intent on converting anyone to his view, he conversed with the man in such a manner that the latter went away convinced that he had himself, by his own reasoning and not under Lenin's pressure, arrived at a new viewpoint. There was little of that subtlety in Trotsky, who could rarely withstand the temptation to remind others of their errors and to insist on his superiority and foresight.

His very foresight, not less real because of its ostentatiousness, was offensive. His restless and inventive mind perpetually startled, disturbed, and irritated. He did not allow his colleagues and subordinates to abandon themselves to the inertia of circumstances and ideas. No sooner had the party decided upon a new policy than he laid bare its 'dialectical contradictions', seized its consequences, anticipated new problems and difficulties, and urged new decisions. He was the born troublemaker. His judgement, even though it turned out to be correct in most cases, inevitably aroused resistance. The rapidity with which his mind worked left others breathless, exhausted, resentful, and estranged.

And yet, almost a stranger in Moscow, in the Kremlin, and within the Old Guard, by Lenin's side he still dominated the stage of the revolution.

.　　.　　.　　.　　.　　.　　.　　.　　.　　.

In April 1922 an incident occurred which did much to cloud relations between Lenin and Trotsky. On 11 April, at a session of the Politbureau, Lenin proposed that Trotsky should be appointed deputy chairman of the Council of People's Commissars. Categorically and somewhat haughtily Trotsky declined to fill this office. The refusal and the manner in which it was made annoyed Lenin; and much was made of this in the new controversies which, added to old animosities, divided the Politbureau.[1]

Lenin had hoped that Trotsky would consent to act as his deputy at the head of the government. He made the proposal a week after Stalin had become the party's General Secretary. Even though the General Secretary was supposed only to give effect to the Politbureau's and the Central Committee's decisions, Stalin's appointment was calculated to enhance discipline in the ranks. Lenin, we know, had already demanded the expulsion of the leaders of the Workers' Opposition; and at the Central Committee he failed by only one vote to obtain for this the necessary two-thirds majority.[2] He expected that Stalin would enforce the ban on organized inner party opposition which the tenth congress had declared in a secret session. In these circumstances it was almost inevitable that the General Secretary should assume wide discretionary powers.

Lenin had had his misgivings about Stalin's appointment; but having brought it about he apparently sought to countervail it by placing Trotsky in a post of comparable influence and

---

[1] *The Archives.*

[2] This was on 9 August 1921—the fact was frequently referred to at the eleventh congress, *11 Syezd RKP (b)*, pp. 605-8 and *passim*.

responsibility at the Council of Commissars. He may have
designed this distribution of offices between Stalin and Trotsky
as a means towards that separation of party and state on the
need for which he had insisted at the Congress. For the separation to be effective it was necessary, so it seemed, that the work
of the government machine should be directed by a man as
strong-minded as the one who would manage the party machine.

In Lenin's scheme, however, Trotsky was not to be the only
vice-Premier. Rykov, who was also chief of the Supreme Council of the National Economy, and Tsurupa, the Commissar of
Supplies, already held the same title. Later Lenin proposed
that Kamenev too should fill a parallel post.[1] Each vice-Premier
supervised certain branches of the administration or groups of
Commissariats. But although nominally Trotsky was to be only
one of three or four vice-Premiers, there can be little doubt
that it was Lenin's intention that he should act as his real
second in command. Without any formal title Trotsky had
acted in this capacity in any case by the sheer force of his initiative in every field of government; and Lenin's proposal was
calculated to regularize and enhance his status.

How anxious Lenin was that Trotsky should occupy the
post can be seen from the fact that he returned to the question
over and over again and that he made the same proposal
several times in the course of nine months. When he first put
it forward in April, he was not yet ill; and the thought of the
succession to his leadership had probably not yet crossed his
mind. But he was overworked and tired. He suffered from
long spells of insomnia and was compelled to try to lighten his
own burden of office. Before the end of May he was struck down
by the first attack of paralysis, and he did not return to work
until October. Yet, on the 11 September, still ill and warned
by doctors to take an absolute rest, he telephoned Stalin asking
that the question of Trotsky's appointment should be placed
again in the most formal and urgent manner before the Politbureau. Finally, early in December, when the problem of the
succession was already causing Lenin grave anxiety, he took up
the matter once again, this time directly with Trotsky and in
private.

Why did Trotsky refuse? His pride may have been hurt by

[1] Lenin, *Sochinenya*, vol. xxxiii, pp. 299–306, 316–18.

an arrangement which would have placed him formally on the same footing as the other vice-Premiers who were only Lenin's inferior assistants. He said that he saw no reason for so many vice-Premiers; and he commented sarcastically on their ill-defined and overlapping functions.[1] He also made a distinction between the substance and the shadow of political influence and held that Lenin had offered him the shadow. All levers of the government were in the hands of the party's Secretariat, i.e. in Stalin's hands. The antagonism between him and Stalin had outlasted the civil war. It was ever present in the differences over policy and the bickerings over appointments that went on at the Politbureau. Trotsky had no doubt that even as Lenin's deputy he would depend at every step on decisions taken by the General Secretariat which selected the Bolshevik personnel for the various government departments and by this alone effectively controlled them. On this point his attitude, like Lenin's, was self-contradictory: he wanted the party, or rather the Old Guard, to be in exclusive command of the government; yet he sought to prevent the party machine from interfering with the government's work. The two things could not be had simultaneously, if only because the Old Guard and the party machine were largely, though not altogether, identical. Having rejected Lenin's proposal, Trotsky at first canvassed his own scheme for an overhaul of the administration; but then he formed the conviction that no such scheme would produce the results desired as long as the powers of the General Secretariat (and of the Orgbureau) were not curtailed.

Personal animosities and administrative disagreements were, as usual, mixed up with wider differences over policy.

The Politbureau's chief concern was now with the conduct of economic affairs. The broad outlines of the New Economic Policy were not under debate. All agreed that war communism had failed and that it had to be replaced by a mixed economy, within which the private and the socialist (i.e. the state-owned) 'sectors' would coexist and in a sense compete with one another. All saw in N.E.P. not merely a temporary expedient but a long-term policy, a policy providing the setting for a gradual transition to socialism. Everybody took it for granted that N.E.P. had a dual purpose: the immediate aim was the revival

[1] See Trotsky's comments to the Politbureau of 18 April 1922 in *The Archives*.

of the economy with the help of private enterprise; and the
basic purpose was to promote the socialist sector and to ensure
its gradual extension over the whole field of the economy. But if
in these general terms the policy commanded common assent,
differences arose when the general principles had to be translated into specific measures. Some Bolshevik leaders saw primarily
the need to encourage private enterprise; while others, without
denying this need, were above all eager to promote the socialist
sector.

In the first years of N.E.P. the prevalent mood was that of
an extreme reaction against war communism. The Bolsheviks
were anxious to convince the country that it need not be afraid
of any relapse into war communism; and they themselves
were convinced that such a relapse would be impermissible
(except in war). Nothing was more important than to save the
economy from utter ruin; and they saw that only the farmer
and the private trader could *begin* to save it. They regarded
therefore no incentive offered to the farmer and the trader as
too liberal. The results were not long in showing themselves.
Already in 1922 the farmers harvested about three-quarters of
a normal pre-war crop. This brought about a radical change
in the country's condition, for in a primitive agricultural country
one good harvest can work wonders. Famine and pestilence
were overcome. But this first success of N.E.P. at once threw
into relief the dangers of the situation. Industry recovered very
slowly. In 1922 it produced only one-quarter of its pre-war
output; but even this slight advance on previous years occurred
mainly in light industry, especially in the textile mills. Heavy
industry remained paralysed. The country was without steel,
coal, and machines. This threatened once again to bring to a
standstill light industry which could not repair or renew its
machinery and lacked fuel. Prices of industrial goods soared
out of the consumers' reach. The rise was due to vast unsatisfied
demand, underemployment of plant, scarcity of raw materials,
and so on; and the situation was made even worse by the
Bolsheviks' lack of experience in industrial management and by
bureaucratic inefficiency. Stagnation in industry threatened to
react adversely upon farming and to break once again the still
tenuous 'link' between town and country. The peasant was
reluctant to sell food when he was unable to buy industrial

goods for his money. Concessions to private farming and trading, necessary though these had been, could not by themselves solve the problem. Nor could 'the market' be expected to take care of it and to resolve it rapidly, through the spontaneous action of supply and demand, without detriment to the government's socialist aspirations.

The government did not see clearly how to deal with the situation. It lived from hand to mouth. It applied palliatives; and the choice of these was dictated by the prevalent reaction against war communism. The Bolshevik leaders had burned their fingers in a reckless attempt to abolish all market economy; and so they were now wary of interfering with the market. Under war communism they had allowed no scruple to hold them back from extracting food and raw materials from the peasant; and so they were now above all anxious to appease the peasant. They hoped that the continued intense demand for consumer goods would keep the wheels of industry turning, and that heavy industry would somehow muddle through to recovery. The same attitude showed itself in financial policy. Under war communism money and credit, despised as relics of the old order, were supposed to be withering away. Then the Commissariat of Finance and the State Bank rediscovered the importance of money and credit and invested their resources in enterprises which were immediately profitable rather than in those that were of national importance. They pumped credit into light industry and neglected heavy industry. Up to a point this reaction against war communism was natural and even useful. But party leaders like Rykov and Sokolnikov, who were in charge of the economic and financial departments, tended to carry the reaction to an extreme.

It should be recalled that no differences over the promulgation of N.E.P. had divided Trotsky from the other leaders. He had himself advocated the principle underlying N.E.P. a year before the Central Committee came to adopt it; and so it was not for nothing that he reproached Lenin privately that the government tackled urgent economic matters with a delay of two years or a year and a half.[1] But having been the first to advocate N.E.P., Trotsky did not succumb to the extreme

[1] Trotsky's statements to the Politbureau of 7 August 1921 and of 22 August 1922 in *The Archives*.

reaction to war communism. He was less inclined than were his colleagues at the Politbureau to believe that further concessions to farmers and traders would suffice to ensure recovery, or that the automatic work of the market would restore the balance between farming and industry and between heavy and light industry. Nor did he share Sokolnikov's and Rykov's fresh enthusiasm for the rediscovered virtues of financial orthodoxy.

These differences were of little or no importance in 1921 and early in 1922, before farming and private trade had got into their stride. But later a major controversy began to develop. Trotsky held that the first successes of N.E.P. necessitated an urgent revision of industrial policy, and that it was imperative to quicken the pace of industrial recovery. The 'boom' in light industry was superficial and narrowly based; and it could not go on for long unless light industry was enabled to repair and renew machinery. (Farming too needed tools to maintain progress.) A concentrated effort was therefore necessary to break the deadlock in heavy industry: the government must work out a 'comprehensive plan' for industry as a whole, instead of relying on the work of the market and the spontaneous play of supply and demand. A schedule of economic priorities should be fixed and heavy industry should have first claim. Resources and manpower should be rationally concentrated in those state-owned concerns which were of basic importance to the national economy, while establishments which could not contribute effectively and rapidly towards recovery should be closed down, even if this exposed their workers to temporary unemployment. Financial policy should be subordinated to the needs of industrial policy and guided by the national interest rather than by profitability. Credits should be directed into heavy industry; and the State Bank should make long-term investments in its re-equipment. Such a reorientation of policy, Trotsky argued, was all the more urgent because of the lack of balance between the private and the socialist sectors. Private business was already making profits, accumulating capital and expanding while the bulk of state-owned industry worked at a loss. The contrast between the two sectors created a threat to the socialist objectives of the government's policy.

These ideas, which thirty and forty years later were to become truisms, seemed far-fetched at first. Even more far-fetched appeared to be Trotsky's insistence on the need for planning. That planning was essential to a socialist economy was a Marxist axiom with which the Bolsheviks were, of course, familiar and which they had always accepted in general terms. Under war communism they imagined that they were in a position to establish immediately a fully fledged planned economy; and Trotsky then met with no opposition when he spoke of the need for a 'single plan' to assure balanced economic reconstruction.[1] Just before the end of war communism, on 22 February 1921, the government decided to form the State Planning Commission, the *Gosplan*. But after the introduction of N.E.P., when all efforts were directed towards reviving the market economy, the idea of planning suffered eclipse. So much had the idea been associated in people's thoughts with war communism that a reminder of it appeared to be out of season. True, just after the promulgation of N.E.P., on 1 April 1921, the State Planning Commission was constituted and Kzhizhanovsky was appointed as its chief. But the new institution led a shadowy existence. Its prerogatives were ill defined; few were eager to define them; and it had no power to devise long-term policy, and to plan or carry plans into operation. It merely advised industrial managements on their day-to-day administrative troubles.[2]

Almost from the first Trotsky criticized this state of affairs. He held that with the transition to N.E.P. the need for planning had become more and not less urgent, and that the government was wrong in treating it as a marginal or merely theoretical issue. Precisely because they lived again under a market economy, he argued, the government must seek to control the market and to equip itself for control. He renewed the demand for a 'single plan', without which, he said, it was impossible to rationalize production, to concentrate resources in heavy industry, and to redress the balance between the various sectors of the economy. Finally, he asked that the prerogatives of the Gosplan be clearly defined so that it should become a fully fledged planning authority, empowered to assess productive capacities, manpower, and stocks of raw materials, to fix targets of production

---

[1] Trotsky, *Sochinenya*, vol. xv, pp. 215–32. Even then, however, Lenin wrote in a short and expressive note to Kzhizhanovsky: 'We are paupers. Starving, destitute paupers. A comprehensive . . . plan for us = "bureaucratic Utopia".' Lenin, *Sochinenya*, vol. xxxv, p. 405.

[2] *Pyat Let Sovietskoi Vlasti*, pp. 150–2.

for years ahead, and to ensure 'the necessary proportionality between various branches of the national economy'. As early as on 3 May 1921, Trotsky was already writing to Lenin: 'Unfortunately, our work continues to be carried out planlessly and without any understanding of the need for a plan. The State Planning Commission represents a more or less planned negation of the necessity to work out a practical and business-like economic plan for the immediate future.'[1] He found no response at the Politbureau. Lenin was against him. In accordance with classical Marxist theory, Lenin held that planning could be effective only in a highly developed and concentrated economy, not in a country with 20-odd million scattered small farms, a disintegrated industry, and barbarously primitive forms of private trade. It was not that Lenin denied the need for long-term development schemes. He himself jointly with Kzhizhanovsky had put forward a scheme for Russia's electrification and had introduced it with the famous dictum that 'Soviets plus electrification equal Socialism'. But he considered the idea of a 'comprehensive' plan, covering the whole of the nationalized industry, as premature and futile. Trotsky rejoined that even Lenin's electrification scheme was suspended in a void as long as it was not based on a comprehensive plan. How, he asked, could electrification be planned when the output of the industries which were to produce the power plants was not? He too was aware that under present conditions the type of planning which classical Marxist theory had expected was impracticable, because that theory presupposed a modern society with highly developed and fully socialized productive forces. But the comprehensive plan he asked for was to embrace only the state-owned industries, not the private sector; and it was not, he thought, too early to apply it. He saw a contradiction between the fact of state-ownership and the government's inclination to let sundry state-owned enterprises run their affairs in an unco-ordinated manner. National ownership, he argued, had transformed the whole of industry into a single concern which could not be run efficiently without a single plan.[2]

[1] Trotsky's letter to Lenin ('Po povodu knizhki I. Shatunovskovo') is in *The Archives*. See also *Leninskii Sbornik*, vol. xx, pp. 208–9. In a note to Zinoviev Lenin remarked: 'Trotsky is in a doubly aggressive mood.'

[2] Trotsky had argued on these lines even on the eve of NEP. See *Sochinenya*, vol. xv, pp. 215–32, 233–5.

This was a bold view at the time. Even bolder was the idea of 'primitive socialist accumulation' which Trotsky began to expound in 1922.[1] This was an adaptation of one of Marx's historical notions to the conditions of a socialist revolution in an under-developed country. Marx had described as the era of primitive accumulation the initial phase in the development of modern capitalism when normal accumulation of capital had hardly begun or was still too feeble to allow industry to expand from its own resources, that is from its own profits. The early bourgeoisie shrank from no violent, 'extra-economic' method in its striving to concentrate in its hands the means of production; and it went on using those methods until capitalist industry was strong and profitable enough to plough large profits back into production and to acquire a self-perpetuating and expanding basis within its own structure. Expropriation of the yeoman peasantry, plunder of colonies, piracy, and later also the underpayment of wages had been the main sources of that primitive accumulation, which in England, the classical country of capitalism, lasted over centuries. Only when this process was relatively far advanced did the era of normal accumulation set in and 'legitimate' profits formed the main, though not the only, basis for large-scale investment and continued industrialization.

What then was primitive socialist accumulation to be? Marxists had never imagined that socialism too might have to go through a phase of development comparable to the primitive accumulation of capitalism. They had always taken it for granted that a socialist economy would rise on the foundations of modern industrial wealth accumulated by bourgeois society and then nationalized. But there had not been enough of that wealth in Russia; and still less was left after the ravages of recent years. Having proclaimed socialism as their objective, the Bolsheviks now found that the material foundations for socialism were lacking in Russia. They had to lay these first. They had, Trotsky argued, to embark upon primitive accumulation which would differ from its predecessors in that it would be carried out on the basis of social ownership.

He had no intention of suggesting that a socialist government should or could adopt the 'bloody and disgraceful' methods of

[1] See his address at the fifth congress of the Communist Youth on 11 October 1922. *Sochinenya*, vol. xxi, pp. 294–317.

exploitation and plunder which Marx had associated with bourgeois primitive accumulation, or that socialism could come into the world as capitalism had come 'dripping from head to foot, from every pore, with blood and dirt'. But intensive and rapid capital formation was necessary. Soviet industry could not yet expand by the normal process of ploughing profits back into production. Most of it still worked at a loss; and even if it did not, it would still be incapable of producing surpluses large enough to sustain rapid industrialization, that *conditio sine qua non* of socialism. The nation's accumulation fund could be increased either at the expense of the earnings of private business and farming or of the nation's wages bill. Only some time later did Trotsky begin to urge heavier taxation on the N.E.P.-men and the wealthier peasants. At present, in 1922, he merely pointed out with much force that the economy was run and could be reconstructed and expanded only at the workers' expense. He said, for instance, at a congress of the Comsomol in October: 'We have taken over a ruined country. The proletariat, the ruling class in our state, is compelled to embark upon a phase which may be described as that of primitive socialist accumulation. We cannot content ourselves with using our pre-1914 industrial plant. This has been destroyed and must be reconstructed step by step by way of a colossal exertion on the part of our labour force.' And again: the working class 'can approach socialism only through the greatest sacrifices, by straining all its strength and giving its blood and nerves . . .'.[1]

His pleas at once aroused resistance. Men of the Workers' Opposition had already said that N.E.P. stood for the New Exploitation of the Proletariat; and the quip had become something of a slogan. Trotsky's argument came as if to illustrate the truth of the charge and to give point to it. Was he not in fact trying to argue the workers into submission to the new exploitation? He retorted that of exploitation it was proper to speak only when one social class was made to toil and slave for the benefit of another. He asked the workers to toil for their own benefit. At the worst, he said, he might be accused of trying to argue them into 'self-exploitation', for he called the workers to make 'sacrifices' and to give their 'blood and nerves' for their own proletarian state and their own socialist industry.[2]

[1] Loc. cit.   [2] Loc. cit.

This was not the first time that Trotsky rested his case on the identification of the working class with the state. In 1920 and 1921 he had argued in the same terms against the autonomy of the trade unions. The workers, he had said, had no interests of their own to defend against their own state. Lenin then replied that the proletarian state invoked by Trotsky was still an abstraction: it was not yet a workers' state proper, it often had to strike a balance between workers and peasants and, worse still, it was bureaucratically deformed. The workers were bound in duty to defend their state, but they should also defend themselves against it.[1] When Trotsky now again claimed that the interests of the working class and of its state were identical he laid himself open to the same criticism. Was it not in the name of an abstract idea that he urged the workers to shoulder the main burden of primitive socialist accumulation? Would not the bureaucracy and perhaps even the kulak and the N.E.P.-man be the chief beneficiaries? And how could primitive socialist accumulation be pursued if the working class refused to bear the brunt? These questions were to loom large in coming years. Immediately Trotsky replied that the policy he advocated could not and should not be imposed upon the workers—only with their consent could it be pursued. The chief difficulty was therefore of an 'educational character': the workers should be made aware of what was necessary and what was demanded of them, for without their willingness and socialist enthusiasm nothing could be achieved.[2] Once again he attempted to strike the heroic chord in the working class, as he had done, with overwhelming success, in 1919 when the White Armies threatened Moscow and Petrograd; and as he had again tried and utterly failed to do in the winter of 1920–1, before the Kronstadt revolt. It should be added that his advocacy of primitive socialist accumulation did not at this stage meet with objections within the Politbureau, although most of its members preferred not to compromise their popularity and face the workers with the frank demand for their 'blood and nerves'.

Such were the main economic ideas which Trotsky expounded in the early years of N.E.P., when he acted in effect as the forerunner of the Soviet planned economy. He was not their sole

[1] *10 Syezd RKP (b)*, pp. 208 ff.; *The Prophet Armed*, pp. 509–10.
[2] *Sochinenya*, vol. xxi, loc. cit.

originator. What he said represented the collective thought of a small circle of theorists and administrators who were close to Trotsky even though some of them did not approve of his disciplinarian attitudes. According to Trotsky himself, Vladimir Smirnov, the leader of the Decemists, who served on the Supreme Council of the National Economy, first coined the term 'primitive socialist accumulation'.[1] Evgenii Preobrazhensky must be regarded as the chief theorist of the idea: his work *The New Economics*, which appeared in 1925, is distinguished by a greater depth of strictly theoretical argument than can be found in Trotsky's writings; and he undoubtedly mooted his theses in 1922–3. Yuri Pyatakov, who was the moving spirit of the Council of the National Economy, and also argued for a single economic plan, was disturbed by the condition of heavy industry, and criticized the credit policy of the Commissariat of Finance and of the State Bank.[2] No doubt, Trotsky made borrowings from these men, and perhaps from others also. But they were too absorbed in theorizing or too immersed in administration to produce more than either abstract treatises or fragmentary empirical conclusions. Trotsky alone transformed their ideas and conclusions into a programme of policy which he defended before the Politbureau and expounded before a nation-wide audience.

Lenin continued to show little enthusiasm for the 'single plan' and for the 'enlargement of the Gosplan's powers'. He described his electrification plan as the 'only serious work on the question' and dismissed the 'idle chatter' about a 'comprehensive' plan. Stalin did likewise; and he did his utmost to widen the breach between Lenin and Trotsky.[3] The lesser leaders, Rykov and Sokolnikov, saw Trotsky's policy as an encroachment upon their own responsibilities. They were sceptical of planning; and they were opposed to investing the Gosplan with wide powers. In their own circle they commented—presently they were to make the charge in public—that Trotsky demanded such wide powers for the Gosplan because he hoped to assume its direction, and that having ceased to be the country's military

---

[1] *12 Syezd RKP (b)*, p. 321; E. A. Preobrazhensky, *Novaya Ekonomika*, vol. i, part 1, p. 57.   [2] *The Trotsky Archives.*

[3] See Stalin's *Sochinenya*, vol. v, pp. 50–51, where, in a letter to Lenin, he describes Trotsky's ideas on planning as those of 'a medieval artisan who imagines himself an Ibsenite hero destined to save Russia . . .'.

dictator he aspired to become its economic master. We do not know whether Trotsky did indeed wish to become the head of the Gosplan. Even if he did, the aspiration was hardly reprehensible. He criticized Kzhizhanovsky, the actual head of Gosplan, as inefficient;[1] but he never put forward his own candidature; and he argued the case on its merits. However, personal ambitions and departmental jealousies again and again intruded. Thus his opponents suggested that an enhanced Gosplan would compete with the Council of Labour and Defence over which Lenin presided, with Trotsky as his deputy. At a session of the Central Committee, on 7 August 1921, Trotsky replied that in his view the Council should remain in charge of high policy, but that Gosplan should translate that policy into specific economic plans and supervise their execution. He failed to carry the Central Committee with him.[2]

Parallel with these controversies there dragged on a conflict over the Rabkrin, the Workers' and Peasants Inspectorate. Stalin had been the chief of Rabkrin, from 1919 till the spring of 1922, when he was appointed Secretary General; but he exercised a strong influence on it even later. The Inspectorate had wide and manifold functions: it was entitled to audit the morals of the civil service; to inspect without warning the work of any Commissariat; to watch over the efficiency of the entire administration, and to prescribe measures for raising it. Lenin intended that Rabkrin should act as a sort of super-commissariat through which the administration, which was not controlled democratically, was to control itself and maintain stern self-discipline. In reality, Stalin transformed the Inspectorate into his private police within the government. As early as 1920 Trotsky attacked Rabkrin, claiming that its methods of inspection were muddled and ineffective, and that all that it did was to throw spanners into the machinery of government. 'You cannot', he said, 'create a special department endowed with all the wisdom of government and able to audit all the other departments. . . . In every branch of government it is well known that whenever the need arises for any change of policy or for any serious reform in organization it is useless to look to Rabkrin for guidance. Rabkrin itself provides striking illustration

---

[1] Lenin referred to this criticism in his letter to the Politbureau of 5 May 1922. See his *Sochinenya*, vol. xxxiii, pp. 316–18.   [2] *The Trotsky Archives*.

of the lack of correspondence between governmental decree and governmental machinery, and is itself becoming a powerful factor of muddle and wantonness.' In any case, what was needed in a body like Rabkrin was a 'broad horizon, a broad view on matters of the state and of the economy, a view much broader than that possessed by those carrying out this work'. He described Rabkrin as the refuge and haven for frustrated misfits who had been rejected by all other commissariats and were 'utterly cut off from any genuine, creative, and constructive work'. He did not even once mention Stalin in whom he saw the super-misfit risen to eminence.[1]

Lenin defended Stalin and Rabkrin. Exasperated by the inefficiency and the corruption of the civil service, he pinned great hopes on the Inspectorate and was irritated by what he considered to be Trotsky's private vendetta.[2] Trotsky argued that the muddle, at least in the economic departments, was the result of faulty organization which, in its turn, reflected the lack of any guiding principle in economic policy. Inspections by Rabkrin could not change this—the remedy could be found in planning and in a reformed Gosplan. Nor could incompetence be cured by the shock treatment and intimidation to which Stalin's commissariat subjected the civil service. In a backward country, with the worst traditions of uncivilized and corrupt government, Trotsky said, the main task was to educate the government personnel systematically and to train it in civilized methods of work.

All these differences considered, Trotsky's refusal to become vice-Premier is less surprising. He could not, without contradicting himself, accept a post in which he would have had to give effect to an economic policy which in his view lacked focus, and to guide an administrative machinery which he held to be faultily constructed. When, in the summer of 1922, Lenin urged him to use the post for a drive against bureaucratic abuses of power, he replied that the worst abuses had their source at the very top of the party hierarchy. He complained that the Politbureau and the Orgbureau meddled intolerably with the affairs of the government and took decisions concerning various commissariats without deigning to consult even the

---

[1] Trotsky, *Sochinenya*, vol. xv, p. 223.
[2] Lenin, *Sochinenya*, vol. xxxiii, loc. cit. and *passim*.

heads of those commissariats. It was therefore vain to struggle against wantonness in the administration as long as this evil flourished unopposed in the party.[1] Lenin did not take Trotsky's hint. He relied on Stalin as the party's General Secretary not less than he relied on him as chief of Rabkrin.

In the summer of 1922 further disagreement arose over the manner in which Moscow controlled the non-Russian republics and provinces of the Soviet Federation. The Bolsheviks had guaranteed to those republics the right of self-determination, which included expressly the right to secede from the Soviet Federation; the guarantee had been enshrined in the 1918 Constitution. At the same time they insisted on strictly centralized government and in practice overruled the autonomy of the non-Russian republics. Early in 1921, it will be remembered, Trotsky protested against the conquest of Georgia, of which Stalin had been the chief prompter. Then Trotsky reconciled himself to the accomplished fact and even defended the conquest in a special pamphlet.[2] Later still, in the spring of 1922, he remained silent when at the eleventh congress eminent Bolsheviks accused Lenin's government of forsaking the principle of self-determination and restoring the 'one and indivisible' Russia of old. Shortly thereafter, however, he himself voiced the same accusation behind the closed doors of the Politbureau; and it was again over Georgia and Stalin's activities there that the conflict came to a head.

As Commissar of Nationalities, Stalin had just ordered the suppression of the Menshevik party in Georgia. When leading Georgian Bolsheviks, Mdivani and Makharadze, protested against this he sought to intimidate them and to quell their protests.[3] His action was up to a point consistent with the general trend of Bolshevik policy, for if it was right to ban the Menshevik party in Moscow, there was no apparent reason why the same should not be done at Tiflis. Trotsky had endorsed the ban in Russia, but attacked its extension to Georgia. He pointed out that the Russian Mensheviks had, because of their counter-

---

[1] See Trotsky's letters to the Politbureau of 22 August 1922, and of 15, 20, and 25 January 1923 in *The Archives*. Also *Moya Zhizn*, vol. ii, p. 216.

[2] *The Prophet Armed*, pp. 474–5.

[3] Mdivani, Makharadze, Ordjonikidze, Yenukidze, Stalin, and Bukharin gave accounts of the conflict in *12 Syezd RKP (b)*, pp. 150–76, 540–65. See also Deutscher, *Stalin, a Political Biography*, pp. 236–46.

revolutionary attitude, discredited themselves while the Georgian ones still enjoyed strong popular support. This was true enough; but the argument might have carried conviction only if the Bolsheviks had still based their rule on proletarian democracy. It sounded somewhat hollow once the view was accepted, as it was by Trotsky, that the Bolsheviks were entitled, in the interest of the revolution, to maintain their political monopoly regardless of whether they did or did not enjoy popular support. It was only a step from the establishment of the single-party system to the persecution of those Georgian Bolsheviks who opposed it, although this was a step from consistency to absurdity. Stalin for the first time now applied repression to members of the Bolshevik party when he tried to intimidate Mdivani and Makharadze. He also gravely compromised the Bolshevik policy towards the non-Russian nationalities, the policy of which he himself had been an inspirer and in the broadmindedness of which the Bolsheviks had taken great pride.

Defending themselves, Mdivani and Makharadze turned against the ultra-centralistic principle of Stalin's policy. What right, they asked, had any Commissariat in Moscow to take decisions concerning the political life of Tiflis? Where was self-determination? Were not the small nationalities being forced back into the Russian empire, 'one and indivisible'? These were pertinent questions. All the more so as at the same time Stalin was preparing a new constitution which was to be much more centralistic than its 1918 predecessor and was to curtail and abrogate the rights of the non-Russian nationalities and to transform the Soviet *Federation* of republics into the Soviet *Union*. Against this constitution, too, the Georgians, the Ukrainians, and others raised protests.

When these protests came before the Politbureau Trotsky upheld them. He was now confirmed in the misgivings which had caused him to oppose in the first instance the annexation of Georgia. He saw in Stalin's behaviour a scandalous and flagrant abuse of power, which carried centralism to a dangerous excess, offended the dignity of the non-Russian nationalities, and suggested to them that 'self-determination' was a fraud. Stalin and Ordjonikidze prepared an indictment of Mdivani and Makharadze and alleged that these 'national deviationists' opposed the introduction of the Soviet currency in Georgia, refused to co-operate with neighbouring Caucasian republics and to share with them scarce provisions, and that they generally acted in a spirit of nationalist selfishness and to the detriment of the Soviet Federation as a whole. Such behaviour, if the charges had been true, could not be tolerated in party members. Trotsky did not believe the charges to be true. Lenin and most members of the Politbureau viewed the conflict as a family quarrel between two sets of Georgian Bolsheviks; and they thought that the most prudent course for the Politbureau was to accept Stalin's views. Stalin was the Politbureau's expert on these matters; and Lenin saw no reason to suspect that Stalin, of all men, the author of the celebrated treatise on *Marxism and the Nationalities*, the party's classical plea for self-determination, would malignantly offend the national dignity of his own countrymen. Again Trotsky appeared to Lenin to be acting from personal animosity or from that 'individualism' which had led him to oppose the Politbureau on so many other questions. One of Lenin's first moves after his return to office, in October 1922, was to rebuke Mdivani and Makharadze and uphold Stalin's authority.

. . . .

As we follow these dissensions in the Politbureau and consider Trotsky's part in them, we are struck by the change which had occurred in Trotsky himself in about a year. In the first half of 1922 Trotsky still spoke primarily as the Bolshevik disciplinarian; in the second he was already in conflict with the disciplinarians. The contrast shows itself in many of his attitudes; but it becomes most apparent when it is recalled that at the beginning of the year he indicted, on behalf of the Politbureau, the Workers' Opposition before the party and the International. Yet towards its end he himself appeared to air views hitherto voiced by that Opposition (and by the Decemists). It was the Workers' Opposition which had first confusedly expressed the discontent of the Bolshevik rank and file with N.E.P. and had spoken of the need to give the policies of N.E.P. a socialist perspective. It was the Workers' Opposition which had first attacked the new bureaucracy, protested against abuses of power, and denounced new privileges. It was that Opposition and the Decemists who had begun the revolt against the excessive powers of the party machine and had clamoured for

the restitution of inner-party democracy. Trotsky at first castigated them and warned them that Bolsheviks must under no
circumstances oppose themselves to the party leaders in terms
of 'we' and 'they'. Yet, in the course of 1922, he appeared to
have adopted most of their ideas and to have taken up an attitude from which he was bound to argue against the majority
of the Politbureau in terms of 'we' and 'they'. It looked, indeed,
as if in the process of taming the Workers' Opposition he had
been converted to its views and become its most eminent
recruit.

In truth he was grappling all this time with a dilemma
which occupied the party as whole—only he grappled with it
more intensely than others. It was the dilemma between authority
and freedom. Trotsky was almost equally sensitive to the claims
of both. As long as the revolution was struggling for bare survival, he put authority first. He centralized the Red Army,
militarized labour, strove to absorb the trade unions in the
state, preached the need for a strong but civilized bureaucracy,
overruled proletarian democracy, and helped to subdue inner-
party opposition. Yet even in this phase the socialist 'libertarian' was alive and awake in him; and through his sternest
calls for discipline there reverberated, like a counterpoint, a
powerful note of socialist freedom. In his most ruthless deeds
and most severe words there still glowed a warm humanity
which distinguished him from most other disciplinarians. In
the very first phase of the revolution he was already pointing
an accusatory finger at the 'new bureaucrat', uneducated,
suspicious, and arrogant, who was a baneful 'ballast', and a
'genuine menace to the cause of communist revolution', the
cause which would 'fully justify itself only when every toiling
man and woman feels that his or her life has become easier,
freer, cleaner, and more dignified'.[1]

The end of armed hostilities sharpened the tension between
authority and freedom within Bolshevism; and within Trotsky
too. The Workers' Opposition and the groups close to it represented a revulsion against authority. What set Trotsky against
them was his deep grasp of the realities of the situation. He
could not easily dismiss those claims of authority which were
rooted in realities. Nor could he keep his peace of mind when

[1] *The Prophet Armed*, p. 427.

he saw that freedom—socialist freedom—was being uprooted. He wrestled with a real dilemma, whereas the Workers' Opposition seized only one of its horns and clung to it. He sought to strike a balance between Bolshevik discipline and proletarian democracy; and the more the balance was tipped in favour of the former, the more was he inclined to uphold the latter. The decisive shifts which upset the balance occurred in the years 1921–3; and in these years he gradually came to put the claims of inner-party democracy against those of discipline.

Yet he did not become a mere 'libertarian', resentful at the encroachments of authority. He remained the Bolshevik *statesman*, as convinced as ever of the need for a centralized state and a strong party leadership, and as mindful as ever of their prerogatives. He attacked the abuse, not the principle, of those prerogatives. Through his most angry broadsides against bureaucracy and his most spirited pleas for inner-party democracy there would still resound a strong disciplinarian counterpoint. Conscious that 'bureaucracy represented a whole epoch, not yet closed, in mankind's development', and that its evils appeared 'in inverse proportion to the enlightenment, the cultural standards, and the political consciousness of the masses',[1] he was careful not to induce the illusion that it was possible to sweep away those evils at a stroke. As yet he did not even turn against bureaucracy at large—he rather appealed to its progressive and enlightened men against its backward and despotic elements, and hoped that the former, together with the advanced workers, would be able to curb, to re-educate, and if need be to eliminate the latter. He had indeed shifted his ground, come closer to the Workers' Opposition and kindred groups, and implicitly acknowledged the rational side of their revulsion against authority; but, unlike them, he was not carried away by the revulsion. He did not simply 'reject' bureaucracy. He still grappled with a real dilemma, but he did it in a different manner than before and from the opposite end.

It is for this reason that it is impossible to pinpoint the change in Trotsky's attitude and to define more precisely what brought it about and when it occurred. No single event brought it about; and there was no single moment at which it was brought about. The policy of the Politbureau drifted over many issues

---

[1] Trotsky, *Sochinenya*, vol. xv, pp. 218–21; *The Prophet Armed*, p. 503.

from a workers' democracy to the totalitarian state. Trotsky's ideas drifted with the drift of Bolshevik policy—but in the opposite direction. He began to protest against the excesses of centralism as these made themselves felt. He began to defend the rights of the small nations as these rights were being violated. He clashed with the party 'apparatus' as the apparatus grew independent of the party and subjected party and state to itself. Because the processes against which he reacted developed piecemeal and in an ambiguous manner, his reactions, too, were piecemeal and vague. At no point did he feel the need for any drastic revision of his views because what he said now, in his anti-bureaucratic phase, he had also said in his disciplinarian phase, although he had said it with less emphasis and in a different context. He passed from one phase into the other almost without noticing it.

Amid the drift of policies one relatively stable issue stood out —the rivalry between Stalin and Trotsky. It had, we remember, intruded itself even in the conduct of the civil war; and it had sprung from an almost instinctive antagonism of temperaments, backgrounds, political inclinations, and personal ambitions. In this rivalry Stalin played the active and offensive part—he was offended by the inferiority of the place he occupied. Only slowly did Trotsky take cognizance of the rivalry; and only reluctantly did he begin to react and to get involved in it. So far the rivalry had remained in the background, where Lenin's strong personality kept it; and it had not assumed any broader significance, for it was not yet identified with any clear conflict of policies and interests. In 1922 this identification began. As manager of the party machine, Stalin, supported for the time being by Lenin, came to represent authority at its extreme, to enforce its claims and to exact obedience. A deep conflict of policies and interests began to take shape, to absorb the personal antagonism, and even to focus itself on it, until the personal antagonism was thus at once overshadowed and yet magnified by the wider conflict.

.   .   .   .

An account of the disagreements in which Trotsky was opposed to Lenin, Stalin, and the majority of the Politbureau, may leave a one-sided impression of his actual position in the

Bolshevik leadership. The biographer is bound to throw into relief the events and situations out of which grew Trotsky's later struggles with Stalin and which were therefore of the greatest consequence to his fortunes. These events and situations did not, however, appear in the same bold relief to contemporaries. Nor were the discords related here of the greatest importance in determining Trotsky's place among the Bolshevik leaders, especially his relationship with Lenin. The controversies were confined to the Politbureau. The party and the country had no inkling of them. The public voice still coupled Trotsky's name with Lenin's; and in the eyes of the world he was one of the chief inspirers of Bolshevik policy. And in truth, his disagreements with Lenin did not, in the balance of their common work, outweigh their solid and close agreement on an incomparably wider range of domestic and foreign issues.

As Commissar of War Trotsky continued to enjoy Lenin's full support. Even after the civil war he had to contend with the 'military opposition' which had challenged his policy in earlier years. Tukhachevsky still sought to win the party's support for his pet idea of an International General Staff of the Red Army. Frunze and Voroshilov, encouraged by Zinoviev and Stalin, still tried to obtain official sanction for their conceptions of 'proletarian strategy' and for their 'offensive military doctrine'. These issues were important enough to be thrashed out at the eleventh congress at a special session held in secret.[1] Trotsky obtained the final disavowal of the demands of his opponents; and he was assisted by the fact that he had Lenin's authority behind him. Lenin had learned to value his military work so much that he accepted almost automatically his judgement in that field. A curious incident may be cited as illustration. After the Kronstadt rising Lenin suggested to Trotsky that the Baltic Navy be scuttled or 'closed down'. The sailors, he held, were unreliable; the navy was useless; it consumed coal, food, and clothing, of which the country was desperately short; and so its disbandment would be a pure gain. Trotsky objected. He was determined to preserve the navy and was confident that he could reorganize it and bring about a change in its morale. The matter was settled in a most informal manner, through

---

[1] Trotsky's speech made at that session is in *Kak Vooruzhalas Revolutsia*, vol. iii, book 2, pp. 244 ff. See *The Prophet Armed*, pp. 484–5.

little private notes which Trotsky and Lenin scribbled to each other during a session of the Politbureau. Lenin accepted Trotsky's assurances, and the navy was saved.[1]

Lenin also repeatedly indicated to the party and the International his regard for Trotsky as interpreter of Marxism; and he lent wholehearted support to the outstanding influence Trotsky exercised on Russia's cultural life. (This aspect of Trotsky's activity is discussed in a later chapter.) Both rejected the ambition of clamorous groups of writers and artists, especially of the *Proletkult*, to sponsor a 'proletarian culture' and 'proletarian literature'. In educational affairs which since the civil war both considered to be of paramount importance, and in all matters relating to the advocacy of Marxism, both counselled caution and tolerance; and both discouraged firmly the crudity of approach, the conceit, and the fanaticism, which influential party members began to exhibit.

Trotsky showed also a highly active and constant initiative in the conduct of foreign policy. Important issues of diplomacy were decided upon by a small committee which consisted of Lenin, Trotsky, and Kamenev, who invited Chicherin, the Foreign Commissar, and often also Radek, to take part in deliberations. The present efforts of Soviet diplomacy were directed towards the consolidation of peace and the establishment of relations with bourgeois Europe. Trotsky, we remember, had used all his influence to secure the final conclusion of peace with Poland in 1921, a peace for which Lenin had not been so eager. He had similarly exerted himself to obtain the Politbureau's consent to the demarcation of frontiers and to the conclusion of peace with the small Baltic republics.[2] As early as 1920 Trotsky had urged Lenin to conciliate Great Britain; but it was only some time later that this advice was acted upon. But his most important initiative in the diplomatic field came early in 1921, when he set afoot a number of bold and highly delicate moves which eventually led to the conclusion of the Rapallo Treaty with Germany, by far the most outstanding feat of Soviet diplomacy in the two decades that lay between the Brest Litovsk Treaty and the Soviet-German agreement of 1939.

---

[1] This happened at the session of 21 March 1921. *The Trotsky Archives*. Some months later Trotsky mentioned the incident in a public speech. *Kak Vooruzhalas Revolutsia*, vol. iii, book 1, p. 81.   [2] See *The Prophet Armed*, pp. 463–70.

As Commissar of War Trotsky was anxious to equip the Red Army with modern weapons. The Soviet armament industry, primitive and run down, could not supply them. Through his agents abroad he purchased munitions wherever he could, even as far as the United States. But the purchases were haphazard and the Red Army was dangerously dependent on foreign supplies. Trotsky was bent on building up with foreign assistance a modern armament industry in Russia. But where, the question arose, could such assistance be obtained? Which bourgeoisie would consent to help in the building up of the military power of a communist government? There was only one country to which he could turn with a prospect of success; and that was Germany. Under the Versailles Treaty Germany had been forbidden to manufacture munitions. Her armament factories, the most modern in Europe, stood idle. Could not their owners be tempted to supply equipment and technological advice, if the enterprise was made sufficiently attractive? At the beginning of 1921 Victor Kopp, the former Menshevik who had once worked for the Viennese *Pravda*, established, on Trotsky's behalf, secret contacts with the great concerns of Krupp, Blohm und Voss, and Albatross Werke. As early on as 7 April 1921 he reported that these concerns were prepared to co-operate and to supply equipment and technological assistance needed for the manufacture in Russia of planes, submarines, artillery, and other munitions. Throughout the year envoys travelled between Moscow and Berlin; and Trotsky kept Lenin and Chicherin informed about every phase. The Politbureau authorized him to pursue the negotiations in the strictest secrecy; and he held their threads in his hands during all these preliminaries to the Rapallo Treaty, until the moment came for the diplomats to act.[1]

As the negotiations proceeded, the scope of transactions widened. Not only the armament industries were idle in Germany. The old and splendid officers' corps was also unemployed. Its members were therefore glad to undertake to instruct Russian soldiers and airmen; and in exchange they were allowed to train secretly in Russia German military cadres, whom they could not train at home. Thus the groundwork was laid for that long co-operation between the Reichswehr and the Red Army

---

[1] Kopp's report and Trotsky's and Lenin's notes are in *The Trotsky Archives*.

which was to outlast Trotsky's tenure of office by a full decade and which contributed greatly to the modernization of the Soviet armed forces before the Second World War.

However, till the spring of 1922 all these moves were still tentative; and there was hesitation in both Moscow and Berlin, for here and there diplomacy still hoped for a *rapprochement* with the powers of the Entente at the forthcoming Genoa conference, the first international gathering to which both Germany and Soviet Russia, hitherto the outcasts of diplomacy, were invited. Only when these hopes failed was the Rapallo Treaty concluded. The Treaty was a 'sober and business-like' bargain rather than a genuine alliance. Anxious to obtain for themselves through give and take as much advantage as possible, the Bolsheviks were as a rule careful not to encourage revisionism and a movement of revenge in the Reich, although they themselves had, as a matter of principle, denounced the Versailles Treaty from the outset, when their government was not even recognized by Germany and when the memories of the Brest Litovsk *Diktat* were still fresh.

Trotsky in particular worked to prevent any entanglement of Soviet policy with German nationalism. After as before Rapallo he sought to improve Russia's relations with France. In the autumn of 1922 he received in the Kremlin Edouard Herriot, who, as leader of the *Cartel de Gauche*, was presently to become French Prime Minister. Herriot describes the visit in detail and recalls the strength of conviction with which Trotsky argued for an improvement of relations between their countries. He assured Herriot that it was only the Entente's blind hostility that had driven Russia to come to terms with Germany first at Brest Litovsk and then at Rapallo; and that the Rapallo Treaty contained no clauses directed against France. He evoked the Jacobin tradition of France and appealed to French statesmen and French opinion for a greater understanding of the Russian Revolution. As he spoke about the affinity of Jacobinism and Bolshevism, Herriot recalls, a detachment of Red Army men marched past singing the *Marseillaise* in French and through the open window the words *Nous saurons mourir pour la liberté* burst into the conference room.[1]

- - - - - - - - - -

[1] E. Herriot, *La Russie nouvelle*, pp. 157–8.

The importance which diplomacy had by now assumed in Soviet affairs was connected with the defeats of communism outside Russia. In Europe the tide of revolution had ebbed, and the Communist International had run aground. Its parties led only a minority of the European working class and they were not in a position to undertake with any chance of success a frontal attack on the bourgeois order. Yet, most Communist parties refused to acknowledge defeat and were inclined to rely on their own strength and to go on staging revolts and coups in the hope that if they tried persistently enough they would carry the majority of the workers with them. A reorientation of the International was overdue; and this was the joint work of Lenin and Trotsky. With regard to the International they acted in a close and intimate partnership which, as far as can be ascertained, was not even once disturbed by the slightest discord.[1]

Neither Trotsky nor Lenin had abandoned their fundamental belief that the October upheaval in Russia had opened an era of international proletarian revolution; and Trotsky was to cling to this conviction throughout the next two decades, to the end of his life. But he now came to realize that the class struggle outside Russia was more complicated and protracted than he and others had at first imagined. He no longer took its outcome for granted; and he was anxious to dispel complacency about this and 'ultra-left' illusions in the International. Thus, in July 1921, he made a striking criticism of those Communists who held that the advent of socialism was 'inevitable'.[2] Such a belief in the predetermined progress of society, he said, was based on a 'mechanistic' misinterpretation of the Marxist approach to history.

Mankind has not always and not invariably moved upwards. . . . It has known in history long spells of stagnation. It has known relapses into barbarism. There have been instances . . . when society, having reached a certain height of development, was incapable of maintaining itself on that height. . . . Mankind can never be at a standstill. Any equilibrium, which it may attain in consequence of struggles between classes and nations, is unstable by its very nature. A society which does not rise must decline. A society from which no

---

[1] Lenin and Trotsky were the only two Soviet leaders elected Honorary Presidents at the third congress of the International. *Tretii Vsemirnyi Kongress Kominterna*, p. 16.　　　　[2] *Pyat Let Kominterna*, pp. 266–305.

class emerges capable of securing its ascendancy disintegrates. The road is then open to barbarism.

Such had been the main cause of the breakdown of the antique civilizations: the upper classes of Rome and Greece had decayed; and the exploited classes, the slaves, had been inherently incapable of revolutionary action and political leadership. This was a warning for our age. The decay of the bourgeois order was undeniable. True, American capitalism was still a dynamic and expansive force, although even in the United States socialism could already develop the nation's resources more rationally and with greater benefit to society than capitalism did. But European capitalism was historically at the end of its tether. It did not significantly develop its productive forces. It had no progressive role to play. It could open up no new vistas. If this were not so all thought of proletarian revolution in our time would be quixotic. But although European capitalism was decaying, the bourgeois order did not and would not collapse by itself. It had to be overthrown, and only the working class could overthrow it in revolutionary action. If the working class were to fail in this, then Oswald Spengler's gloomy prediction of the *Untergang des Abendlandes* would come true. History confronted the workers with a challenge as if saying to them: 'You ought to know that if you do not overthrow the bourgeoisie, you will perish under the ruins of civilization. Try and carry out your task!'[1]

Meanwhile European capitalism had withstood the shocks of world war and post-war crises. The possessing classes of Western Europe had learned their lessons from the Russian revolution: they did not allow themselves to be taken by surprise as Tsardom did; and they mobilized all their resources and strategic ideas. The appearance of fascism—Trotsky said this in 1922, the year of Mussolini's march on Rome—was a symptom of that mobilization; and there was the danger, he added, that 'a German Mussolini' might also rise to power.[2]

All this was of grave omen to the further course of socialist revolution. The whole development, with its peculiar sequence of phases unforeseen by earlier Marxists, might put socialism at a disadvantage. Proletarian revolution would have produced

---

[1] Loc. cit.   [2] Op. cit., p. 563.

the best results if it had occurred first in the United States or, as a second best, in Britain, against the background of highly developed productive resources. Instead the revolution had won in Russia, where it found only limited possibilities to demonstrate its advantages. It would find itself placed under even worse handicaps in the countries of Asia and Africa which were more backward than Russia. This led Trotsky to make the melancholy remark that '*History seems to be unwinding her skein from the other end*', that is from the countries which are least mature.[1]

He did not cease to hope that 'the skein' would still unwind from the Western, the European, end as well. The delays of revolution, the mobilization of counter-revolution, the prospect of a stalemate in the class struggle, and of the decadence of European civilization were for him not certainties to be accepted fatalistically, but dangers to be acted against and averted. The chances were still overwhelmingly in favour of revolution; but much depended on the attitude of the Communist parties. It was their duty to lead European society out of the impasse. They had to struggle for leadership. They could succeed in this only if they became militant and conscious parties, versed in the strategy and tactics of revolution, and accustomed to concert their efforts under strict international discipline. They were bound to fail if they remained only a radical variety of the old Social Democratic parties, if they cherished illusions about bourgeois parliamentarianism, and if they worked only within the framework of their national politics. But they would fail just as surely if, reacting against the Social Democratic tradition, they were to become narrow, self-centred sects, rigid in outlook and tactics; if they contented themselves with purely negative and arid boycott of the institutions of bourgeois society instead of promoting the revolutionary idea even from within those institutions; and if they went on trying to storm the bastions of capitalism without paying due regard to the circumstances and the balance of forces.

The Communist parties were not immediately confronted with revolutionary opportunities. Their job was to gather strength and to win over the majority of the workers without whose support no revolution could ever succeed.

Together with Lenin, Trotsky worked out the tactics of the

[1] Op. cit., pp. 429–30.

'united front'.[1] The gist of it was this: the Communist parties, still too weak to overthrow the established order, should be the most active participants in the 'day-to-day' struggles of the workers for higher wages, shorter hours, and democratic freedoms. They should not change the idea of socialism into the small coin of trade unionism and parliamentary reform, but carry into the struggle for 'partial demands' their own revolutionary spirit and purpose. They should make the workers realize how tenuous were all the gains they could win under capitalism and so rally them, even through the fight for such gains, for the last battle. The Social Democrats directed the struggle for 'partial demands' in such a way as to contain the workers' militant energy within the framework of capitalism; and they used reform as the diversion from revolution. The Communists, on the contrary, should use it as the spring-board of revolution.

But since Communists had to fight for partial gains and reforms they had some common ground, however narrow, with the Social Democrats and the moderate trade unionists. They should try to concert action with them within a united front. This should remove at least one dangerous consequence of the fundamental and irremediable cleavage between reformism and communism: it should overcome the division of the working class and prevent the dispersal of its energies. Marching separately, Communists and reformists should strike jointly at the bourgeoisie whenever they were threatened by it or could wrest concessions from it. Common action should extend to parliaments and elections, in which Communists must be prepared to support Social Democrats. But the main arena of the united front lay outside parliaments, in trade unions, in industry, and 'in the street'. The Communists had to pursue a double objective: they should seek to secure the immediate success of the united front; and at the same time assert their own viewpoint within the united front in order to wean Social Democratic workers from reformist habits of mind and to develop in them a revolutionary consciousness.

Lenin had expounded these ideas as early as 1920 in the

---

[1] Trotsky presented the 'Report on the World Crisis and the Tasks of the International' at the second session of the congress on 23 June 1921. Radek presented the 'Report on Tactics', in place of Zinoviev who was inclined towards the 'ultra-left' opposition. *Tretii Vsemirnyi Kongress Kominterna.*

Infantile Disease of 'Leftishness' in Communism, where he dwelt on the harm done to communism by unreasoning ultra-radical sectarians. The need for a firm and formal disavowal of 'ultra-radicalism' became pressing after the German March rising of 1921. It was then that Lenin placed proposals for the united front before the Executive of the International. He met with strong opposition from Zinoviev, Bukharin, Bela Kun, and others. For a moment it seemed that the ultra-radicals would prevail. It was only after animated debates in the course of which Lenin and Trotsky jointly faced the opposition that the Executive was persuaded to authorize the policy of the united front and to instruct both Lenin and Trotsky to expound it at the forthcoming congress of the International.[1]

At the congress, in July 1921, the ultra-radicals made a stand. They exercised a strong influence on the German, Italian, and Dutch parties and they drew their strength from a powerful emotional current in the whole International. The Communist parties had come into existence in a desperate struggle against the leaders of the old Socialist parties whom they blamed for supporting the 'imperialist slaughter' of 1914–18, for the subsequent suppression of revolution in Europe, for the assassination of Rosa Luxemburg and Karl Liebknecht, and for an ambiguous attitude towards European intervention in Russia. No wonder that many Communists were bewildered and indignant when they now heard Lenin and Trotsky urging them to acknowledge defeat, be it temporarily, and to co-operate with the hated 'social imperialists' and 'social traitors'. This to the ultra-radicals was surrender or even betrayal. At the congress, as earlier on the Executive, Trotsky and Lenin had to use all their influence and eloquence to prevent the opposition from

---

[1] Alfred Rosmer gives an informative account of these days in *Moscou sous Lenine*, pp. 172–88. Radek, *Pyat Let Kominterna*, vol. ii, preface. At the Executive Lenin made a speech in which he declared his full solidarity with Trotsky and strongly attacked Bela Kun, the spokesman of the ultra-left, repeatedly describing Kun as a 'fool'. The full text of the speech, which I read many years ago, was not available at the time of writing. Trotsky published excerpts from it in his *Bulleten Oppozitsii* (December 1932). Lenin said: 'I have come here in order to protest against Bela Kun's speech who came out against comrade Trotsky instead of defending him—which he should have done if he had wished to act as a genuine Marxist. . . . Comrade Trotsky was a thousand times right. . . . I have considered it my duty to support in all essentials everything comrade Trotsky has said. . . .' Lenin backed Trotsky also against Cachin and Frossard who at the congress represented the extreme right wing. (Ibid.)

gaining the upper hand—they even threatened to split the International if it backed the ultra-radicals.

The congress voted for the united front. But it did so with mental reservations and without a clear grasp of the issues involved. Lenin and Trotsky had set the Communist parties the dual task of fighting arm in arm with the reformists against the bourgeoisie and of wresting from the reformists influence over the working class. The idea of the united front embodied the whole tactical experience of the Bolsheviks who had indeed fought first against Tsardom, then the Cadets, and then Kornilov, in a sort of a united front with the Mensheviks and Social Revolutionaries until, in the end, they gained ascendancy over the latter too. The Bolshevik success was secured not merely by the resourcefulness of the Bolshevik leaders, but by the breakdown of a whole social order and by the subsequent shift *from right to left* typical of all classical revolution. Could such tactics, even if no other tactics were, from the communist viewpoint, realistic, be applied outside Russia with comparable chance of success? In Europe the old order had regained a measure of stability which produced a confused but distinct shift *from left to right*. This alone tended to secure the ascendancy of the reformists within any united front. Nor was there among European Communists a single leader with a mastery of tactics comparable to Lenin's or Trotsky's. And so the European Communists were to prove incapable of applying the united front in both its aspects. Some took to heart their duty to co-operate in all earnestness with the Social Democrats. Others were above all eager to discredit the Social Democrats. Some saw the united front as a serious endeavour to unify the working class in the struggle for partial demands. Others saw it merely as a clever trick. Still others wavered between the opposed views. And so the International began to split into right and left wings and intermediate and extreme groups, 'centrists' and 'ultra-lefts'.

At the congress Trotsky and Lenin contended mainly with opposition from the ultra-radicals; and so at times they appeared to encourage the right wing. Trotsky in particular spoke scathingly and disdainfully about the ultra-radicals, for instance, Arkadi Maslov and Ruth Fischer, the leaders of the communist organization of Berlin, describing them as empty-headed emotionalists, who had little in common with Marxism and might be expected to switch over to the most unprincipled opportunism.[1] He was enthusiastically applauded by all the moderate elements at the congress; and the applause rose to an ovation when on behalf of the majority of the delegates Klara Zetkin, the famous veteran of German communism, paid him a solemn and stirring tribute.[2]

At the next, the fourth, congress Lenin, already ill, spoke only briefly and with great difficulty; and Trotsky came to the fore as the chief expounder of the International's strategy and tactics. He advocated once again the united front. He went a step farther and urged the Communist parties to support, on conditions, Social Democratic governments and even, under special circumstances, in pre-revolutionary situations, when such coalitions could pave the way for proletarian dictatorship, to participate in them.[3] The opposition was outraged. From the first day of its existence the International had declared it an axiom of its policy that a Communist party must never enter any coalition government: its task was to destroy the bourgeois state machine, not to try and capture it from within. However, the congress accepted the tactical innovation; and the Communist parties were instructed to watch for opportunities to form government coalitions with the Social Democrats. This decision was to assume a crucial importance in the crisis of German communism in the autumn of 1923.

Such were the tactical efforts through which Trotsky (and Lenin) still hoped to 'unwind the skein of revolution' from its 'proper', that is from its European, end.

.   .   .   .

Throughout the summer of 1922 the disagreements in the Politbureau over domestic issues dragged on inconclusively. The dissension between Lenin and Trotsky persisted. On 11 September from his retreat in Gorki, outside Moscow, Lenin made contact with Stalin and asked him to place before the Politbureau once again and with the utmost urgency a motion proposing Trotsky's appointment as deputy Premier. Stalin communicated the motion by telephone to those members and

[1] Trotsky, *Pyat Let Kominterna*, pp. 288 ff.
[2] *Tretii Vsemirnyi Kongres Kominterna*, p. 58.
[3] See Trotsky's report on the fourth congress in his *Pyat Let Kominterna*.

alternate members of the Politbureau who were present in Moscow. He himself and Rykov voted for the appointment; Kalinin declared that he had no objection, while Tomsky and Kamenev abstained. No one voted against. Trotsky once again refused the post.[1] Since Lenin had insisted that the appointment was urgent because Rykov was about to take leave, Trotsky replied that he, too, was on the point of taking his holiday and that his hands were, anyhow, full of work for the forthcoming congress of the International. These were irrelevant excuses, because Lenin had not intended the appointment to be only a stopgap for the holiday season. Without waiting for the Politbureau's decision, Trotsky left Moscow. On 14 September the Politbureau met and Stalin put before it a resolution which was highly damaging to Trotsky; it censured him in effect for dereliction of duty.[2] The circumstances of the case indicate that Lenin must have prompted Stalin to frame this resolution or that Stalin at least had his consent for it.

Less than a month later an unexpected event put an end to the sparrings between Lenin and Trotsky. At the beginning of October the Central Committee adopted certain decisions concerning the monopoly of foreign trade. The Soviet government had reserved for itself the exclusive right to engage in trade with foreign countries; and it had centralized all foreign commercial transactions. This was a decisive measure of 'socialist protectionism'—the term was coined by Trotsky[3]—designed to defend the weak Soviet economy from hostile pressures and unpredictable fluctuations of the world market. The monopoly also prevented private business from overlapping into foreign trade, exporting essential goods, importing inessential ones, and disrupting even further the country's economic balance. The new decisions of the Central Committee, taken in Trotsky's and Lenin's absence, did not go so far as to admit private business to foreign trade; but they did loosen central control over Soviet trade agencies abroad. This might have enabled individual state concerns working on foreign markets to act independently with a view primarily to their sectional advantage and thus to make a breach in 'socialist protectionism'. In time private business might have benefited from the breach.[4]

---

[1] *The Trotsky Archives.*   [2] Ibid.   [3] Preobrazhensky, op. cit., p. 79.
[4] Lenin, *Sochinenya*, vol. xxxiii, pp. 338–40.

Lenin at once objected to the decision, describing it as a grave threat to the Soviet economy. He was alarmed, irritated, and—paralysed. In brief moments, snatched from his doctors and nurses, he dictated notes and memoranda, protests, and exhortations; but he could not intervene personally with the Central Committee. Then, to his relief, he learned that Trotsky adopted a view identical with his. In the course of nearly two months the issue hung in the balance. On 13 December Lenin wrote to Trotsky: 'I earnestly beg you to take it upon yourself to defend at the forthcoming plenary session [of the Central Committee] our common view about the imperative need to preserve and reinforce the monopoly of foreign trade.' Trotsky readily agreed. But having repeatedly warned Lenin and the Politbureau that their policy encouraged the administration to submit passively to the uncontrolled forces of the market economy, he pointed out that the Central Committee's latest decision showed that his warnings had been all too justified. Once again he urged the need for co-ordination and planning and for vesting wide powers in Gosplan. Lenin still tried to shelve the issue of Gosplan and entreated Trotsky to concentrate on the trade monopoly. 'I think we have arrived at a full agreement', he wrote to Trotsky again, 'and I am asking you to announce our solidarity at the plenary session.' Should they both be outvoted there, then Trotsky should announce that they would both go to any length to nullify the vote: they would both attack the Central Committee in public.[1]

There was no need for them to resort to such drastic action. Contrary to Lenin's fears, when the Central Committee came to review the issue in the second half of December, Trotsky easily persuaded it to reverse its decision. Lenin was all exultation. 'We have captured the position without firing a shot . . .', he commented to Trotsky in a note written 'with Professor Forster's permission',[2] 'I propose that we do not stop but press on with the attack. . . .'[3]

[1] See the correspondence between Lenin and Trotsky of 12–27 December 1922 in *The Trotsky Archives*, and Trotsky, *The Stalin School of Falsification*, pp. 58–63.

[2] Professor Forster was one of Lenin's doctors.

[3] I wrote the first two chapters of this volume in 1954, basing the documentation largely on *The Trotsky Archives*. Only two years later, after Khrushchev's disclosures at the twentieth congress of the Soviet Communist party, were some of these important documents published in Moscow for the first time; and they have

The incident brought the two men closer together than they had been for some time. In the next few days Lenin reflected further over the criticisms of economic policy Trotsky had made in the last two years. He communicated the result of his reflections to the Politbureau in a letter of 27 December:

> Comrade Trotsky, it seems, advanced this idea [about the Gosplan's prerogatives] long ago. I opposed it . . . but having attentively reconsidered it I find that there is an essential and sound idea here: Gosplan does stand somewhat apart from our legislative institutions . . . although it possesses the best possible data for a correct judgment of [economic] matters. . . . In this, I think, one could and should go some way to meet Comrade Trotsky. . . .[1]

He realized that this would be a disappointment to members of the Politbureau—hence his apologetic undertone. The Politbureau was indeed annoyed by his sudden conversion and, despite Trotsky's protests, it resolved not to publish Lenin's remarks.[2]

In the last weeks and days of the year Lenin went a very long way to 'meet Comrade Trotsky' on further issues which had separated them. At the beginning of December he once again urged Trotsky to accept the post of vice-Premier.[3] This time he did it in a private talk, not amid the formalities of Politbureau proceedings. The question of the succession was already uppermost in his mind—presently he was to write his will. But he gave no hint of this to Trotsky. Instead he spoke in a tone of grave anxiety about the abuses of power which he saw were getting worse and worse and about the need to curb them. Trotsky did not this time reject the offer outright. He repeated that a drive against bureaucratic abuses in the government would yield little or no result as long as such abuses were

since been included in a special volume (vol. xxxvi) added to the fourth edition of Lenin's *Works*. Comparing the texts I have not had the need to alter a single comma in the quotations taken from *The Trotsky Archives*. Even now, however, only a fraction of the Lenin correspondence which these *Archives* contain, not to speak of other documents, has been published.

[1] *The Archives*; Lenin, *Sochinenya*, vol. xxxvi, pp. 548–9. Lenin in fact completely accepted Trotsky's basic idea, but not his allegation about Kzhizhanovsky's incompetence as head of Gosplan.

[2] *The Archives*. Stalin observed evasively: 'I suppose there is no necessity to print this, especially as we do not have Lenin's authorization.'

[3] Trotsky, *Moya Zhizn*, vol. ii, pp. 215–17.

tolerated in the party's leading bodies. Lenin replied that he was ready for a 'bloc' with Trotsky, that is for joint action against bureaucracy in the party as well as in the state. There was no need for either of them to mention names. Such action could be directed only against Stalin. They did not have the time to pursue the matter and to discuss any plan of action. A few days afterwards Lenin suffered another stroke.

In their last conversation Lenin gave Trotsky no indication that he had also pondered anew the other major issue over which they had differed: Stalin's policy in Georgia. On this, too, he was at last going to 'meet Comrade Trotsky'. He was in the mood of a man who, with one foot in the grave, uneasily looks back on his life's work and is seized by a poignant awareness of its flaws. Some months earlier, at the eleventh congress, he said that often he had the uncanny sensation which a driver has when he suddenly feels that his vehicle is not moving in the direction in which he steers it. Powerful forces diverted the Soviet state from its proper road: the semi-barbarous peasant individualism of Russia, pressure from capitalist surroundings, and above all, the deep-seated native traditions of uncivilized absolutist government.[1] After every spell of illness, when he returned to watch anew the movements of the state machine, Lenin's alarm grew; and with pathetic determination he struggled to grip the steering wheel in his paralysed hands.

The 'vehicle', he discovered, had run into the rut—oh, how familiar—of Great Russian chauvinism. In the second half of December he re-examined the circumstances of the conflict with the Georgian Bolsheviks, the conflict in which he had sided with Stalin. He carefully collected, sifted, and collated the facts. He learned about the brutality with which Stalin and Ordjoni-kidze, Stalin's subordinate, had behaved in Tiflis; he found that the accusations they levelled against the Georgian 'deviationists' were false, and he grew angry with himself for having allowed Stalin to abuse his confidence and to cloud his judgement.

In this mood, on 23 and 25 December, Lenin dictated that letter to his followers which became in effect his last will and testament. He intended to offer the party guidance about those who would presently be called upon to lead it. He characterized briefly the men of the leading team, so that the party should

[1] Lenin, *Sochinenya*, vol. xxxiii, pp. 235–76.

know which, in his view, were the merits and the faults of each. He contained his emotion and weighed his words so as to convey a judgement based on observation of many years and not a view formed on the spur of the moment.

The party, he wrote, should beware of the danger of a split in which Stalin and Trotsky 'the two most eminent leaders of the present Central Committee', would confront each other as the chief antagonists. Their antagonism reflected as yet no basic conflict of class interest or principle: it still was, he suggested, merely a clash of personalities. Trotsky was 'the most able' of all the party leaders; but he was possessed of 'excessive self-confidence', a 'disposition to be too much attracted by the purely administrative aspect of affairs', and an inclination to oppose himself individualistically to the Central Committee. In a Bolshevik leader these were, of course, important faults, impairing his capacity for teamwork and his judgement. Yet, Lenin added, the party ought not to hold against Trotsky his pre-revolutionary disagreements with Bolshevism. The warning implied that the disagreements had long since been lived down; but Lenin was aware that this was not necessarily the view taken by his disciples.

About Stalin he had only this to say: 'Having become General Secretary, Stalin has concentrated immeasurable power in his hands; and I am not sure that he will always know how to use that power with sufficient caution.' The warning was suggestive but inconclusive. Lenin refrained from offering explicit advice and stating personal preferences. He seemed to place somewhat stronger emphasis on Trotsky's faults than on Stalin's, if only because with Trotsky's qualities he dealt in greater detail. Soon, however, he had afterthoughts; and on 4 January 1923 he wrote that brief and pregnant postscript in which he stated that Stalin's rudeness was already 'becoming unbearable in the office of the General Secretary' and in which he advised his followers to 'remove Stalin' from that office and to appoint to it 'another man . . . more patient, more loyal, more polite, more attentive to comrades, less capricious, etc.'. If this were not to be done, the conflict between Stalin and Trotsky would grow more bitter with dangerous consequences to the party as a whole.[1] Lenin had no doubt that his advice to 'remove' Stalin could only establish Trotsky in the leadership.

[1] Lenin, *Sochinenya*, vol. xxxvi, pp. 545–6.

The understatements of the will and even of the postscript give no idea of the full force of Lenin's fresh fury against Stalin and of his fixed resolve to discredit him once and for all. It was between 25 December and 1 January that Lenin formed this resolution. The Congress of the Soviets had just assembled at which Stalin proclaimed the *Union* of Soviet Socialist Republics in place of the *Federation* established under the 1918 constitution.[1] Having supported this constitutional change, Lenin now suspected that it would do away completely with the autonomy of the non-Russian Republics and indeed re-establish Russia 'one and indivisible'. He formed the opinion that Stalin had used the need for centralized government to screen the oppression of the small nationalities. The suspicion hardened into a certainty when Lenin had a new insight into Stalin's character: he saw him as churlish, sly, and false. On 30 December, just when Stalin was proclaiming the Union, Lenin, cheating once again his doctors and his health, began to dictate a series of notes about policy towards the small nations. This was in effect his last message on the subject; and it was full of heart-searching, passionate remorse, and holy anger.[2]

He wrote that he felt 'strongly guilty before the workers of Russia for not having intervened vigorously and drastically enough in this notorious issue . . .'. Illness had prevented him from doing so, even though he had confided his fears and doubts to Zinoviev. But only now, after he had heard Dzerzhinsky's report on Georgia, did it become quite clear to him 'in what sort of a swamp' the party had landed. All that had happened in Georgia and elsewhere was being justified on the ground that the government must possess a single and integrated administrative machine or 'apparatus'. 'Where do such statements emanate from?' Lenin asked. 'Do they not come from that same Russian apparatus . . . [we had] borrowed from Tsardom and only just covered with a Soviet veneer?' To the small nations 'freedom of secession' from the Union was becoming an empty promise. They were in fact exposed to 'the irruption of that truly Russian man, the Great Russian chauvinist, who is essentially a scoundrel and an oppressor as is the typical

[1] Stalin, *Sochinenya*, vol. v, pp. 145–59.
[2] Lenin, op. cit., pp. 553–9. See also L. A. Fotieva's memoirs published in *Voprosy Istorii KPSS*, Nr. 4, 1957.

Russian bureaucrat'. It was high time to defend the non-Russian nationalities from that 'truly Russian *dzerzhymorda* [the great brutish bully of Gogol's satire]. . . . The rashness of Stalin's administrative zeal and his spite have played a fatal role. I fear that Dzerzhinsky too . . . has distinguished himself by his truly Russian state of mind (it is well known that Russified aliens are always much more Russian than the Russians themselves).'

On New Year's Eve Lenin continued:

> . . . internationalism on the part of a . . . so called great nation (great only through its acts of oppression, great only in the sense in which the bully may claim to be great)—internationalism on the part of such a nation should consist not merely in respecting formal equality between nations. It is necessary to create such [real] equality as would reduce . . . the actual inequality which arises in life. The Georgian who treats this aspect of the matter with contempt and charges others with being 'social chauvinists' (that Georgian who himself is not merely a genuine social-chauvinist but a coarse brutish bully on behalf of a Great Power) that Georgian is offending against the interests of proletarian class solidarity. . . . Nothing hampers the growth and consolidation of such solidarity as much as does injustice towards smaller nationalities. . . . That is why it is better to show too much conciliation and softness towards national minorities, rather than too little.

The rights of the Georgians, Ukrainians, and others were more important than the need for administrative centralization which Stalin evoked in order to justify 'a quasi-imperialist attitude towards oppressed nationalities'. If need be, Lenin concluded, the new constitution sponsored by Stalin, together with the new centralistic organization of government, would have to be scrapped altogether.

Having expressed himself with so much anguish and merciless bluntness, Lenin apparently intended to turn the matter over in his thoughts and to consider what course of action to take. For over two months he did not communicate his notes to any member of the Politbureau.

.   .   .   .

The upheaval in Lenin's mind which caused him to reverse so many of his crucial policies may appear even more startling and more sudden than the change which had occurred in

Trotsky in 1921 and 1922. It, too, resulted from the intense conflict between the dream and the power of the revolution, a conflict which was going on in Lenin's mind, and not only in his. In its dream the Bolshevik party saw itself as a disciplined yet inwardly free and dedicated body of revolutionaries, immune from corruption by power. It saw itself committed to observe proletarian democracy and to respect the freedom of the small nations, for without this there could be no genuine advance to socialism. In pursuit of their dream the Bolsheviks had built up an immense and centralized machine of power to which they then gradually surrendered more and more of their dream: proletarian democracy, the rights of the small nations, and finally their own freedom. They could not dispense with power if they were to strive for the fulfilment of their ideals; but now their power came to oppress and overshadow their ideals. The gravest dilemmas arose; and also a deep cleavage between those who clung to the dream and those who clung to the power.

The cleavage was not clear-cut, because dream and power were up to a point inseparable. It was from attachment to the revolution that the Bolsheviks had mounted and operated the machine of power, which now functioned according to its own laws and by its own momentum, and which demanded from them all their attachment. Consequently those who clung to the dream were by no means inclined to smash the machine of power; and those who identified themselves with power did not altogether abandon the dream. The same men who at one moment stood for one aspect of Bolshevism at the next moment rushed to embrace its opposite aspect. Nobody had in 1920-1 gone farther than Trotsky in demanding that every interest and aspiration should be wholly subordinated to the 'iron dictatorship'. Yet he was the first of the Bolshevik chiefs to turn against the machine of that dictatorship when it began to devour the dream. When subsequently Trotsky became involved in the struggle over the succession to Lenin, many of those who heard him invoke the revolution's ideals doubted his sincerity, and wondered whether he did not use them merely as pretexts in the contest for power. Lenin stood above any such suspicion. He was the party's undisputed leader; and he had and could have had no ulterior motive when in the last weeks of his activity he confessed with a sense of guilt that he had not sufficiently resisted the new oppression of the weak by the strong, and when he used his last ounce of strength to strike a blow at the over-centralized machine of power. He invoked the purpose of the revolution for its own sake, from a deep, disinterested, and remorseful devotion to it. And when at last, a dying man, his mind ablaze, he moved to retrieve the revolution from its heavy encumbrance, it was to Trotsky that he turned as his ally.

CHAPTER II

The Anathema

FROM the beginning of the civil war the Politbureau acted as the party's brain and supreme authority although the party statutes contained no provision even for its existence. The annual congresses elected only a Central Committee which was endowed with the widest powers of determining policy and managing the organization and was accountable to the next congress. The Central Committee elected the Politbureau. At first, the Politbureau was to take decisions only on urgent matters arising during the weekly or fortnightly intervals between the sessions of the Central Committee. Then, as the scope of the affairs with which that Committee had to deal widened, including more and more of the business of government, and as the members of the Committee became increasingly absorbed in manifold departmental responsibilities and were often absent from Moscow, the Central Committee gradually and informally delegated some of its prerogatives to the Politbureau. The Central Committee once consisted only of a dozen or so members; but then it became too big and cumbersome to act effectively. In 1922 it met only once in two months, while the members of the Politbureau worked in close day-to-day contact. In their work they adhered strictly to democratic procedure. Where differences of opinion were marked, they decided by a simple majority. It was within this framework, as *primus inter pares*, that Lenin exercised supreme power.[1]

From December 1922 the problem of the succession to Lenin was uppermost in the Politbureau's mind. Yet in principle the problem could not even exist. With or without Lenin it was the Politbureau as a body (and through it the Central Committee) which was supposed to rule the party; and the will of the Politbureau was what its majority willed. The question therefore became not who would succeed Lenin, but how the alignments in the Politbureau would shape themselves without Lenin, and

[1] *KPSS v Rezolutsyakh*, vol. i, pp. 525, 576–7, 657–8.

what sort of a majority would form itself to provide stable leadership. Stability of leadership had so far rested, at least in part,
on Lenin's unchallenged authority and on his powers of persuasion and tactical skill, which, as a rule, allowed him to
secure in each matter as it arose majority votes for his proposals.
Lenin had no need to form for this purpose any special faction
of his own within the Politbureau. The change which occurred
either in December 1922, or in January 1923 when Lenin
finally ceased to take part in the Politbureau's work, was the
creation of a special faction the sole purpose of which was
to prevent Trotsky from having a majority which would enable him to take Lenin's place. That faction was the triumvirate of Stalin, Zinoviev, and Kamenev.

The motives which prompted Stalin to set his face against
Trotsky are clear enough. Their antagonism dated back to the
early Tsaritsyn battles of 1918;[1] and recently Trotsky's wound-
ing criticisms of the Commissar of Rabkrin and of the General
Secretary had exacerbated it. In December 1922 or in the following January Stalin could have had no knowledge of the
'bloc' against him which Lenin and Trotsky mooted, of Lenin's
resolve to see him removed from the General Secretariat, or of
the attack which Lenin was preparing against his policy in
Georgia and his 'Great Russian chauvinism'. But he sensed
danger.[2] He saw Lenin and Trotsky acting in unison over the
trade monopoly and then over Gosplan. He heard Lenin in-
veighing against bureaucratic misrule; and he probably knew
from Zinoviev that Lenin was disturbed by events in Georgia.
As General Secretary, Stalin had already gained enormous
power: the Secretariat (and the Organization Bureau) had
taken over from the Politbureau most of its executive functions
and left to it decisions on high policy. Nominally, however,
the Politbureau exercised control over the Secretariat and the
Orgbureau; and it could prolong or refuse to prolong Stalin's
tenure of office. Stalin was convinced that he could expect no
good for himself from a Politbureau swayed by Trotsky. At this
stage he was anxious merely to preserve the influence he had
acquired rather than to take Lenin's place. He was aware
that the party saw in him only the supreme technician and

[1] *The Prophet Armed*, pp. 423–6.
[2] See Fotieva, '*Iz Vospominanii o Lenine*' in *Voprosy Istorii KPSS*, Nr. 4, 1957.

manipulator of its machine, but not a policy-maker and an expounder of Marxism such as it would expect Lenin's successor to be. No doubt Stalin's ambition was stung by this lack of appreciation, but his caution induced him to make allowance for it.

Next to Lenin and Trotsky, Zinoviev was by far the most popular member of the Politbureau. He was President of the Communist International; and in these years, when the Russian party had not yet come to use the International as a mere tool, but considered itself to be under its moral authority, the Presidency of the International was the most exalted position for any Bolshevik to occupy. Zinoviev was also the head of the Northern Commune, the Soviet of Petrograd. He was an agitator and speaker of tremendous power; and he was almost constantly before the party's eyes as one of the revolution's giants, an embodiment of Bolshevik virtue, indomitable and implacable. This popular image of his personality did not correspond to his real character, which was complex and shaky. His temper alternated between bursts of feverish energy and bouts of apathy, between flights of confidence and spells of dejection. He was usually attracted by bold ideas and policies which it needed the utmost courage and steadfastness to pursue. Yet his will was weak, vacillating, and even cowardly.[1] He was superb at picking Lenin's brain and acting as Lenin's loud and stormy mouthpiece; but he had no strong mind of his own. He was capable of the loftiest sentiments. In his best moments, in his idealistic vein, he impressed his listeners with such force that in a single speech lasting three hours and made in a foreign language, arguing against the most brilliant and authoritative men of European socialism, he persuaded a divided and hesitant congress of the German Independent Socialist party to join the Communist International.[2] His grip on the imagination of

[1] In a letter to Ivan Smirnov (written at Alma Ata in 1928) Trotsky relates this 'short talk' he had had with Lenin soon after the October Revolution: 'I told Lenin: "Who surprises me is Zinoviev. As to Kamenev, I have known him close enough to see where in him the revolutionary ends and the opportunist begins. But I did not know Zinoviev personally [before 1917]; and from accounts of him and appearances I imagined that this was the man who would stop at nothing and be afraid of nothing." To this Vladimir Ilyich replied: "If he is not afraid, it only means that there is nothing to be afraid of . . .".' *The Archives.*

[2] See *Protokoll über die Verhandlungen des Ausserordentlichen Parteitags zu Halle,* and Zinoviev, *Zwölf Tage in Deutschland.*

Russian crowds is described by eye-witnesses as 'demonic'.[1] Yet from the loftiest sentiments he could stoop at once to the meanest tricks and the demagogue's cheapest jokes. In the course of the many years which he had spent by Lenin's side in Western Europe, his quick mind had absorbed a considerable mass of knowledge about the world; yet it remained unrefined and unpolished. His temper was warm and affectionate; yet it was also savage and brutal. Genuinely attached to the principle of internationalism and a man of 'world outlook', he was at the same time a parochial politician inclined to settle the greatest issues by horse trading and petty manœuvre. He had risen to an undreamt-of height; and, devoured by ambition, he strove to rise even higher; but he laboured under inner uncertainty and doubt in himself.

It was Zinoviev's great pride that he had been Lenin's closest disciple in the ten years between 1907 and 1917, the years of reaction, isolation, and despair, when they were both struggling to keep the party in being and to prepare it for the great day, and when, at the time of the Zimmerwald and Kientahl conferences, together they launched on the world the idea of the Third International. But it was Zinoviev's great shame, or so he himself and his comrades thought, that he had failed at his test in October 1917, when he opposed the insurrection and Lenin branded him as 'strike-breaker of the revolution'. Between this shame and that pride his whole political life was torn. He did his best to get over the memories of 1917; and he was helped in this by Lenin, who even in his will begged the party not to remind Zinoviev and Kamenev of their 'historic error'. By 1923 most party members had almost forgotten the grave incident or were not inclined to delve into the past. The Old Guard preferred to let bygones be bygones, if only because the cleavage on the eve of the October Revolution had run right across it and many of its members were then on Zinoviev's side. All the more did the historians and the legend-mongers of the Old Guard turn the limelights on the earlier period, the one in which Zinoviev's great pride resided. If any man could in Lenin's absence speak for the Old Guard then it was surely Zinoviev.

It was unthinkable that he should now accept Trotsky's

---

[1] This is how Heinrich Brandler and Angelica Balabanoff, among others, have described it.

leadership. Not only was his memory crowded with the many incidents of their pre-revolutionary feud when, encouraged by Lenin, he had often vehemently inveighed against Trotsky.[1] Not only had his great shame been connected with the event on which rested Trotsky's chief title to glory, the October insurrection. Ever since 1917 he had been opposed to Trotsky at almost every crucial turn of Bolshevik policy. He was the most extreme advocate of the peace of Brest Litovsk; and he vaguely encouraged the military opposition to Trotsky during the civil war. In the spring of 1919 Trotsky arrived at Petrograd to organize its defences against Yudenich's offensive after Zinoviev, the city's official leader, had thrown up his hands in panic. During the Kronstadt rising Trotsky blamed Zinoviev for having needlessly provoked it. On the other hand, Zinoviev was one of Trotsky's most vocal critics in the debate over militarization of labour and trade unions.[2] Later, at the Politbureau, he cast his vote against Trotsky over economic policy and Gosplan only to find himself defeated when Lenin 'went over' to Trotsky. Even at the Executive of the International he was again defeated by Trotsky when the latter, together with Lenin, forced through the policy of the United Front. No wonder his attitude towards Trotsky was one of sneaking admiration mingled with envy and that sense of inferiority which Trotsky induced in so many members of the Old Guard.

Zinoviev's attitude was as a rule shared by Kamenev. The political partnership of these two men was so close that the Bolsheviks regarded them as their Castor and Pollux. Paradoxically, however, it was not the likeness but the contrast of their minds and temperaments that made of them political twins. Kamenev, although he headed the party organization of Moscow City, was far less popular than Zinoviev but far more respected in the inner circle of the leaders. Less self-confident on the public platform, not given to oratorical flourishes and heroic postures, he possessed a stronger and more cultivated intellect and a steadier character; but he lacked Zinoviev's fervour and imagination. He was a man of ideas rather than slogans. Unlike Zinoviev, he was as a rule attracted by moderate ideas and policies; but the strength of his Marxist convictions

[1] Zinoviev, *Sochinenya*, vols. i, ii, and v; and *Gegen den Strom*.
[2] *The Prophet Armed*, chapters x–xiii.

inhibited him in his moderation—his theoretical thinking was at loggerheads with his political inclination. His conciliatory character suited him well for the part of the negotiator; and in the early days Lenin often used him as the party's chief representative in contacts with other parties, especially when Lenin was anxious for agreement. (In inner-party controversy, too, Kamenev acted as the edge-blunter and the seeker for common ground between opposed viewpoints.) But his moderation repeatedly brought him into conflict with Lenin. During the 'treason' trial of the Bolshevik deputies to the Duma, early in the First World War, Kamenev declared from the dock that he was no adherent of Lenin's 'revolutionary defeatism'; in March and April 1917, before Lenin's return to Russia, he steered the party towards conciliation with the Mensheviks; and in October he was an opponent of the insurrection. Yet it was not courage that he lacked. Nor was he a mere trimmer. Cool and reserved, free from excessive vanity and ambition, he hid behind his phlegmatic appearance an infinite loyalty to the party. His character showed itself on the very day of the October Revolution: having publicly opposed the insurrection, he appeared at the insurgents' headquarters right at the very beginning, put himself at their disposal and wholeheartedly co-operated with them, thus assuming responsibility for the policy he had opposed and courting all the political and personal risks involved.[1]

What attracted him so strongly to Zinoviev was probably the very contrast of their characters. In each of them impulses were active which should have driven them wide apart; but in each strong inhibitions were also at work which kept their conflicting impulses in check, with the result that the two men usually met half-way between the opposed extremes towards which they gravitated.

Kamenev felt none of Zinoviev's and Stalin's intense hostility towards Trotsky, his former brother-in-law; and he might have put up with his leadership more easily than they. It was from sheer devotion to the Old Guard and friendship to Zinoviev that he turned against Trotsky. Whatever his private inclinations and tastes, he was extremely sensitive to the mood which prevailed among the Old Bolsheviks and by it he was swayed. When that mood went against Trotsky, Kamenev, full of mis-

[1] *Protokoly Tsentralnovo Komiteta*, pp. 141–3; *The Prophet Armed*, p. 307.

givings and heartbroken, went with it. He did not and could not hope to gain anything for himself by joining the triumvirate: he had no ambition to become Lenin's successor. But he supported and encouraged the restless ambition of his political twin, in part because he was convinced that it was harmless, that Zinoviev could not take Lenin's place anyhow, and that the triumvirs would in fact rule the party collectively; and partly, because in his moderation Kamenev was genuinely afraid of Trotsky's dominant and imperious personality and of his risky ideas and policies.

Zinoviev, Stalin, and Kamenev, however they differed in their characters and motives, were flesh and blood of the Old Guard; and between them they seemed to embody every aspect of the party's life and tradition. In Zinoviev were found the *élan* and the popular appeal of Bolshevism; in Kamenev its more serious doctrinal aspirations and its sophistication; and in Stalin the self-assurance and the practical sense of its solid and battle-hardened caucus. When they joined hands to debar Trotsky from the leadership they expressed a distrust and instinctive aversion felt by many members of the Old Guard. As yet they had no intention of eliminating him from the party, or even from its leading bodies. They acknowledged his merits. They wished him to occupy a prominent place in the Politbureau. But they did not consider him worthy of occupying Lenin's place; and they were horrified at the thought that, if nothing was undertaken against him, he might do so.

The triumvirs pledged themselves to concert their moves and act in unison.[1] In doing so they automatically swayed the Politbureau. In Lenin's absence the Politbureau consisted of only six members: the triumvirs, Trotsky, Tomsky, and Bukharin. Even if Trotsky had won over Tomsky and Bukharin, the vote would still have been divided equally. But as long as he, Bukharin, and Tomsky, formed no faction and voted each in his own way, it was enough that one of them should vote with the triumvirs, or abstain, to give them a majority. The triumvirs knew beforehand that Tomsky would not make common cause with Trotsky. An upright worker, a veteran Bolshevik,

---

[1] Stalin made the first public admission of the existence of the triumvirate at the twelfth congress in April 1923. See his *Sochinenya*, vol. v, p. 227, and also my *Stalin*, pp. 257–8.

and a trade-union leader in the first instance, Tomsky was the most modest member of the Politbureau. He was eager to defend, within limits and with caution, the demands and wage claims of the workers; and so in 1920 he was the first to oppose Trotsky over the militarization of labour and to raise a storm when Trotsky threatened to 'shake up' the trade unions. Trotsky criticized him harshly as an old-fashioned type of the trade unionist who from pre-revolutionary habit encouraged the 'consumptionist' attitude in the workers and showed no understanding for the 'productionist' outlook of the socialist state. For some time Tomsky led the trade unions in virtual revolt against the party. He was deposed from their Central Council and sent 'on an assignment', which was a barely veiled form of exile, to Turkestan. After the promulgation of N.E.P. he returned to the Kremlin and was promoted to membership of the Politbureau. But the wound inflicted on him rankled; and his attitude reflected the hostility towards Trotsky, the militarizer of labour, which many Bolshevik trade unionists had felt since 1920.

Bukharin was the only member of the Politbureau who was still friendly towards Trotsky. In his early thirties, yet an 'old' Bolshevik, he was the party's leading theorist, brilliant and profoundly educated. Lenin criticized his inclination to scholasticism and the doctrinaire angularity of his ideas. These ideas, however, exercised a strong influence even on Lenin who often adopted them and gave them a more realistic and supple expression.[1] Bukharin's was indeed an angular mind, fascinated more by the logical neatness of abstract propositions than by confused and confusing realities. Yet angularity of intellect was combined in him with an artistic sensitivity and impulsiveness, a delicacy of character, and a gay, at times almost schoolboyish, sense of humour. His rigidly deductive logic and his striving for abstraction and symmetry induced him to take up extreme positions: for years he had been the leader of the 'left Communists'—and by a process of radical reversal he was to become the leader of the party's right wing.

Bukharin had been in conflict with Trotsky as often as in agreement. During the Brest Litovsk crisis he led the war party

[1] Bukharin's intellectual relationship with Lenin will be discussed in my *Life of Lenin*.

and opposed the 'shameful peace'. During the civil war he was in sympathy with those who opposed the discipline and the centralistic organization which Trotsky gave to the Red Army. Then in the debate over trade unions he drew close to Trotsky. Like Trotsky, and even more passionately, he defended the rights of the non-Russian nationalities and stood up for the Georgian 'deviationists'. But whether he saw eye to eye with Trotsky or not, he was attracted to him by a strong affection and was spellbound by his personality.[1] Trotsky describes how in 1922, when he himself was laid up with a minor illness, Bukharin visited him and told him about Lenin's first stroke of paralysis.

At that time Bukharin was attached to me in his characteristic manner, half-hysterically, half-childishly. He finished his account of Lenin's illness and dropped down on my bed and muttered, as he gripped me through the blanket: 'Don't you fall ill too, I implore you, don't. . . . There are two men of whose death I always think with horror, Lenin and you.'

Another time he sobbed on Trotsky's shoulder: 'What are they doing with the party, they are turning it into a gutter.'[2] But with only this one friend in the Politbureau, Trotsky could not do much: Bukharin's sobs and sighs were of little assistance to him when he was confronted by the triumvirs.

Apart from these full members of the Politbureau, there were two alternate members: Rykov, chief of the Supreme Council of National Economy, and Kalinin, nominal Head of the State. Both were 'moderate' Bolsheviks. Both were of peasant origin and both retained much of the muzhik's character and outlook. In both, receptiveness to the moods of rural Russia, to the peasantry's hopes and fears, and also to some of its prejudices was stronger than in perhaps any other leader. Both embodied the element of nativeness in the party—'genuine Russianness'—and all that it implied: a distinct anti-intellectual bias, a distrust of the European element, a pride in social roots, and a certain stolidity of outlook. All this predisposed them against Trotsky. The peasantry, we know, cherished the regained freedom of private property and trade and was afraid of nothing more than of a

---

[1] 'Trotsky, the brilliant and heroic tribune of the October insurrection, the tireless and fiery preacher of revolution . . .' wrote Bukharin in his account of the events of 1917.   [2] Trotsky, *Moya Zhizn*, vol. ii, p. 207.

relapse into war communism. Of that fear Rykov and Kalinin were the mouthpieces within the party. More than anyone else they sensed a danger of such a relapse in Trotsky's ideas on planning. When Trotsky spoke of the lack of any guiding idea in the Supreme Council of the National Economy and of its inclination towards a Soviet variety of *laissez faire* he had Rykov in mind. Rykov, for his part, saw in Trotsky's scheme for a new Gosplan an encroachment upon his own prerogatives and more than that—an encroachment upon the basic principle of N.E.P. He was now the first to level against Trotsky the charge of hostilitity towards the peasant, the charge which was to resound through all the campaigns against Trotsky in coming years.[1]

Kalinin, on the contrary, had a deep respect for Trotsky and a friendly feeling, which he was to express even at the height of the drive against Trotskyism. The circumstance that in 1919 it was Trotsky who sponsored Kalinin's candidature for the office of Head of State, because of Kalinin's exceptional appeal to the peasants, had perhaps something to do with this.[2] Yet, when Rykov began to speak of Trotsky's hostility towards the peasantry, Kalinin was undoubtedly impressed. He had no strong views about Trotsky's proposals for policy, of which, in any case, he understood little; but he concluded, without rancour, that nothing could be safer and sounder than to keep in check Trotsky's influence, an influence which might endanger the 'alliance between workers and peasants'.

Two other men, Dzerzhinsky and Molotov, were at this time closely associated with the Politbureau, although they were not members. Dzerzhinsky, chief of Cheka and G.P.U., was the only one in this group of leaders who did not belong to the Old Guard. He had come from the Social Democratic party of the Kingdom of Poland and Lithuania, the party founded by Rosa Luxemburg; and he had adhered to the Bolsheviks only in 1917, about the same time as Trotsky. His original party had, under Rosa Luxemburg's inspiration, adopted towards the Bolsheviks an attitude indistinguishable from Trotsky's: it was usually critical of both Bolsheviks and Mensheviks; and it was the only party in the Socialist International to agree with

---

[1] *13 Konferentsya RKP*, pp. 6–7; *8 Vserossiiski Syezd Sovetov*, pp. 100–2.
[2] Trotsky, *Sochinenya*, vol. xvii, book 2, p. 542.

Trotsky's theory of permanent revolution. Dzerzhinsky, even after he had joined the Bolsheviks, remained opposed to Lenin over the self-determination of the non-Russian nationalities; and, again following Luxemburg, he argued that socialism should overcome, not encourage, separatist tendencies among the small nations. Paradoxically, this internationalist reasoning led him, the Pole of noble origin, to back Stalin's ultra-centralistic policy and to act *vis-à-vis* the Georgians as a spokesman of the new 'indivisible' Russia.

Dzerzhinsky's views, however, had not hitherto counted for much within the party. Important as the revolution's chief security officer, he was not a political leader. When the Bolsheviks decided to set up the Extraordinary Commission for the Struggle against Counter-revolution, as their political police were first called, they looked for a man with absolutely clean hands to do the 'dirty work'; and they found such a man in Dzerzhinsky. He was incorruptible, selfless, and intrepid—a soul of deep poetic sensibility, constantly stirred to compassion for the weak and the suffering.[1] At the same time his devotion to his cause was so intense that it made him a fanatic who would shrink from no act of terror as long as he was convinced that it was necessary for the cause. Living in permanent tension between his lofty idealism and the butchery which was his daily job, high-strung, his life force burning itself out like a flame, he was regarded by his comrades as the strange 'saint of the revolution' of the Savonarola breed. It was his misfortune that his incorruptible character was not allied to a strong and discriminating mind. His need was to serve the cause; and he came to identify the cause with the party of his adoption and then to identify that party with its leaders, with Lenin and Trotsky until lately, and now with the triumvirs behind whom he saw the Old Guard. Not being himself one of the Old Guard, he was all the more eager to promote its interest; and so he became more Bolshevik than the old Bolsheviks themselves just as he was, according to Lenin, more Great Russian than the Russians themselves.

For sheer lack of colour Molotov forms a striking contrast to Dzerzhinsky. In his late twenties, he already occupied a high

---

[1] Dzerzhinsky's private correspondence, published in *Z Pola Walki* and other Polish periodicals, gives a good insight into his character.

position in the hierarchy: he had been secretary to the Central Committee before Stalin became General Secretary, and then he served under Stalin as his chief aide. Even at this stage his narrowness and slow-mindedness were already bywords in Bolshevik circles; he appeared to be devoid of any political talent and incapable of any initiative. He usually spoke at party conferences as *rapporteur* on a second- or third-rate point; and his speech was always as dull as dishwater. The descendant of an intellectual family, a relative of Scriabin, the great musician, he seemed the very opposite of the intellectual—a man without ideas of his own. He could not have been altogether without his spark—the spark had shown itself in 1917—but it was now quite extinguished.

Molotov was the almost perfect example of the revolutionary turned official; and he owed his promotion to the completeness of this conversion. He possessed a few peculiar virtues which helped him along: infinite patience, imperturbable endurance, meekness towards superiors, and a tireless, almost mechanical industry which in the eyes of his superiors compensated for his mediocrity and incompetence. Very early he attached himself like a shadow to Stalin; and very early, too, he conceived an intense dislike, mingled with fear, of Trotsky. The story is told that Trotsky once appeared at the Secretariat, dissatisfied with something that had been done there, and all but pointing at Molotov taunted the dull-witted bureaucrats of the Secretariat. 'Comrade Trotsky', Molotov stammered out, 'Comrade Trotsky, not everyone can be a genius.'[1]

.   .   .   .

Thus even before the beginning of the struggle for the succession Trotsky stood almost alone in the Politbureau. He had the first inkling of a concerted action against him in the early weeks of 1923—a full year before Lenin's death—when at sessions of the Politbureau he found himself attacked by Stalin with quite unwonted ferocity and venom.[2] Stalin assailed him for his persistence in declining to be vice-Premier. He questioned Trotsky's motives and insinuated that Trotsky refused to respond to the call of duty because in his craving for power he would not content himself with being one of Lenin's deputies.

---

[1] Bajanov, *Avec Staline dans le Kremlin*, p. 139.   [2] *The Trotsky Archives.*

Then he heaped on Trotsky accusations of pessimism, bad faith, and even of defeatism, all on the flimsiest of grounds. Thus, to show up Trotsky's 'defeatism' he made much play of a remark Trotsky had once made to Lenin in private, saying that the 'cuckoo would soon sound the death knell for the Soviet Republic'.[1]

Stalin had several purposes in mind. He still reckoned with the possibility of Lenin's return to office; and so he took up the issue of the appointment Lenin had proposed hoping that he might drive this wedge between Lenin and Trotsky. He knew that nothing could embarrass Trotsky more than the insinuation that he craved to inherit Lenin's position. The calculation was shrewd. Trotsky was touched to the quick. He had sounder reasons than Stalin had to hope for Lenin's return, which would bring into action their 'bloc'. Even apart from this, he was so confident of his own position in the party and the country and of his superiority to his adversaries, that he had no inclination to fight for the succession. He did not try to recruit partners and associates; and it did not even occur to him to manœuvre for position. Yet Stalin's charges and insinuations were such that it was as absurd for Trotsky to refute them as it was dangerous to ignore them. Their effect was to drag him down and to extract from him those denials and excuses of which it is said *qui s'excuse s'accuse*. Once a man in a position comparable to Trotsky's is charged with craving for power, no denial on his part can dispel the suspicion aroused, unless he resigns all office on the spot, goes out into the wilderness, and ceases even to voice his views. This Trotsky was, of course, not prepared to do. Time and again he had explained that he could not see what useful role he could perform as one of the vice-Premiers whose functions overlapped; and that the division of labour in the government was faulty because 'every Commissar was doing too many jobs and every job was done by too many Commissars'. He now added that as vice-Premier he would have no machinery through which to work and no real influence. 'My appointment to such a work would, in my view, efface me politically.' He denied the imputation of pessimism and defeatism: he had indeed made the remark about the 'cuckoo sounding the death knell for the Soviet Republic' when he tried

[1] In Slavonic folk-lore the cuckoo is a bird of omen.

to impress on Lenin the ruinous effects of economic waste and red tape; but his purpose—was there any need to say it?—was to remedy those ills, not to sow panic.[1] To such irrelevancies had the bickering in the Politbureau sunk; and it dragged on for weeks during which Trotsky, waiting for Lenin's return, held his fire.

He had some reason for waiting. The medical reports on Lenin's health were encouraging. Even from his sick-bed Lenin dealt blow after blow at Stalin with a relentless resolve which surprised Trotsky. It was only proper, Trotsky held, that he should leave to Lenin the initiative in this matter. At the beginning of February Lenin produced *inter alia* a severe criticism of Rabkrin and communicated it to the Politbureau. Although Stalin had already withdrawn from Rabkrin, Lenin's attack affected him personally, because Lenin left no doubt that he considered the Commissariat to have been an utter failure during Stalin's tenure of office. He spoke of the vices of the Commissariat in almost the same terms that Trotsky had used: 'lack of culture', 'muddle', 'bureaucratic misrule and wantonness', &c.; and he inserted barbed remarks against 'bureaucracy in the party as well'. He concluded with proposals for an overhaul of Rabkrin, a reduction of its staff, and for the setting up of a Central Control Commission, which was to take over many of Rabkrin's functions. For several weeks Trotsky demanded that Lenin's criticism be published, but the Politbureau refused.[2]

At the same time Trotsky submitted a scheme for a radical reorganization of the Central Committee and of its various agencies; and he supported this by a critical survey of the party's condition. The Central Committee, he said with emphasis, had lost touch with the lower ranks and had become transformed into a self-sufficient bureaucratic machine. This was the issue over which the controversy was to burst into the open next autumn; but already in January and February Trotsky posed it before the Politbureau with even greater bluntness than that which he was to allow himself in the public debate later. In some details, such as the size of the Central

---

[1] See the papers of January 1923 in *The Archives*.

[2] Lenin, *Sochinenya*, vol. xxxiii, pp. 440 ff. Trotsky's letter to all members of the Central Committee of 23 February 1923 in *The Archives*. (See also Fotieva's memoirs in *Voprosy Istorii KPSS*, 4, 1957.)

Committee and its relationship with the Central Control Commission, his scheme differed from Lenin's. The triumvirs made the most of these differences, saying that Trotsky not only snubbed Lenin by refusing to become his deputy but also tried to divert the party from Lenin's ideas of organization. At this stage the upper ranks of the hierarchy were becoming initiated into the Politbureau dispute; and nothing could do more to damage Trotsky's position in their eyes as Lenin's presumed successor than a whispering campaign in which he was depicted as resisting Lenin on almost every issue. The words of the triumvirs were calculated to feed such a campaign. Their charges were recorded in the Politbureau minutes and opened for inspection to members of the Central Committee who were not slow in divulging their secrets to friends and subordinates.

The campaign had been afoot for some time already when Trotsky first reacted to it. On 23 February 1923 he addressed a letter to the Central Committee in which he said: 'Some members . . . have expressed the opinion that Comrade Lenin's scheme aims at preserving the party's unity while the purpose of my project is to create a split.' This insinuation was concocted and canvassed by a clique which in fact concealed Lenin's writings from party members. He disclosed what had happened at the Politbureau: 'While the majority . . . held it impossible even to publish Lenin's letter, I . . . not only insisted on publication, but defended the essential ideas of the letter, or, to put it more accurately, those of its ideas which seemed essential to me.' 'I reserve', he concluded, 'the right to expose these facts before the entire party, should this become necessary, in order to refute an insinuation [the authors of which] have enjoyed all too great an impunity because I have almost never reacted to insinuation.'[1] The occasion for the 'exposure' was to be the twelfth party congress convened for April. The threat was characteristic of Trotsky: he felt that he was by the unwritten code of inner party loyalty bound to give his antagonists due notice of any move against them which he might contemplate. He thereby deprived himself of the advantage of surprise and gave them time to parry the blow—this was the exact opposite of Stalin's tactics. Trotsky did not even intend to carry out his threat, however. His aim was merely to curb Stalin and to gain

[1] *The Archives.*

time while waiting for Lenin's recovery. He obtained one immediate result: on 4 March *Pravda* at last published Lenin's attack on Rabkrin.

On 5 March, while he, too, was confined to bed with illness, Trotsky received from Lenin a message of the utmost importance and urgency.[1] Lenin begged him to speak out in defence of the Georgian 'deviationists' at the forthcoming session of the Central Committee. This was Trotsky's first contact with Lenin since their talk about the 'bloc' in December, and the first intimation he had of Lenin's changed attitude in the Georgian affair. 'At present', Lenin wrote, 'their case [i.e. the case of the "deviationists"] is under *"prosecution"* by Stalin and Dzerzhinsky, and I cannot rely on Stalin's and Dzerzhinsky's impartiality. Quite the contrary. If you would agree to undertake the defence, my mind would be at rest.' Lenin attached a copy of his notes on Stalin's policy towards the nationalities (which are summarized in the previous chapter). These notes for the first time gave Trotsky a full idea of the relentlessness with which Lenin intended to press home the attack—by comparison Lenin's criticism of Rabkrin seemed mild. Lenin's secretaries added that Lenin had prepared, to use his own word, a 'bombshell' against Stalin to be exploded at the congress. Moreover, in a last moment of an exhausting tension of mind and will he urged Trotsky to show no weakness or vacillation, to trust no 'rotten compromise' Stalin might propose, and, last but not least, to give Stalin and his associates no warning of the attack. The next day he himself sent a message to the Georgian 'deviationists', conveying his warm sympathy and promising to speak up. About the same time Trotsky learned from Kamenev that Lenin had written a letter to Stalin threatening to 'break off all personal relations'.[2] Stalin had behaved in an offensive manner towards Krupskaya when she was collecting information for Lenin on the Georgian affair; and when Lenin learned about this, he could hardly contain his indignation. He decided, Krupskaya told Kamenev, 'to crush Stalin politically'.

What a moment of moral satisfaction and triumph this was

---

[1] *Moya Zhizn*, vol. ii, pp. 220–1; *The Stalin School of Falsification*, pp. 69–70.

[2] This letter was read out by Khrushchev at the twentieth congress and is included in the text of his speech published in the U.S.A. and Great Britain, but not in vol. xxxvi of Lenin's *Sochinenya*—nor in *Kommunist*, No. 9, 1956. Fotieva only hints at the existence of this letter.

for Trotsky. As on so many previous occasions Lenin at last acknowledged that Trotsky had been right all along. As so often before, Trotsky's bold foresight had condemned him for a time to political solitude and had caused dissension between him and Lenin; and just as events had vindicated him and led Lenin to conclusions identical with his, first over Gosplan, then over Rabkrin and 'party bureaucracy', so now they vindicated him over Georgia. Trotsky was confident that the triumvirate. was ruined and Stalin beaten. He was the victor and could dictate his terms. His adversaries thought likewise. When on their behalf Kamenev came to see Trotsky on 6 March, he was crestfallen, ready for chastisement, and anxious to mollify Trotsky.[1]

Not much mollifying was needed. Trotsky's revenge was to display magnanimity and forgiveness. Forgetting Lenin's warning, he jumped at a 'rotten compromise'. Lenin intended to demote Stalin and Dzerzhinsky and even to expel from the party 'for at least two years' Ordjonikidze (once his favourite disciple) because of the latter's brutal behaviour at Tiflis. Trotsky at once reassured Kamenev that he himself would propose no such severe reprisals. 'I am', he said, 'against removing Stalin and against expelling Ordjonikidze and displacing Dzerzhinsky . . . but I do agree with Lenin in substance.'[2] All he asked of Stalin was that he should mend his ways: let him behave loyally towards his colleagues; let him apologize to Krupskaya; and let him stop bullying the Georgians. Stalin had just prepared 'theses', to be submitted to the party congress, on policy towards the non-Russian nationalities—he was to address the congress on this point as the Central Committee's *rapporteur*. Anxious to justify his own behaviour, he had placed strong emphasis on the condemnation of 'local nationalisms'. Trotsky proposed that Stalin should reword his resolution, insert a denunciation of Great Russian chauvinism and of Russia 'one and indivisible', and give the Georgians and Ukrainians a firm assurance that henceforth their rights would be respected. This was all he demanded of Stalin—no breast-beating and no personal apologies. On these terms he was prepared to let Stalin continue as General Secretary.

On these terms Stalin was, of course, ready to surrender or at

---

[1] Trotsky, *Moya Zhizn*, vol. ii, pp. 223–4.   [2] Loc. cit.

least to feign surrender. To find himself threatened with political
ruin, to feel Lenin's anger bursting over his head, and at this very
moment to see Trotsky stretching out to him a forgiving hand
was a quirk of fortune for which he could not but be grateful.
He accepted Trotsky's terms at once. He rephrased his 'theses'
and inserted all of Trotsky's amendments. As to the other 'conditions', well, all the offence he had given and all the hurts he
had caused had sprung, he said, from misunderstandings and he
was only too anxious to clear these up.

While Kamenev was still acting as go-between, Lenin succumbed to another stroke. He was to survive it by ten months,
but paralysed, speechless most of the time, and suffering from
spells of unconsciousness, the torment of which was all the
greater because in the intervals he was acutely and helplessly
aware of the intrigue in the background. The news of Lenin's
relapse at once relieved the triumvirs. A few days after they had
meekly submitted to Trotsky, they were once again working
with redoubled energy but greater discretion to eliminate him
from the succession. He still felt on top. He did not abandon the
hope that Lenin would recover. In any case, he had in his
hands Lenin's messages and manuscripts; and if he were to
come out with these at the congress, especially with the notes on
the Georgian affair, the party would have not a shadow of
doubt where Lenin stood. Surely, he concluded, the triumvirs
must know this and, fearing exposure, they must adhere to the
compromise.

The triumvirs knew that Trotsky had promised Lenin to take
up the case of the Georgian deviationists and to acquaint the
congress with Lenin's views. (Kamenev had already read the
notes on Georgia.) Stalin's chief preoccupation was now to
prevail upon Trotsky not to act on this promise. Had he, Stalin,
not done everything Trotsky demanded of him? He had, indeed; and so Trotsky consented to submit Lenin's notes to the
Politbureau and to leave it to the Politbureau to decide whether
or in what form they should be communicated to the congress.
The Politbureau resolved that the notes should in no case be
published, and that only chosen delegates should be acquainted
in strict confidence with their content. This was not how Lenin
had expected Trotsky to behave when he urged him to remain
adamant, to address the congress with complete bluntness, and

to allow no patching up of differences. But all these urgings and warnings were lost on Trotsky, who in his magnanimous mood helped the triumvirs to conceal from the world Lenin's death-bed confession of shame and guilt at the revival of the Tsarist spirit in the Bolshevik state. Lenin's notes on policy towards the non-Russian nationalities were to remain unknown to the party for thirty-three years.[1]

Hindsight makes Trotsky's behaviour appear incredibly foolish. This was the moment when his adversaries were taking up positions; and every one of his steps was as if calculated to smooth their way. Years later he remarked wistfully that if he had spoken up at the twelfth congress, with Lenin's authority behind him, he would probably have defeated Stalin there and then, but that in the long run Stalin might still have won.[2] The truth is that Trotsky refrained from attacking Stalin because he felt secure. No contemporary, and he least of all, saw in the Stalin of 1923 the menacing and towering figure he was to become. It seemed to Trotsky almost a bad joke that Stalin, the wilful and sly but shabby and inarticulate man in the back-ground, should be his rival. He was not going to be bothered about him, he was not going to stoop to him or even to Zino-viev; and, above all, he was not going to give the party impression that he, too, participated in the undignified game played by Lenin's disciples over Lenin's still empty coffin. Trotsky's conduct was as awkward and as preposterous as must be the behaviour of any character from high drama sud-denly involved in low farce.

Of farce there was indeed no lack. When the Politbureau met on the eve of the congress, Stalin proposed that Trotsky should address the congress as the Central Committee's political *rapporteur*, that is in the role hitherto always reserved for Lenin. Trotsky refused saying that as General Secretary Stalin should be *rapporteur ex officio*. Stalin, all modesty and meekness, replied: 'No, the party would not understand it . . . the report must be made by the most popular member of the Central Committee.'[3] The 'most popular member', who only a few weeks earlier had been charged with craving for power, now leaned over back-wards to show that the charge was baseless; and so he made it all

---

[1] They were first published in *Kommunist* in June 1956.
[2] *Moya Zhizn*, vol. ii, p. 219.   [3] Trotsky, *Stalin*, p. 366.

the easier for the triumvirs to overthrow him. The Politbureau
decided that Zinoviev should deliver the address the party had
been accustomed to hear from Lenin.

When the twelfth congress at last assembled, in the middle of
April, its opening provided an occasion for a spontaneous display of homage to Trotsky. As usual, the chairman read out the
greetings to the congress which poured in from party cells,
trade unions, and groups of workers and students all over the
country. In almost every message tributes were paid to Lenin
and Trotsky. Only now and then did the greetings refer to
Zinoviev and Kamenev, and Stalin's name was hardly mentioned. The reading of the messages went on during several
sessions; and it left no doubt whom, if the party had now been
asked to choose, it would have chosen as Lenin's successor.[1]

The triumvirs were surprised and annoyed; but they had
little to fear. Lenin was not there to explode his 'bombshell';
and Trotsky, having promised not to explode it either, honoured
his promise. He did not give the congress even the slightest hint
of any disagreement between him and the triumvirs; and he
kept himself well to the background. In the meantime the
triumvirs acted behind the scenes. Their agents initiated delegates into the crisis in the leadership and turned against Trotsky
even the homage just paid to him. They did their best to impress upon provincial delegates the dangers which they alleged
were inherent in Trotsky's extraordinary popularity: had not
Bonaparte, the 'grave-digger' of the French Revolution, risen
to power on such acclaim? Could the imperious and ambitious
Trotsky be trusted not to abuse his popularity? Was not, in
Lenin's absence, the 'collective leadership' of smaller men, but
men whom the party knew and trusted, preferable to his pre-
eminence? Such questions, uttered in worried whispers, made
many a delegate apprehensive. The Bolsheviks had been accustomed to look back to the great French precedent and to
think in historical analogies. Occasionally, they cast round for
that unpredictable character among their leaders, the potential
Danton or the would-be Bonaparte, who might spring a dangerous
surprise upon their revolution. Among all the leaders none
seemed to have as much affinity with Danton as Trotsky; and
none, it also seemed, would the mask of a Bonaparte fit as well

[1] *12 Syezd RKP (b)*, pp. 89, 488, 496, 502–3.

as him. In the eyes of many an old Bolshevik Trotsky's pre-eminence was a liability; and on reflection it seemed, indeed, safer that the party should be run by a team of less brilliant but reliable comrades.[1]

The triumvirs behaved with studious modesty. They declared that the only claim they had on the party's confidence was that they were Lenin's loyal and tested disciples. It was at this congress that Zinoviev and Kamenev initiated the exalted glorification of Lenin which was later to become a state cult.[2] No doubt the exaltation was in part sincere: this was the first Bolshevik congress without Lenin; and the party already felt bereft. The triumvirs played upon this mood, knowing that the glorification of Lenin would reflect glory upon those whom the party had known as his oldest disciples. Yet they had to work hard to convince the congress that they spoke with Lenin's voice. Delegates were uneasy. They received Zinoviev with sullen silence, when he came forward as *rapporteur*. His exaggerated and even ridiculous expressions of adoration of Lenin disgusted the sophisticated and the critically minded; but these were in a minority, and they did not protest lest they be misunderstood.

The triumvirs followed this up by calls for discipline, unity, and unanimity. When the party was leaderless it would have to close its ranks. 'Every criticism of the party line', Zinoviev exclaimed, 'even a so-called "left" criticism, is now objectively a

[1] A critic of my *Stalin*, where I mentioned this whispering campaign (p. 273), writes: 'That he [Trotsky] was viewed by some communists as a potential Bonaparte is a discovery made only quite recently by writers like Mr. Deutscher. . . . It was not appreciated at the time.' (G. L. Arnold in *Twentieth Century*, July 1951.) It is not always that a writer can give chapter and verse for a 'whispering campaign'; and in *Stalin* I referred to this particular campaign on the basis of what I had heard about it in Moscow, when the memory of it was still fairly fresh. In the meantime Alfred Rosmer, who in 1923 was in Moscow as member of the Executive of the Comintern and was extremely well informed about matters concerning Trotsky's person, has published his memoirs; and this is what he says: 'But now [in 1923] a rumour into which one ran everywhere indicated a well-prepared manœuvre . . .: "Trotsky imagines himself a Bonaparte", or "Trotsky wants to act a Bonaparte". The rumour circulated in every corner of the country. Communists arriving in Moscow came to tell me about it; they understood that something was afoot against Trotsky and urged me: "You should warn him about it." ' Rosmer, *Moscou sous Lénine*, p. 283. References to this 'whispering campaign' occur also in contemporary literature. In Eastman's *Since Lenin Died* there is a whole chapter about it under the title 'The anti-Bonaparte fraction'.

[2] See Kamenev's and Zinoviev's opening speeches in *12 Syezd RKP (b)*.

Menshevik criticism.'[1] He flung this warning at Kollontai, Shlyapnikov, and their followers; and, working himself up as he went, he told them that they were even more obnoxious than the Mensheviks. Ostensibly directed only against the Workers' Opposition, his words carried wider implications: they intimated to every potential critic with what sort of a denunciation he would be met. The maxim that *every* criticism was to be regarded *a priori* as a Menshevik heresy was novel—nothing like it had been pronounced before. Yet the maxim could be deduced from the argument Zinoviev had presented at the previous congress, when he said that in consequence of their political monopoly the Bolsheviks found that there were two or more potential parties within their party and that one of these consisted of the 'unconscious Mensheviks'. Concerned only with the immediate circumstances of the struggle for power and flushed with self-confidence, Zinoviev now went a step farther and described every opponent of the leading group as a virtual mouthpiece for those 'unconscious' and inarticulate Mensheviks. It followed that the leaders, whoever they were, had the right and even the duty to suppress opponents within the party as they had suppressed the real Mensheviks. In this way Zinoviev came to formulate what was to be the canon of Bolshevik self-suppression.

This call for discipline and the new view of unity did not pass unchallenged. The members of the Workers' Opposition and other dissenters mounted the platform to denounce the triumvirate and demand its disbandment. Lutovinov, a prominent party worker, protested against the 'papal infallibility' and immunity from criticism Zinoviev had claimed for the Politbureau.[2] Kossior, another old Bolshevik, maintained that the party was ruled by a clique, that the General Secretariat persecuted critics, that Stalin had during his first year in that office demoted and victimized the leaders of such important organizations as those of the Urals and of Petrograd; and that the talk about collective leadership was a fraud. Amid uproar Kossior demanded that the congress should revoke the 1921 ban on inner party groupings.[3]

---

[1] *12 Syezd RKP (b)*, pp. 46–47.   [2] Ibid., pp. 105–6.

[3] Ibid., pp. 92–95. Another speaker referred to an anonymous leaflet circulated at the time of the congress and demanding the removal of the triumvirate from the Central Committee. He suggested that the Workers' Opposition was responsible for the leaflet. Ibid., p. 136.

The triumvirs, however, dominated the congress: Kamenev presided over it, Zinoviev enunciated policy, and Stalin manipulated the party machine. They made no bones about their partnership any longer: in reply to the challenge from the Workers' Opposition they defiantly acknowledged the existence of the triumvirate.[1] But within the triumvirate a shift was making itself felt: Zinoviev was losing his position as senior triumvir. He had overreached himself, antagonized many delegates, and drawn upon himself most of the attacks from the floor. Stalin's more discreet conduct gained him credit. The eyes of the delegates turned appreciatively on him when Nogin, an old influential and moderate member of the Central Committee, made his eulogy, praising the unobtrusive but vital work of direction he had done at the General Secretariat. 'Essentially', Nogin said, 'the Central Committee constitutes that basic apparatus which sets in motion all political activity in our country. The Bureau of the Secretariat is the most important part of the apparatus.'[2] Even some of the malcontents appealed from Zinoviev's flamboyance and demagoguery to Stalin's common sense.

Stalin's position was further enhanced in the debate over policy towards non-Russian nationalities, the debate which might have brought his undoing. The Georgians had come to Moscow expecting to get that strong support Lenin had promised them.[3] They did not obtain it. Rakovsky, who was the head of the Ukrainian government but had not enough influence in Moscow, took up their case. Was Moscow out to russify the small nationalities as the Tsarist gendarmes had done? he asked.[4] The Georgians were perplexed and confounded when they heard Stalin himself speaking with righteous indignation against the bullying of non-Russian nationalities, and when they found that their own denunciations of Great Russian chauvinism were inserted into the text of Stalin's 'theses'. This spectacle, the result of Trotsky's compromise with Stalin, seemed to them a mockery of all their complaints and protests. In vain did they demand that at least Lenin's notes should be read out. The members of the Politbureau were enigmatically reticent. Only one of them, Bukharin, broke the conspiracy of silence and in a great and stirrings peech—this was to be the swan-song

[1] Stalin, *Sochinenya*, vol. v, p. 227.   [2] *12 Syezd RKP (b)*, p. 63.
[3] Ibid., pp. 150-1.   [4] Ibid., pp. 528-34.

of Bukharin the leader of Left Communism—he defended the small nationalities and exposed Stalin's pretences. He exclaimed that Stalin's disavowal of Great Russian chauvinism was sheer hypocrisy and that the atmosphere at the congress, where the party's *élite* was assembled, proved it: every word uttered from the platform against Georgian or Ukrainian nationalism aroused stormy applause, while even the mildest allusion to Great Russian chauvinism was received with irony or with icy silence.[1] It was with icy silence that the delegates received Bukharin's own speech. Stalin, emboldened by the attitude of the congress, could now permit himself to play down the meaning and import of Lenin's attack on his policy and to rout the 'deviationists'.

Trotsky followed the proceedings impassively or absented himself. He observed scrupulously the terms of his compromise with the triumvirs and the principle of the Politbureau's 'Cabinet solidarity'. This principle did not prevent Zinoviev from treating Trotsky to allusive pinpricks about his 'obsession with planning'.[2] Trotsky did not react. He showed a blank face to the speakers of the Workers' Opposition when they demanded the disbandment of the triumvirate and attacked the General Secretariat. He gave not a nod of encouragement to the disheartened Georgians; and when the debate over nationalities opened he left the assembly, excusing himself on the ground that he would be busy preparing his own report to the congress.[3]

When at last, on 20 April, Trotsky addressed the congress he turned away from the issues that had aroused so much heat and passion and spoke strictly on economic policy.[4] This, no doubt, was a great subject and the one in which he saw the key to all other problems; and at last he had the opportunity to present fully and before a nation-wide audience the ideas he had so far developed only loosely or only within the closed circle of the leaders. It was part of his bargain with the triumvirs that he was authorized to present his views as a statement of official policy, although the Politbureau agreed with his views no more than he

---

[1] *12 Syezd RKP (b)*, pp. 561–5.   [2] Ibid., pp. 45–46.

[3] Ibid., p. 577. Yet only a month later Trotsky once again attacked in *Pravda* Stalin's policy in Georgia without mentioning Stalin. He wrote that if Great Russian chauvinism were to have its way in the Caucasus, then the Soviet invasion of the Caucasus would turn out to have been 'the greatest crime'. *Sochinenya*, vol. xxi, pp. 317–26.   [4] *12 Syezd RKP (b)*, pp. 282–322.

agreed with Stalin's policy towards the non-Russian nationalities. He attached the greatest importance to his being able to launch his economic policy as the party's official 'line'; and this in his eyes probably justified in part his concessions to the triumvirs. And in fact no member of the Politbureau contradicted him openly while the congress debated his address.

He appealed to the party to master the country's economic destiny and to tackle the great and difficult task of primitive socialist accumulation. He surveyed the experience of two years of the New Economic Policy and redefined its principles. The twin purpose of N.E.P., he argued, was to develop Russia's economic resources and to direct that development into socialist channels. The rise in industrial production was still slow; it lagged behind the recovery of private farming. Thus a discrepancy arose between the two sectors of the economy; and it was reflected in the 'scissors' that opened between high industrial and low agricultural prices. (This metaphorical term which Trotsky coined soon entered the economists' idiom all over the world.)[1] Since the peasants could not afford to buy industrial goods and had no real incentive for selling their produce, the 'scissors' threatened to cut once again the economic ties between town and country and to destroy the political alliance between worker and peasant. The 'scissors' should be closed by lowering industrial prices rather than by raising agricultural ones. It was necessary to rationalize, modernize, and concentrate industry; and this required planning.

Planning was his main theme. He did not, as his adversaries later claimed, advocate that N.E.P. should be abandoned in favour of planning. He urged the party to pass from the 'retreat' to a socialist offensive within the framework of N.E.P. 'The New Economic Policy', he said, 'is the arena which we ourselves have set up for the struggle between ourselves and private capital. We have set it up, we have legalized it, and within it we intend to wage the struggle seriously and for a long time.'[2] Lenin had said that N.E.P. had been conceived 'seriously and for long'; and the opponents of planning often quoted the saying. 'Yes, seriously and for a long time', Trotsky retorted, 'but not for ever. We have introduced N.E.P. in order to defeat it on its own ground and largely by its own methods. In what way?

[1] Ibid., pp. 292–3.   [2] Ibid., p. 285.

By making effective use of the laws of the market economy . . .
and also by intervening through our state-owned industry in the
play of those laws and by systematically broadening the scope
of planning. *Eventually* we shall extend planning to the entire
scope of the market, thereby absorbing and abolishing the
market.'[1]

Bolshevik views on the relationship between planning and a
market economy were still extremely vague. Most Bolsheviks
considered N.E.P. almost incompatible with planning. They
saw in N.E.P. an act of appeasement of private property to
which they had been driven by weakness. They thought that
the need for such appeasement would remain for years, and so
it was necessary to stress the stability of N.E.P. and to strengthen
the peasants' and the merchants' confidence in it. Only in a
more or less remote future would the party be able to withdraw
the concessions it had made to private property and to abolish
N.E.P.; and only then would it be possible to set up a planned
economy. This view was to underlie Stalin's policies throughout
the decade, in the course of which he first resisted planning in the
name of N.E.P. and then for the sake of planning decreed the
'abolition' of N.E.P., 'liquidated' private trade, and destroyed
private farming.

In Trotsky's conception N.E.P. was designed not merely to
appease private property. It had set the framework for long-term
co-operation, competition, and struggle between the socialist
and the private sectors of the economy. Co-operation and
struggle appeared to him dialectically opposed aspects of a
single process. Consequently he called on the party to protect
and expand the socialist sector, even while it conciliated and
helped to develop the private sector. Socialist planning would
not one day supersede N.E.P. at a stroke. Planning should
develop within the mixed economy until the socialist sector had
by its growing preponderance gradually absorbed, transformed,
or eliminated the private sector and outgrown the framework of
N.E.P. There was thus in Trotsky's scheme of things no room
for any sudden 'abolition' of N.E.P., for prohibition of private
trade by decree and for the violent destruction of private farming, just as there was no room for any administrative proclamation of the 'transition to socialism'. This difference between

---

[1] *12 Syezd RKP (b)*, p. 331.

Trotsky's and Stalin's approach was to show itself most strikingly only at the turn of the decade. Immediately, however, because of his insistence on the need for an offensive socialist policy, Trotsky seemed to many people basically opposed to N.E.P.

There is no need to go here into the economic detail of Trotsky's argument or into the case he made for primitive socialist accumulation—his ideas on this are summarized in the previous chapter. Suffice it to say that his address and the 'Theses' he presented are among the most crucial documents on Soviet economic history; and that he drew there a perspective on the Soviet economy for several decades ahead, the decades during which the evolution of the Soviet Union was to be determined by the processes of forced capital formation in an underdeveloped but largely nationalized economy. The Marxist historian may indeed describe and analyse those decades, the Stalinist decades, as the era of primitive socialist accumulation; and he may do so in terms borrowed from Trotsky's exposition of the idea in 1923.[1]

But whatever were the historic merits of Trotsky's performance at the twelfth congress and of whatever interest that performance may be to any study of Marxist ideas, it did not improve Trotsky's position for the struggle which awaited him. His central idea was, on the whole, beyond the understanding of his audience. The congress was as usual impressed, but this time it was impressed by the *élan* of his speech rather than by its content. The few implications of his thought that the mass of delegates could grasp were such as to arouse apprehension and even suspicion. Some could not help wondering whether he was not, after all, calling on the party to abandon N.E.P. and to return to the disastrous policies of war communism. When he demanded that industrial output should be concentrated in a small number of large and efficient concerns, the question arose what would happen to the workers who lost employment through the closing down of inefficient factories. When he argued that the working class would have to shoulder the main burden of industrial reconstruction, he made not the slightest attempt to soften the harsh impact of his words. On the contrary, he gave

---

[1] In later years Trotsky himself rarely, if ever, spoke of 'primitive socialist accumulation'.

his thought an overemphasis which was bound to startle and
shock many workers. 'There may be moments', he said, 'when
the government pays you no wages, or when it pays you only
half your wage and when you, the worker, have to lend [the
other half] to the state.'[1] It was in this way, by 'taking away
half the worker's wage', that Stalin later promoted accumulation; but then he told the workers that the state paid them two
or three times the wages they had earned before. When Trotsky
put this issue before the congress with all his bluntness and
merciless honesty, the workers were struck by his mercilessness
rather than by his honesty. Is he again telling us, they could not
help reflecting, as he told us when he formed the Labour Armies,
that we must take the producer's, not the consumer's view?
Nothing would be easier for the agents of the triumvirs than to
confirm the workers in this suspicion.

And how, asked others, would Trotsky's policy affect the
peasantry? Would it not drive the party to a collision with the
muzhik? It would, Rykov and Sokolnikov had already said at
the Politbureau and at the Central Committee. A significant
incident at the congress gave new point to the question. In the
debate Krasin, Trotsky's old comrade, addressed himself directly
to Trotsky and asked whether he had thought out to the
end the implications of primitive socialist accumulation? Early
capitalism, Krasin pointed out, did not merely underpay workers
or rely on the entrepreneur's 'abstinence' to promote accumulation. It exploited colonies; it 'pillaged entire continents'; it
destroyed the yeomanry of England; it ruined the cottage
weavers of India and on their bones, which 'whitened the plains
of India', rose the modern textile industry. Did Trotsky carry
the analogy to its logical conclusion?[2]

Krasin put the question without hostile intent. He approached
it from his particular angle: as Commissar of Foreign Trade he
had tried to persuade the Central Committee of the need for
more foreign trade—and of the need to make more concessions
to foreign capital. He wished to impress on the congress that
since as Bolsheviks they could not expropriate peasants and
plunder colonies—everyone took this for granted—they must
seek to attract foreign loans; and that foreign capital might
help Russia to proceed with primitive accumulation and to

[1] *12 Syezd RKP (b)*, p. 315.   [2] Ibid., pp. 351–2.

avoid the horrors that had accompanied such accumulation in the West. The Bolsheviks, however, had found out by now that they had little chance of attracting foreign credits on acceptable terms; and so the question which Krasin posed retained its full force: where would the resources needed for rapid accumulation come from? When Krasin spoke of the plunder of the peasantry and the 'white bones' of the Hindu cottage weavers, Trotsky jumped to his feet to protest that he had 'proposed nothing of the sort'.[1] This was true enough. Still, did not the logic of his attitude lead, after all, to 'plundering the peasantry'? That Trotsky jumped to his feet to deny it indicates that he felt a cloud of suspicion not yet larger than a man's hand gathering over his head.

Having said so much that was likely to antagonize the workers and to stir in the party the fear of a collision with the peasantry, Trotsky then incurred the enmity of the industrial managers and administrators. He could not help saying the most unpopular things once he was convinced that what he had to say was of vital importance and that it was his duty to say it. And so he drew the picture of the condition of industry in such dark colours and flayed the new economic bureaucracy so pitilessly for complacency, conceit, and inefficiency that it smarted under his lash and sought to work off the grudge. Trotsky, the managers replied, saw the economy in such dark colours and was so displeased with their work because he would content himself with nothing less than the Utopia of a planned economy.[2]

Thus slowly but inexorably the circumstances which eventually led to Trotsky's defeat began to unfold and agglomerate. He missed the opportunity of confounding the triumvirs and discrediting Stalin. He let down his allies. He failed to act as Lenin's mouthpiece with the resolution Lenin had expected of him. He failed to support before the entire party the Georgians and the Ukrainians for whom he had stood up in the Politbureau. He kept silent when the cry for inner-party democracy rose from the floor. He expounded economic ideas the historic portent of which escaped his audience but which his adversaries could easily twist so as to impress presently upon workers, peasants, and bureaucrats alike that Trotsky was not their

[1] Loc. cit.   [2] Ibid., pp. 322–50 and *passim*.

well-wisher, and that every social class and group ought to tremble at the mere thought that he might become Lenin's successor. At the same time the triumvirs sought assiduously to please everybody, promising something to every social class and group, pandering to every kind of complacency, and flattering every imaginable conceit.

Finally, Trotsky directly strengthened the triumvirs when he declared his 'unshaken' solidarity with the Politbureau and the Central Committee and called the rank and file to exercise 'at this critical juncture' the strictest self-restraint and the utmost vigilance. Speaking on a motion appealing for unity and discipline in Lenin's absence, he stated: 'I shall not be the last in our midst to defend [this motion], to put it into effect, and to fight ruthlessly against all who may try to infringe it.'[1] 'If in the present mood', he went on, 'the party warns you emphatically about things which seem dangerous to it, the party is right, even if it exaggerates, because what might not be dangerous in other circumstances must appear doubly and trebly suspect at present.' In this state of alarm and heightened suspicion the triumvirs would, of course, find it easy to assert themselves and stifle opposition. Trotsky shared their anxiety over the shock to which Lenin's death might expose the party; and in his eagerness to strengthen the party he weakened his own position in it. No doubt he counted on the triumvirs' loyalty. Little though he thought of them, he treated them as comrades whom he expected to behave towards him with a certain propriety. He did not imagine that they would turn his unselfish gestures to their immediate and private advantage.

The enlarged Central Committee elected at the twelfth congress reappointed Stalin as General Secretary. Trotsky made no attempt to prevent this—at any rate, he did not propose any other candidate, as he knew Lenin would have done. In Lenin's absence he had no chance of displacing Stalin anyhow. The triumvirs swayed the Politbureau and through it the Central Committee as before. They also dominated the new Central Control Commission elected to act as the party's supreme disciplinary Court. The man appointed to preside over it was Kuibyshev, Stalin's close associate.

[1] *12 Syezd RKP (b)*, p. 320.

The triumvirs had no reason to precipitate a showdown with Trotsky. He offered no provocation; and they were not yet sure how the party would behave if the conflict came into the open. Yet Stalin lost no time in setting the stage. He used his wide powers of appointment to eliminate from important posts, in the centre and in the provinces, members who might be expected to follow Trotsky; and he filled the vacancies with adherents of the triumvirate or preferably of himself. He took great care to justify the promotions and demotions on the apparent merits of each case; and he was greatly assisted by the rule, which Lenin had established, that appointments should be made with reference to the number of years a member had served the party. This rule automatically favoured the Old Guard, especially its caucus.

It was in the course of this year, the year 1923, that Stalin, making full use of this system of patronage, imperceptibly became the party's master. The officials whom he nominated as regional or local secretaries knew that their positions and confirmation in office did not depend on the members of the organization on the spot but on the General Secretariat. Naturally they listened much more attentively to the tune called by the General Secretary than to views expressed in local party branches. The phalanx of these secretaries now came to 'substitute' itself for the party, and even for the Old Guard of which they formed an important section. The more they grew accustomed to act uniformly under the orders of the General Secretariat, the more it was the latter which virtually substituted itself for the party as a whole. In theory the party was still governed by the Central Committee and by decisions of party congresses. But henceforth a party congress could only be a sham: as a rule only nominees of the General Secretariat had any chance of being elected as delegates.

Trotsky watched this change in the party, grasped its significance, but could do nothing to arrest it. There was only one way in which he might have tried to counteract it: by appealing openly to the rank and file and calling on them to resist the impositions of the General Secretariat. But as Stalin was backed by the Politbureau and the majority of the Central Committee, this would have been incitement against the newly elected and regularly constituted leadership. No single member of the

Politbureau, not even one enjoying the highest authority, could risk such a step. Least of all could Trotsky risk it now, after he had concealed from the party his differences with the triumvirs, after he had solemnly declared his full solidarity with them, and pledged himself to act as the most zealous and vigilant guardian of discipline. If he were to try to arouse the party against the triumvirs, he would appear to act hypocritically, from a private grudge, or from the ambition to take Lenin's place.

For the time being he could resist Stalin only within the Politbureau and the Central Committee. But there he was isolated and his words counted for little. Even Bukharin was inclining ever more towards the triumvirs. (Among the forty members of the new Central Committee Trotsky had no more than three political friends: Rakovsky, Radek, and Pyatakov.) The sessions of the Politbureau held in his presence were becoming mere formalities: all the cards were stacked against him; and the real Politbureau worked in his absence. Thus shortly after the twelfth congress he began to pay the penalty of procrastination. He was already the political prisoner of the triumvirs. Unable to achieve anything against them within the leading party bodies, and unable to undertake any action against them from without, he could only bide his time and wait for some event to open a new prospect.

.   .   .   .

In the summer of 1923 Moscow and Petrograd were suddenly shaken by a political fever. Throughout July and August there was a great deal of industrial unrest. Workers felt that they were made to carry too much of the burden of industrial re-covery. Their wages were a mere pittance; and often they did not receive even this. Industrial managers, running concerns at a loss and deprived of state subsidies and credits, had been unable to pay the men, had been in arrears to them for long months, and resorted to painful frauds and tricks to cut wage bills. Trade unions, reluctant to disturb the industrial revival, refused to press claims. Finally, 'wild' strikes broke out in many factories, spread, and were accompanied by violent explosions of discontent. The trade unions were caught by surprise; and so were the party leaders. The threat of a general strike was in the air; and the movement seemed on the point of turning into a political revolt. Not since the Kronstadt rising had there been so much tension in the working class and so much alarm in ruling circles.

The shock was all the more severe because it was unexpected. The ruling circles had viewed the economic situation with smugness and had boasted of continuous improvement. They had not received timely signals of the approaching trouble; or, if any warning had reached them, they ignored it. Rudely awakened, they began to look for the culprits who had incited the workers. Lower down, in the party branches, the commotion led people to inquire more seriously how it was that more than two years after the promulgation of N.E.P. there was still so much bitter discontent. What was the worth, they asked, of the official progress reports? Had not the party leaders been too complacent and had they not lost contact with the working class? There was not much use in looking for culprits if these questions remained unanswered.

The culprits were not easy to find. The agitation for strikes could not be traced to any source such as the remnants of the anti-Bolshevik parties—these, thoroughly suppressed, had been inactive. Official suspicion turned on the Workers' Opposition. But the leaders of the latter, too, had been surprised by the strikes. Intimidated by constant threats of expulsion, the Workers' Opposition had lain low and was breaking up. Its splinter groups, however, had to some extent been involved in the strike agitation, which was spontaneous in the main. The most important of these was the Workers' Group, led by three labourers, Myasnikov, Kuznetsov, and Moiseev, all party members at least since 1905. In April and May, immediately after the twelfth congress, they circulated a manifesto denouncing the New Exploitation of the Proletariat and urging the workers to fight for Soviet democracy.[1] In May Myasnikov was arrested. But his followers went on propagating his views. When the strikes broke out they wondered whether they should not go to the factories with the call for a general strike. They were still arguing about this when the G.P.U. arrested them, about twenty persons in all.[2]

[1] The manifesto was published by German sympathizers of the group in Berlin in 1924. *Das Manifest der Arbeitergruppe der Russischen Kommunistischen Partei.*

[2] V. Sorin, *Rabochaya Gruppa*, pp. 97–112. The group apparently had 200 members in Moscow.

The discovery that this and similar groups, like the Workers' Truth, had been active in the factories caused among party leaders a dismay which seemed quite out of proportion to its cause. But small as these groups were, they had many contacts in party and trade unions. Rank-and-file Bolsheviks listened to their arguments with open or sneaking sympathy. As the trade unions did not voice, and as the party paid all too little attention to, the workers' grievances, small political sects, had they not been stopped, might have rapidly acquired a broad influence and placed themselves at the head of the discontent. The instigators of the Kronstadt revolt had not been more numerous or influential; and where there is much inflammable material a few sparks may produce a conflagration. The party leaders sought to stamp out the sparks. They determined to suppress the Workers' Group and the Workers' Truth on the ground that the members of these organizations no longer considered themselves bound by party discipline and conducted half-clandestinely an agitation against the government. Dzerzhinsky was charged with the business of suppression. As he investigated the activities of the presumed culprits, he found that even party members of unquestioned loyalty regarded them as comrades and refused to testify against them. He then turned to the Politbureau and asked it to declare that it was the duty of any party member to denounce to the G.P.U. people who inside the party engaged in aggressive action against the official leaders.

The issue came before the Politbureau just after Trotsky had had several clashes with the triumvirs which had envenomed their relations; and Dzerzhinsky's demand was more than he could stomach. He was not at all eager to defend the Workers' Group and kindred sets of dissenters. He did not protest when their adherents were thrown into prison. Although he held that much of their discontent was justified and that many of their criticisms were well founded, he had no sympathy with their crude and anarchic tub-thumping. Nor was he inclined to countenance industrial unrest. He did not see how the government could meet the workers' demands when industrial output was still negligible: it was no use paying higher wages when wages could buy no goods. He saw that the strikes by delaying recovery only made matters worse; and he refused to seek popularity by bandying promises that could not be honoured or by exploiting grievances. Instead, he urged again the long over-due change in economic policy. Nor was he at all eager to sup-port the demand for Soviet democracy in that extreme form in which the Workers' Opposition and its splinter groups had raised it. But he took exception to the manner in which the triumvirs and Dzerzhinsky proposed to deal with the trouble and to the obstinacy with which they dwelt on the symptoms of the discontent instead of turning to the underlying cause. When he saw that the Politbureau was on the point of ordering party members to spy upon and to denounce one another, he was seized with disgust.

Dzerzhinsky's demand had raised a delicate issue, because the attitude of the Bolsheviks to the G.P.U. had in it nothing of that haughty distaste with which the good bourgeois democrat normally views any political police. The G.P.U. was the 'sword of the revolution'; and every Bolshevik had been proud to assist it in work directed against the revolution's enemies. But after the civil war, when the reaction against the terror set in, many of those who had volunteered to serve in the G.P.U. were glad to leave its ranks. 'Only saints or scoundrels can serve in the G.P.U.', Dzerzhinsky complained to Radek and Brandler about this time, 'but now the saints are running away from me, and I am left with the scoundrels.'[1] Yet this debased G.P.U. was still the guardian of the Bolshevik monopoly of power. Hitherto it had defended it only against external enemies, White Guards, Mensheviks, Social Revolutionaries, and Anar-chists. The question was whether the G.P.U. should also defend the monopoly against its supposed Bolshevik enemies? If so, then it could not do it otherwise than by operating within the party itself.

Trotsky did not tell the Politbureau plainly that it should reject Dzerzhinsky's demand. He evaded the question and dwelt on the underlying issue. 'It would seem', he wrote in a letter to the Central Committee on 8 October 1923, 'that to in-form the party organization of the fact that its branches are being used by elements hostile to it is an obligation of members so elementary that it ought not to be necessary to introduce a special resolution to that effect six years after the October Revolution. The very demand for such a resolution is an

[1] This has been related to the writer by Brandler.

extremely startling symptom alongside of others no less clear.
. . .'[1] It pointed to the gulf that now separated the leaders
from the rank and file, the gulf which had grown especially
wide since the twelfth congress and was deepened by Stalin's
system of patronage.

When Trotsky stated this, the triumvirs reminded him that
he himself had under war communism ruled the trade unions
through his nominees. He replied that even at the height of the
civil war 'the system of appointment within the party did not
have one-tenth of the extent that it has now. Appointment of
the secretaries of provincial committees is now the rule. That
creates for the Secretary a position essentially independent of
the local organization. . . .' Trotsky did not explicitly question
the General Secretary's prerogatives—he merely urged him to
make moderate and prudent use of them. He confessed that at
the last congress, when he listened to the pleas made there for
proletarian democracy, many of these 'seemed to me exaggerated and to a considerable extent demagoguish, because a
fully developed workers' democracy is incompatible with the
régime of dictatorship'. However, the party ought not to go on
living under the high pressure of civil war discipline. This 'ought
to give place to a livelier and broader party responsibility. The
present régime . . . is much further from any workers' democracy
than was the régime of the fiercest period of war communism.'
'Secretarial selection' was responsible for 'unheard-of bureau-
cratization of the party apparatus.' The hierarchy of secretaries
'created party opinion', discouraged members from expressing
or even possessing views of their own, and addressed the rank
and file only in words of command and summons. No wonder
that discontent which could not 'dissipate itself through open
exchange of opinions at party meetings and through the exercise of influence upon the party organization by the mass of
members . . . accumulated in secret and gave rise to strains and
stresses'.[2]

Trotsky also renewed his attack on the triumvirs' economic
policy. The ferment within the party, he argued, was intensified
by industrial unrest; and this had been brought about by lack
of economic foresight. He had found out by now that the only
gain the triumvirs had allowed him to score at the twelfth

[1] Max Eastman, *Since Lenin Died*, pp. 142–3.   [2] Ibid.

congress, the gain for the sake of which he had yielded so much ground to them, was spurious: the congress had adopted his resolutions on industrial policy, but these had remained a dead letter. Now as before the economic administration bungled and muddled. Nothing had been done to make Gosplan the guiding centre of the economy. The Politbureau set up a number of committees to investigate symptoms of the crisis instead of going to its root. Trotsky himself had been invited to serve on a committee which was to inquire into prices; but he refused to do so. He had no wish, he declared, to participate in an activity designed to dodge issues and to postpone decisions.

Just before Trotsky made these criticisms he had his clashes with the triumvirs, already mentioned. Some of these occurred in deliberations over the situation in Germany, where, Trotsky held, the turmoil provoked by the French occupation of the Ruhr offered the German communists a unique chance. Other collisions developed when the triumvirs proposed changes in the Military Revolutionary Council over which Trotsky presided. Zinoviev was bent on introducing into that Council either Stalin himself or at least Voroshilov and Lashevich. It is not quite clear what induced him to make this proposal, and whether he acted in agreement with Stalin from an anxiety to gain for the triumvirs a decisive share in the control of military affairs; or whether he was already engaged in a subtle move against Stalin designed to oust him from the General Secretariat.[1] Enough that when Zinoviev tabled his motion, Trotsky, hurt and indignant, declared that he was resigning in protest from every office he held, the Commissariat of War, the Military Revolutionary Council, the Politbureau, and the Central Committee. He asked to be sent abroad 'as a soldier of the revolution' to help the German Communist party to prepare its revolution. The idea had not come out of the blue. The leader of the German party, Heinrich Brandler, had just arrived in Moscow; and doubting his own and his comrades' capacity to lead an insurrection, had inquired in all earnestness from Trotsky and Zinoviev whether Trotsky could not come *incognito* to Berlin or Saxony to take charge of revolutionary operations.[2] The idea stirred Trotsky; and the danger of the mission excited his courage. Disillusioned by the turn events had taken in

[1] See further, p. 241.   [2] The source for this statement is Brandler.

Russia, disgusted with the Politbureau cabal, and perhaps already tired of it, he asked for the assignment. To contribute once more to the victory of a fighting revolution suited him better than to taste the maggoty fruit of a victorious one.

The triumvirs could not let him go. In Germany he might have become doubly dangerous. If he went, succeeded, and returned in triumph, he would have dwarfed them as the acknowledged leader of both the Russian and the German revolutions. But if something untoward were to happen to him, if he were to fall into the hands of the class enemy or to die fighting, the party would suspect that they had sent him on a hopeless mission to get rid of him; and neither Stalin nor his partners could as yet risk such a suspicion. They could not permit Trotsky to win either the laurels of a new revolutionary victory or even the martyr's crown. They got out of the difficulty by turning the painful scene into a farce. Zinoviev replied that he himself, the President of the Communist International, would go to Germany 'as a soldier of the revolution' instead of Trotsky. Then Stalin intervened, and with a display of bonhomie and common sense said that the Politbureau could not possibly dispense with the services of either of its two most eminent and well-beloved members. Nor could it accept Trotsky's resignation from the Commissariat of War and the Central Committee, which would create a scandal of the first magnitude. As for himself, he, Stalin, would be content to remain excluded from the Military Revolutionary Committee if this could restore harmony. The Politbureau accepted Stalin's 'solution'; and Trotsky, feeling the grotesqueness of the situation, left the hall in the middle of the meeting 'banging the door behind him.'[1]

Such was the state of affairs in the Politbureau just before Dzerzhinsky made his proposal and Trotsky wrote the letter of 8 October, in which he confronted the triumvirs with a definite

---

[1] Stalin's ex-secretary thus underlines the grotesqueness of the incident: 'The scene took place in the Throne Hall. The door of the Hall is enormous and massive. Trotsky ran towards it, pulled it with all his strength, but it was dead slow to open. Some doors do not lend themselves to banging. In his fury, however, he failed to notice this; and he made yet another violent effort to close the door. Alas, the door was as slow to close as it was to open. Thus, instead of witnessing a dramatic gesture, indicating a historic break, we watched a sorry and helpless figure struggling with a door. . . .' Bajanov, *Avec Staline dans le Kremlin*, pp. 76–77.

PLATE V

(*a*) Lenin during his illness

(*b*) Lenin in his family circle with his wife Krupskaya next to him and his sister Elizarova

(*a*) Stalin during the struggle for the
succession

Radio Times Hulton Picture Library

(*b*) Zinoviev

Radio Times Hulton Picture Library

(*c*) Kamenev

THE TRIUMVIRATE

challenge. The latter were not yet unduly disturbed, because he did not carry the controversy into the open: his letter was addressed only to the members of the Central Committee who were entitled to know the Politbureau's secrets.

However, a week later, on 15 October, forty-six prominent party members issued a solemn statement directed against the official leadership and criticizing its policy in terms almost identical with those Trotsky had employed. They declared that the country was threatened with economic ruin because the 'majority of the Politbureau' had no policy and did not see the need of purposeful direction and planning of industry. They did not demand any definite change in the leadership; they only urged the Politbureau to awaken to its task. They, too, protested against the rule of the hierarchy of secretaries and against the stifling of discussion, alleging that the regular party congresses and conferences, packed by nominees, had ceased to be representative. Then, going farther than Trotsky, the Forty Six demanded that the ban on inner party groupings should be abolished or relaxed because it served one faction as a screen for its dictatorship over the party, drove disgruntled members to form clandestine groups, and strained their loyalty towards the party. 'The inner party struggle is waged all the more savagely the more it is waged in silence and secrecy.' Finally, the signatories of the statement asked that the Central Committee should call an emergency conference to review the situation.[1]

The Forty Six echoed Trotsky's criticisms so faithfully that the triumvirs could not but suspect that he was their direct inspirer, if not the organizer of their protest.[2] They assumed that the Forty Six had come together to form a solid faction. Trotsky's attitude was in fact more reserved than the triumvirs believed. True, among the Forty Six were his close political friends: Yuri Pyatakov, the most able and enlightened of the industrial administrators, Evgenii Preobrazhensky, the economist and former Secretary of the Central Committee, Lev Sosnovsky, *Pravda*'s gifted contributor, Ivan Smirnov, the victor over Kolchak, Antonov-Ovseenko, hero of the October insurrection, now chief political commissar of the Red Army, Muralov,

---

[1] *The Trotsky Archives.*

[2] Trotsky's responsibility for the action of the Forty Six was in the centre of the debate at the thirteenth party conference in January 1924.

commander of the Moscow garrison, and others. To these men Trotsky had confided his thoughts and anxieties; and he kept some of them informed even of his intimate talks with Lenin.[1] They formed the leading circle of the so-called 1923 Opposition and represented the 'Trotskyist' element in it. But the Forty Six were not a uniform group. There were among them also adherents of the Workers' Opposition and Decemists like V. Smirnov, Sapronov, Kossior, Bubnov, and Ossinsky, whose views diverged from those of the Trotskyists. Many of the signatories appended to the common statement strong reservations on special points or plain expressions of dissent. The statement dwelt with equal emphasis on two issues: economic planning and inner-party democracy. But some signatories were primarily interested in the former while others took the latter more to heart. Men like Preobrazhensky and Pyatakov demanded freedom of criticism and debate primarily because they were opposed to specific economic policies and hoped through debate to convert others to their views; while members like Sapronov and Sosnovsky were in opposition chiefly because they cherished inner-party freedom for its own sake. The former voiced the aspirations of the advanced and educated *élite* of the Bolshevik bureaucracy itself while the latter expressed a revulsion against bureaucracy at large. Far from forming a solid faction, the Forty Six were a loose coalition of groups and individuals united only on a vaguely common denominator of discontents and strivings.

Whether or to what extent Trotsky should be regarded as the direct sponsor of this coalition is not certain. He himself denied this while his adversaries claimed that his denial was a *ruse de guerre* to which he resorted in order to avoid the blame for organizing a faction.[2] However, they offered no specific proof; and the Forty Six did not act as a coherent faction with a distinct line of conduct and discipline. Even many years after Trotsky's death those who had stood close to him claimed that he observed the rules of discipline so strictly that he could not have acted as the sponsor of this particular demonstration of protest. In the light of all that is known about Trotsky's conduct in such matters, this may be accepted as true. However, it is

[1] *Moya Zhizn*, vol. ii, p. 215.
[2] *13 Konferentsya RKP (b)*, pp. 46, 92–102, 104–13; *13 Syezd RKP (b)*, pp. 156 ff.

doubtful that he had, as is also claimed, no foreknowledge of the action of the Forty Six or that he was surprised by it. Preobrazhensky, Muralov, or Antonov-Ovseenko undoubtedly kept him informed about what they were doing, and would not have done what they were doing without some encouragement from him. And so even if Trotsky was not formally responsible for their action, he must be regarded as its actual prompter.

The Forty Six addressed their protest to the Central Committee with the request that the Committee should, in accordance with long-established custom, bring it to the party's knowledge. The triumvirs refused the request. Moreover, they threatened to apply disciplinary sanctions if the signatories themselves were to circulate the document among party members. At the same time agents of the Central Committee were sent to the cells to denounce the authors of the unpublished protest. Then a special enlarged session of the Central Committee was held to deal with the statement of the Forty Six and with Trotsky's letter of 8 October.[1] Replying to Trotsky, the triumvirs repeated the charges which Stalin had brought against him at the Politbureau meetings of January and February. Trotsky, they alleged, was actuated by lust of power, and, sticking to the maxim 'all or nothing', he refused not merely to serve as Lenin's deputy, but even to attend to his normal duties. Then they enumerated all the issues over which he had in recent years dissented from Lenin; but they passed over in silence the fact that on nearly all these issues Lenin had in the end found himself in agreement with Trotsky. The Central Committee endorsed the charges and censured Trotsky. It also reprimanded the Forty Six, qualifying their joint protest as an infringement of the 1921 ban on factions. As to Trotsky, it did not charge him plainly with organizing the faction but held him morally responsible for the offence of which it found the Forty Six guilty.

The condemnation threw into relief the vicious circle in which any incipient opposition found itself under the disciplinary rules of 1921. The Forty Six had come forward precisely in order to demand that those rules be revoked or relaxed. But it was enough of them to speak up for a revision of the rules to lay themselves open to the accusation that they had already violated them. The ban on inner-party groupings was self-perpetuating

[1] *KPSS v Rezolutsyakh*, vol. i, pp. 766–8.

and irreversible: under it no movement for its revision could be set afoot. It established within the party that barrack discipline which may be meat for an army but is poison for a political organization—the discipline which allows a single man to vent a grievance but treats the joint expression of the same grievance by several men as mutiny.

The triumvirs could not easily suppress this particular 'mutiny'. The mutineers were not ordinary rankers—they were forty-six generals of revolution. Every one of them had held important positions in government and party. Most had a heroic civil-war record. Many had been members of the Central Committee. Some had joined the Bolsheviks in 1917, together with Trotsky; others had been Bolsheviks since 1904. Their protest could not be concealed. By denouncing it to the cells and calling upon the cells to join in the denunciation but refusing to show the condemned statement, the triumvirs aroused intense suspicion. The party was astir with alarming rumours. The triumvirs had to open at least a safety valve. On 7 November, the sixth anniversary of the revolution, Zinoviev made a solemn statement promising to restore democracy within the party. As a token, *Pravda* and other newspapers opened their pages for discussion and invited members to write frankly on all issues which troubled them.

To initiate a debate after 'three years of silence' was a risky undertaking.[1] The triumvirs knew it. They opened the discussion in Moscow and delayed it in the provinces. But no sooner had they lifted a safety valve than they were hit by pressure of unsuspected force. Moscow's party cells were in revolt. They received official leaders with hostility and acclaimed spokesmen of the opposition. At some meetings in large factories the triumvirs themselves were met with derision and were heavily outvoted.[2] The discussion was at once focused on the statement of the Forty Six who were now free to expound their views to the rank and file. Pyatakov was their most aggressive and effective spokesman; wherever he went he easily obtained large majorities for bluntly worded resolutions. Antonov-Ovseenko addressed

---

[1] At the thirteenth party conference Radek spoke of 'three years of silence' which preceded the discussion. *13 Konferentsya RKP (b)*, pp. 135-7.

[2] This was admitted by Rykov. Ibid., pp. 83–91. See also Preobrazhensky's description of the crisis in the party. Ibid., pp. 104–13.

the party organizations of the garrison; and shortly after the debate had begun at least one-third of those organizations had sided with the opposition. The Central Committee of the Communist Youth and most of the Comsomol cells in Moscow did likewise. The universities were seized with excitement; and a large majority of student cells declared enthusiastic support for the Forty Six. The leaders of the opposition were in a buoyant mood. According to one version, they were so self-confident that they discussed among themselves in what proportion they would be willing to share with the triumvirs control over the party machine.

The triumvirs took fright. When they saw which way the vote in the garrison cells went, they resolved that these cells must not be permitted to proceed with the vote. They at once dismissed Antonov-Ovseenko from his post as the Red Army's chief political commissar, alleging that he had threatened the Central Committee that the armed forces would stand up 'like one man' for Trotsky, 'the leader, organizer, and inspirer of the revolution's victories'.[1] Antonov-Ovseenko had not in fact threatened any military revolt. What he had meant and said was that the military party cells were 'like one man' behind Trotsky. This, no doubt, was an impulsive overstatement, but it was not very far from the truth. Nor had Antonov-Ovseenko acted illegitimately in carrying the discussion to the military cells. These had the same right as had the civilian cells to take part in any debate and to vote on policy; and they had never before been denied this right. But whether Antonov's behaviour was or was not beyond reproach—Trotsky held that he might have exercised more prudence in a delicate situation—the triumvirs decided that they could not leave him at the head of the army's political department. Demotions of other critics followed. The General Secretariat, violating the statutes, disbanded the Central Committee of the Comsomol and replaced it by nominees.[2] Disciplinary reprisals were applied to other supporters of the opposition as well, and every imaginable device was used to obstruct the further progress of controversy.

All this, however, did not relieve the tension. The triumvirs then decided to confound the opposition by taking a leaf from its book. They framed a special resolution bluntly denouncing

[1] Ibid., p. 124.　　　　　[2] *14 Syezd VKP (b)*, p. 459.

the 'bureaucratic régime within the party' in terms which sounded like a plagiarism from Trotsky and the Forty Six; and they proclaimed the opening of a New Course which was to guarantee full freedom of expression and criticism for party members.

Throughout November, when Moscow was all excitement, Trotsky did not take part in the public controversy. An accident of ill health reduced him to silence. Late in October, during a week-end hunting trip to the marshy country outside Moscow, he had contracted a malarial infection; and he was bedridden with fever during these decisive months. It is curious to note how such accidents—first Lenin's illness and then his own—contributed to the trend of events which was more solidly determined by the basic factors of the situation. 'One can foresee a revolution or a war', Trotsky remarks in *My Life*, 'but it is impossible to foresee the consequences of an autumn shooting trip for wild ducks.'[1] It was certainly no mean disadvantage to Trotsky that at this crucial stage the use of his live voice and direct appeal to an audience was denied him.

Those were hard days [his wife writes], days of tense fighting for Lev Davidovich at the Politbureau against the rest of its members. He was alone and ill and had to fight them all. Because of his illness the Politbureau held its meetings in our apartment; I sat in the adjoining bedroom and heard his speeches. He spoke with his whole being; it seemed as if with every such speech he lost some of his health—he spoke with so much 'blood'. And in reply I heard cold and indifferent answers. . . . After each of these meetings L.D.'s temperature rose. He came out of his study soaked through, and undressed and went to bed. His linen and clothes had to be dried as if he had been drenched in a rain storm.[2]

When the triumvirs decided to confound the opposition by a resounding proclamation of the New Course, they were anxious that Trotsky should endorse the proclamation. They asked him to put his signature next to theirs under the text they had plagiarized from him. He could not refuse without giving the party the impression that it was he who stood in the way of its freedom; and he hoped that the formal inauguration of a public debate would at least enable him to bring into the open the issues over which he had wrestled with the triumvirs in the

[1] *Moya Zhizn*, vol. ii, pp. 234 ff.   [2] Op. cit., vol. ii, p. 240.

secrecy of the Politbureau. Yet he could not but suspect that he was being asked to endorse an empty promise. Only a few weeks later one of the leaders of the opposition compared this pro-clamation to the October Manifesto of 1905, that promise of constitutional freedoms which the last Tsar had made in a moment of weakness and which he withdrew as soon as his strength returned.[1] In October 1905 the young Trotsky, when he appeared for the first time before the revolutionary crowds of St. Petersburg, crumpled in his hand the Tsar's Manifesto and warned the people: 'To-day it has been given us and to-morrow it will be taken away and torn into pieces as I am now tearing it into pieces, this paper-liberty, before your very eyes.'[2] Now, in 1923, he could not go out to the crowds and tear to pieces the 'new October Manifesto' in front of them. It was to be pro-claimed in the name of the Politbureau of which he was a mem-ber; and he was striving to reform, not to subvert, the established government. So, when the Politbureau brought the motion on the New Course to his bedside, he could only seek to introduce amendments designed to make the promise of inner-party free-dom as plain and emphatic as possible and thereby to commit the triumvirs. The Politbureau accepted all his amendments; and on 5 December it voted unanimously for the motion.[3] Yet, although he had voted in favour, Trotsky could not help repeat-ing after a fashion his gesture of 1905.

He did this in a few brief articles which he wrote for *Pravda* and which later appeared in his pamphlet *The New Course*.[4] These articles contain in a nutshell most of the ideas which at once became the hallmark of 'Trotskyism'. He began with an essay which appeared on 4 December, the day before the Polit-bureau voted on the New Course. This was a somewhat cryptic attack on 'officialdom' in his own department, the army, 'and —elsewhere'. The vices of officialdom, he wrote, show them-selves when people 'cease to think things through; when they smugly employ conventional phrases without reflecting on what they mean; when they give the customary orders without asking

[1] See Sapronov's speech in *13 Konferentsya RKP (b)*, pp. 131–3.
[2] *The Prophet Armed*, pp. 128–9.
[3] The text appeared in *Pravda* on 7 December 1923.
[4] The quotations in the following pages are from the American edition of the pamphlet; but the text of the translation has been occasionally rephrased after comparison with the original.

if they are rational; when they take fright at every new word, every criticism, every initiative, every sign of independence....'[1] The 'soul-uplifting' lie was the daily bread of officialdom. It could be found in histories of the Red Army and of the civil war, where truth was sacrificed to bureaucratic legend. 'To read it, you would think that there are only heroes in our ranks; that every soldier burns with the desire to fight; that the enemy is always superior in numbers; that all our orders are reasonable and appropriate to the occasion; that the execution is always brilliant; and so on.' The edifying effect of such legends is itself a legend. The Red soldier would listen to them as 'his father listened to the *Lives of Saints*: just as magnificent and uplifting, but not true to life'.

Supreme heroism, in military art as in revolution, consists of truthfulness and a sense of responsibility. We speak for truthfulness not from the standpoint of the abstract moralist who teaches that man must never lie or deceive his neighbour. Such idealistic talk is sheer hypocrisy in a class society where there are antagonistic interests, struggle, and war. Military art in particular includes, as it must, *ruse*, dissimulation, surprise, and deception. But it is one thing to deceive the enemy consciously and deliberately and to do so in the name of a cause for which life itself is given; and another—to spread injurious false information and assurances that 'all goes well' . . . from a spirit of sheer sycophancy.

Then he drew a parallel between army and party, especially between their attitudes towards tradition. The young communist stood in the same relation to the Old Guard as that in which the military subaltern stood to his superiors. In both the party and the army the young enter a ready-made organization which their elders had to build from scratch. Here and there tradition is therefore of 'vast importance'—without it there can be no steady progress.

But tradition is not a rigid canon or an official manual; it cannot be learned by heart or accepted as gospel; not everything the old generation says can be believed merely 'upon its word of honour'. On the contrary, tradition must, so to speak, be conquered by internal travail; it must be worked out by oneself in a critical manner and in that way assimilated. Otherwise the whole structure will be built on sand. I have already spoken of the representatives of the

[1] L. Trotsky, *The New Course*, pp. 99–105.

'Old Guard' . . . who impart tradition to the young in the manner of Famusov [a character from classical Russian comedy]: 'Learn by looking at the elders: at us, for example, or at our deceased uncle.' But neither from the uncle nor from his nephews is there anything worth while learning.

It is incontestable that our old cadres which have rendered immortal services to the revolution enjoy very great authority in the eyes of the young military men. And that is excellent because it assures the indissoluble bond between the higher and lower commands and their link with the ranks. But on one condition: that the authority of the old does not efface the personality of the young and most certainly that it does not terrorize them. . . . Any man trained merely to say 'Yes, Sir' is a nobody. Of such people the old satirist Saltykov said: 'They keep saying yes, yes, yes, till they get you in a mess.'[1]

This was Trotsky's first attack on the Old Guard. But it was couched in terms so general and allusive that very few grasped its meaning. The party and the country still had no inkling of his differences with the Politbureau and held him to be responsible for official policy. So much was this the case that when the Forty Six, addressing the cells, claimed that they had Trotsky's support, Stalin could reply that they had no right to do so because Trotsky, far from agreeing with the opposition, was one of the most determined disciplinarians among the leaders.[2] This, it seems, was the last straw which broke Trotsky's patience. On 8 December he wrote an Open Letter to party meetings in which he made clear his position.[3] He described the New Course as a historic turning-point; but he warned the rank and file that some of the leaders were already having second thoughts and trying to nullify the New Course in practice. It was, he said, the party's task and duty to free itself from the tyranny of its own machine. The rank and file must rely solely upon themselves, their own understanding, and their own initiative and courage. True, the party could not dispense with its machine; and the machine had to work in a centralized manner. But it must be the party's tool, and not its master; and the needs of centralism must be harmonized and balanced with the demands of democracy. 'During this last period there was no such balance.'

---

[1] Ibid., p. 104.   [2] Stalin, *Sochinenya*, vol. v, pp. 369-70.
[3] *The New Course*, pp. 89-98.

'The idea or at the very least the feeling that bureaucratism is threatening to get the party into a blind alley has become pretty general. Voices have been raised to point out the danger. The resolution on the New Course is the first official expression of the change that has taken place in the party. It will be effective only to the degree that the party, that is its 400,000 members, want to make it effective and succeed in this.' Some leaders, afraid of this, were already arguing that the mass of members was not mature enough to enable the party to govern itself democratically. But it was precisely the bureaucratic tutelage that prevented the mass from growing politically mature. It was right 'to make astringent demands upon those who want to enter the party and stay in it'; but once they have been admitted they must be free to exercise all rights reserved for members. He then explicitly appealed to the young to assert themselves and not to regard the Old Guard's authority as absolute. 'It is only by constant active collaboration with the young, within the framework of democracy, that the Old Guard can preserve the Old Guard as a revolutionary factor.' Otherwise it will ossify and degenerate into a bureaucracy.

This was the first time that Trotsky confronted the Old Guard with the charge, still strongly qualified, of 'bureaucratic degeneration'. He supported the charge by a telling analogy; he recalled the process by which the Old Guard of the Second International had become transformed from a revolutionary into a reformist force and surrendered its greatness and historic mission to its own party machines. But Bolshevism was threatened not only by a divorce between generations. Even more menacing was the divorce between the party and the working class. Only 15 or 16 per cent. of the entire membership consisted of factory workers. He demanded 'an increasingly large flow into the party of working-class elements'; and he concluded his Letter with this tempestuous war cry:

Away with passive obedience, with mechanical levelling by the authorities, with suppression of personality, with servility, and with careerism! A Bolshevik is not merely a disciplined man: he is a man who in each case and on each question forges a firm opinion of his own and defends it courageously and independently not only against his enemies but inside his own party. To-day perhaps he will be in a minority . . . he will submit . . . but this does not always signify that he is in the wrong. Perhaps he has seen or has understood a new task or the necessity of a turn earlier than others have done. He will persistently raise the question a second, a third, a tenth time, if need be. Thereby he will render his party a service helping it to meet the new task fully armed, or to carry out the necessary turn without organic upheaval and without factional convulsions.[1]

This was the crux of the matter. He put forward the idea of a party which allowed the freedom of various trends of thought in its midst as long as these were compatible with its programme; and he opposed this idea to the conception of the monolithic party which the triumvirs had already advanced as belonging to the essence of Bolshevism. Of course, the party must not be 'chopped up into factions'; but 'factionalism' was only an extreme and morbid reaction against the excessive centralism and the domineering attitude of the bureaucracy. It could not be uprooted as long as its cause persisted. And so it was necessary to 'renew the party apparatus', to 'replace the mummified bureaucrats by fresh elements who are in close touch with the life of the party as a whole', and, above all, to remove from the leading posts 'those who, at the first word of criticism, of objection, or of protest, brandish the thunderbolts of penalties . . . the New Course must begin by making every one feel that from now on nobody will dare terrorize the party'.

Thus, after a delay of nearly nine months, he threw at last, alone, the bombshell he had hoped to explode together with Lenin at the twelfth congress. The delay was fatal. Stalin had already carried out the overhaul of the party machine and had placed his and to a lesser extent Zinoviev's subordinates at every sensitive spot, in every branch of the organization. By insinuation, obloquy, and stage-whisper he had prepared them thoroughly for the expected clash with Trotsky. And now he moved the phalanx of his secretaries into action.

When Trotsky's Letter was read out at the party meetings, pandemonium broke loose. Many received the Letter as the message for which they had long waited, the inspiring call from the great revolutionary who had at last turned his back upon the Pharisees and placed himself once again at the head of the humble and the humiliated. Even members of the opposition groups against whom he had only recently acted as counsel for

[1] *The New Course*, p. 94.

the prosecution responded with fervour, and acknowledged that even in his severity towards them he had been guided only by pure and high-minded motives. 'We address ourselves to you, Comrade Trotsky', wrote one of them, 'as to the leader of the Russian Communist Party and of the Communist International whose revolutionary thought has remained alien to caste exclusiveness and narrow-mindedness.' 'I approach you, Comrade Trotsky', wrote another, 'as one of the leaders of Soviet Russia to whom considerations of political revenge are alien.'[1] But many a Bolshevik was stunned by the sombre picture of the party he had drawn and by his harsh language; and some were outraged by what they considered to be an unprovoked insult to the party, if not a stab in its back. Everywhere, the secretaries led and organized this latter section of Bolshevik opinion, exacerbated it, excited it to the utmost, and gave to it a weight which was out of proportion to its real strength, placing at its disposal all means of expression, most of the time reserved for debates at meetings, and most of the discussion columns in leading newspapers and in local bulletins and sheets which played an enormous part in forming opinion in the provinces.

At the branch meetings the opposition's adherents often overwhelmed the party machine by their numbers and articulateness. But when the branch meetings with all their sound and fury were over, it was the secretaries who spoke on behalf of the branches, who handled the resolutions adopted, and who decided whether to suppress them or not, and, if not, how much currency to give them. Once a secretary had been confronted with the unmanageable temper of one meeting he prepared carefully for the next meeting, packed it with his men, and ruled out or silenced the opposition.

The debate was to be concluded by the holding of the thirteenth party conference. The preparations for the conference were also in the hands of the secretaries. The election of delegates was indirect and proceeded through several stages. At every stage the secretaries checked how many of the sympathizers of the opposition were elected; and they saw to it that they were eliminated at the next stage. It was never disclosed how many votes were cast for the opposition in the primary

[1] Yaroslavsky quoted these letters at the thirteenth party conference with the intention to discredit Trotsky. *13 Konferentsya RKP (b)*, p. 125.

cells of Moscow. The Forty Six claimed, without meeting with denial, that at the regional conference, which was the tier above the primary cells, they had obtained not less than 36 per cent. of the vote; yet at the *gubernia* conference, the next tier, that percentage dwindled to 18. The opposition concluded that, if its representation had been whittled down in the same proportion all the way from the primary to the final elections, then the opposition had behind it the great majority of the Moscow organization.[1] This was almost certainly true, but the secretaries were on top of the majority.

The triumvirs were anxious to bring the contest to a speedy conclusion. They replied to Trotsky's Letter with a deafening barrage of counter-accusations. It was, they said, disloyal on Trotsky's part to vote with the whole of the Politbureau for the New Course and then to cast aspersion upon the Politbureau's intentions. It was criminal to incite the young against the Old Guard, the repository of revolutionary virtue and tradition. It was wicked of him to try and turn the mass of the party against the machine, for every good old Bolshevik was aware how much importance the party had always attached to its machine and with how much care and devotion it had surrounded it. He equivocated over the ban on factions: he knew that the ban was essential to the party's unity and did not dare to demand plainly that it be revoked; but he sought to sap it surreptitiously. He played false when he described the party régime as bureaucratic; and he played with fire when he aroused an exaggerated and dangerous appetite for democracy in the masses. He pretended to speak for the workers, but played up to the students and the intelligentsia, that is to the petty bourgeois gallery. He spoke about the rights and responsibility of the rank and file only to cover up his own irresponsibility, *folie de grandeur*, and frustrated dictatorial ambition. His hatred of the party machine, his contumelious attitude towards the Old Guard, his reckless individualism, his disrespect for Bolshevik tradition, yes, and his notorious 'underestimation' of the peasantry—all this clearly indicated that at heart he had remained something of a stranger in the party, an alien to Leninism, an unreformed semi-Menshevik. Agreeing to become the mouthpiece of all the disparate opposition groups, he had set himself up as the chief, even if

[1] Ibid., pp. 131–3.

unconscious, agent of all the petty bourgeois elements which pressed upon the party from all sides, seeking to breach its unity and to inject into it their own moods, prejudices, and pretensions.[1]

.   .   .   .

In the long history of inner-party oppositions none had been weighed down by so heavy a load of accusations and none had been ground down so remorselessly by the party machine as was the 1923 Opposition. By comparison the Workers' Opposition had been treated fairly, almost generously; and the oppositions which had been active before 1921 had as a rule enjoyed unrestricted freedom of expression and organization. What accounted for the vehemence and fury with which the party machine now bore down upon its chief critic?

The triumvirs were not able to meet Trotsky on his own ground, in fair argument. His attack was all too dangerous: his Open Letter and his few articles on the New Course rang out like powerful bells arousing alarm, anger, and militancy. Yet, the triumvirs resorted not merely to falsification and suppression. They also exposed and made the most of the weaknesses and inconsistencies, real or apparent, in Trotsky's attitude. Throughout he took his stand on the Bolshevik monopoly of power; and much more persuasively than the triumvirs did he call on the party to guard it as the sole guarantee of the revolution's survival; and he reaffirmed his own desire to defend it and to consolidate it. He objected only to the monopoly of power which the Old Guard obtained within the party and exercised through the machine. It was not difficult for his adversaries to demonstrate that the latter was the necessary sequel to the former, and that the party could maintain its monopoly only by delegating it to the Old Guard. Trotsky argued that the 400,000 members should be trusted to exercise their judgement and allowed to have their full share in shaping policy. Why then, his adversaries asked, had the party, under Lenin's inspiration and with Trotsky's consent, denied the mass of members that trust in recent years? Was it not because the party had been infiltrated by alien elements, ex-Mensheviks, turncoats, and even N.E.P. men? Had not even some authentic Bolsheviks become

[1] See, for instance, Stalin's replies in *Sochinenya*, vol. v, pp. 383–7, and vol. vi, pp. 5–40.

alienated from their comrades and corrupted by power and privilege? Trotsky stated that the purge in which hundreds of thousands had been expelled should have sufficiently purified the party and restored its integrity. But had not Lenin and the Central Committee stated repeatedly that this was not so? Had they not foreshadowed new and periodical purges? Had they all not agreed with Zinoviev that it was inevitable that the party should, because of its monopoly, comprise 'unconscious Mensheviks' and 'unconscious Social Revolutionaries'? No single purge could eliminate these alien elements, let alone the immature ones. Expelled, they were bound to reappear: they entered the party in good faith and bad with every group of new entrants. After it had been found necessary to expel one third of the membership in one year, how could 'the party' trust the judgement of the mass and allow it to exercise full rights?

Trotsky protested against the irrational self-suppression of Bolshevism which, however, followed ineluctably from the suppression by Bolshevism of all its enemies. If free competition of political trends within the party were to be tolerated, would that not allow the 'unconscious Mensheviks' to become articulate, to form a definite body of opinion, and to split the party? The monolithic system kept the heterogeneous mass unconscious of its heterogeneity and inarticulate; and thus it mechanically assured unity. Some of the more sophisticated adherents of the triumvirs saw that the dangers to which Trotsky pointed were real enough: the Old Guard might degenerate; and the monolithic system was bound to breed discontent and arouse sporadic revulsion which might also lead to schisms. But the party had to face perils whichever road it chose. Under monolithic control at least no schismatic movement could spread as easily as it could in a democratically ruled organization. The party machine would spot it in time, nip it in the bud, and keep the rest of the party more or less immune.

In other words, the party was in danger of losing its pro-letarian-socialist outlook, in danger of 'degeneration', no matter whether it entrusted its future to the mass of members or to the Old Guard. The predicament arose from the fact that the majority of the nation did not share the socialist outlook, that the working class was still disintegrated, and that, the revolution having failed to spread to the West, Russia had to fall back, materially and spiritually, on her own resources. The possibility of 'degeneration' was inherent in this situation; and what remained to be determined was whether its chief source lay in the heterogeneous mass of members or in the Old Guard. It was only natural that the Old Guard, or rather its majority, should trust its own socialist tradition and character infinitely more than it trusted the judgement and the political instincts of the 400,000 nominal party members. True, Trotsky did not ask the Old Guard to efface itself—he urged it to maintain its authority by democratic methods. But the Old Guard did not feel—and it was probably right in this—that it could do this. It was afraid of taking the risk; and it had a vested interest in preserving its acquired political privileges.

The reform within the party which Trotsky advocated could be upheld as the first act in the restoration of those free Soviet institutions which the party had sought to establish in 1917, as the beginning, that is, of a return to a workers' democracy and of the gradual dismantling of the single-party system. This idea was not far from Trotsky's mind;[1] but he did not voice it—either because he took it for granted but did not believe that the time had come to question and to weaken the single-party system; or because he did not wish to lay himself open to fresh and damaging charges and to complicate the controversy needlessly. Probably both these motives played their part. In effect, however, he claimed for the Bolsheviks a twofold privilege: the monopoly of freedom as well as the monopoly of power. These two privileges were incompatible. If the Bolsheviks wished to preserve their power they had to sacrifice their freedom.

There was a further weakness in Trotsky's attitude. He urged the party to preserve its proletarian-socialist outlook. At the same time he pointed out that workers from the bench formed only a small minority—one-sixth —of the party's membership. The majority consisted of industrial managers, civil servants, army officers, commissars, party officials, &c. (Some of these were of proletarian origin, but they were becoming more and more assimilated to the professional bureaucracy which the Soviets had inherited from Tsardom.) It was thus precisely under the rule of inner-party democracy that the influence of the

[1] See Trotsky's remarks on the secret vote in the USSR in a 'Letter to Friends' of 21 October 1928. *The Archives*.

workers was bound to be negligible and that the bureaucratic elements were bound to retain the upper hand. Trotsky therefore urged the party to recruit more workers and to 'strengthen its proletarian cells'. But he also insisted that the party should proceed cautiously and regulate carefully the admission of new members from the working class lest it be swamped by a politically raw and uncivilized mass.[1] This state of affairs appeared highly paradoxical from whichever angle it was approached. The application of democratic rules could not render the party democratic because it could only strengthen its bureaucracy; and the party could not become more enlightened and socialist in spirit by opening the doors widely to the working class.

In what then did the party's proletarian outlook consist? It would be easy to conclude that the Bolshevik leaders, including Trotsky, dealt in a mythology which bore no relation to the party's social composition and to its real attitude to the working classes. Inner Bolshevik controversy was indeed conducted, at least in part, in quasi-mythological terms, reflecting that substitutism which had led the party (and then the Old Guard) to consider itself as the *locum tenens* of the working class. Neither side to the controversy could frankly and fully admit the substitution. Neither could say that they were condemned to pursue the proletarian ideal of socialism without the support of the proletariat—such an avowal would have been incompatible with the whole tradition of Marxism and Bolshevism. They had to construct elaborate arguments and a peculiar and ambiguous idiom with its own conventions designed to veil and to explain away this sad state of affairs. The triumvirs were the worse sinners in this respect: and the mythology of substitutism finally congealed into the rigid cults of latter-day Stalinism. But even Trotsky, while he sought to reverse in part the process of substitution and struggled to tear to shreds the thickening fabric of the new mythology, could not help being entangled in it.[2]

[1] *The New Course*, pp. 20–21.

[2] Thus, referring to analogies between Bolshevism and Jacobinism, made by Mensheviks and liberals, as 'superficial and inconsistent', Trotsky wrote that the fall of the Jacobins had been caused by the social immaturity of their following and that the situation of the Bolsheviks was 'incomparably more favourable' in this respect. 'The proletariat forms the nucleus as well as the left wing of the [Russian] revolution. . . . *The proletariat is politically so strong* that while permitting, within limits, a new bourgeoisie to form itself . . . *it enables the peasantry to participate . . . directly . . . in the exercize of state power.*' Ibid., p. 40. (My italics, I. D.)

In truth, the Bolshevik bureaucracy was already the only organized and politically active force in society and state alike. It had appropriated the political power which had slipped from the hands of the working class; and it stood above all social classes and was *politically* independent of them all. And yet the party's socialist outlook was not a mere myth. It was not only that the Bolshevik bureaucracy subjectively saw itself as the exponent of socialism and that it cultivated, in its own manner, the tradition of proletarian revolution. Objectively, too, by the force of circumstances, it had to work as the chief agent and promoter of the country's development towards collectivism. What ultimately governed the behaviour and the policies of the bureaucracy was the fact that it was in charge of the publicly owned industrial resources of the Soviet Union. It represented the interests of the 'socialist sector' of the economy against those of the 'private sector', rather than the specific interests of any social class; and only to the extent to which the general interest of the 'socialist sector' coincided with the general or 'historic' interest of the working class, could the Bolshevik bureaucracy claim to act on behalf of that class.

The 'socialist sector' had its own claims and its own logic of development. Its first claim was that it should be made secure against wholesale restoration of capitalism and even against a partial but massive reintrusion of private enterprise. Its logic of development required planning and coordination of all the publicly owned branches of the economy and their rapid expansion. The alternative was contraction and decay. Expansion had to proceed, at least in part, at the cost of the 'private sector', through the absorption of its resources. This was to lead to conflict between the state and private property; and in this conflict the Bolshevik bureaucracy could not but side ultimately with the 'socialist sector'. True, even then it could not achieve socialism; for that presupposed economic abundance, high popular standards of living, of education, and of general civilization, the disappearance of striking social contrasts, the cessation of domination of man by man, and a spiritual climate corresponding to this general transformation of society. But to the Marxist the nationalized economy was the essential prerequisite of socialism, its genuine foundation. It was quite conceivable that even on that foundation the edifice of socialism might not rise; but it was unthinkable that it should rise without it. It was this foundation of socialism that the Bolshevik bureaucracy could not but defend.

At the point our narrative has reached, in the years 1923-4, the Bolshevik bureaucracy was only dimly aware of the nature of the interest to which it was tied. It was embarrassed and puzzled, as it were, by its own unprecedented command over the nation's industrial resources; and it did not quite know how to exercise it. It regarded uneasily, even fearfully, the property-loving peasantry; and it was momentarily even inclined to give more weight to the claims of the latter than to those of the 'socialist sector'. Only after a series of shocks and internal struggles was the Bolshevik bureaucracy to be driven to identify itself exclusively and irrevocably with the 'socialist sector' and its needs.

It was Trotsky's peculiar fate that even while he declared war on the political pretensions and the arrogance of the bureaucracy, he had to try and awaken it to its 'historic mission'. His advocacy of primitive socialist accumulation aimed at this. Yet such accumulation, in the circumstances under which it was to take place, could hardly be reconciled with the workers' democracy. The workers could not be expected to surrender voluntarily 'half their wages' to the state, as Trotsky urged them to do, in order to promote national investment. The state could take 'half their wages' only by force; and to do this it had to deprive them of every means of protest and to destroy the last vestiges of a workers' democracy. The two aspects of the programme which Trotsky expounded in 1923 were to prove incompatible in the near future; and therein lay the fundamental weakness of his position. The bureaucracy raged furiously against one part of his programme, the one which claimed a workers' democracy; but after much resistance, hesitation, and delay, it was to carry out the other part which spoke of primitive socialist accumulation.

.   .   .   .

At the turn of the year, while preparations for the thirteenth conference and the drive against the opposition were in full swing, Trotsky's health deteriorated. His fever persisted and he suffered from physical exhaustion and depression. He began to be overcome by a sense of approaching defeat. The campaign against him with its relentless barrage of accusations, distortions, and tricks, still seemed to him almost unreal; yet it induced in him a feeling of helplessness. He could only argue his case, but his argument was drowned in the hubbub. (Even the publication of *The New Course* was delayed by the public presses so that the pamphlet could not reach the cells before the opening of the thirteenth conference.) His mood alternated between tension and apathy. And so when his doctors told him to leave frostbound Moscow—the winter was exceptionally severe that year —and to take a cure on the Caucasian coast of the Black Sea, this was for him an opportunity of escaping the oppressive atmosphere of the capital.[1]

He was getting ready for the journey when, on 16 January 1924, the thirteenth conference was opened. The triumvirs prepared a resolution blusteringly denouncing Trotsky and the Forty Six as guilty of a 'petty bourgeois deviation from Leninism'. The proceedings were taken up almost wholly by this question. In Trotsky's absence, Pyatakov, Preobrazhensky, V. Smirnov, and Radek argued the opposition's case. The triumvirs and their adherents replied with much venom; and their replies filled the newspapers. The outcome was a foregone conclusion. So thoroughly had the General Secretariat manipulated the elections that only three votes were cast against the motion condemning Trotsky. Even in the light of the accounts of the opposition's influence which Zinoviev's and Stalin's adherents gave at the conference, this vote was so ludicrously false that it should have had the effect of a bad and insolent joke.[2] But the triumvirs deliberately disregarded all the proprieties of normal political behaviour. Their purpose was to impress upon the party that they would stop at nothing and that all resistance was useless. The cells now knew that no matter how much they

[1] A bulletin on Trotsky's health, signed by Semashko, the Commissar of Health, and five Kremlin doctors, spoke of influenza, catarrh in the upper respiratory organs, enlargement of bronchial glands, persistent fever (not exceeding 38° C.), loss of weight and appetite, and reduced capacity for work. The doctors considered it necessary to release the patient from all duties, and to advise him to leave Moscow and take a 'climatic cure for at least two months'. The bulletin, signed on 21 December 1923, appeared in *Pravda* on 8 January 1924.

[2] According to Rykov, Pyatakov obtained a majority vote for the Opposition's motions at all party cells in Moscow which he addressed. (*13 Konferentsya RKP (b)*, pp. 83–91.) Yaroslavsky stated that one-third of the military party cells of Moscow had voted for the Opposition before the discussion in the garrison was stopped, and that the majority of student cells had done the same. Ibid., pp. 123–6.

stormed or protested, they had no chance of making the slightest imprint on official decisions. This alone was enough to show up the impotence of the opposition and to spread despondency in its ranks.

On 18 January, without waiting for the verdict, Trotsky set off on a slow journey to the south. Three days later his train halted at Tiflis. There, while the train was being shunted, he received a code message from Stalin informing him of Lenin's death. The blow hit Trotsky as if it had come suddenly—to the end Lenin's doctors, and Trotsky even more than they, had believed that they would save Lenin's life. With difficulty he jotted down for the newspapers a brief message mourning the deceased leader. 'Lenin is no more. These words fall upon our mind as heavily as a giant rock falls into the sea.'[1] The last flicker of the hope that Lenin would return, undo the triumvirs' work, and tear up their denunciatory resolutions, was extinguished.

For a moment Trotsky wondered whether he should not return to Moscow.[2] He got in touch with Stalin and asked for advice. Stalin told him that he would not be back in time for the funeral next day and counselled him to stay and proceed with the cure. In fact, Lenin's funeral took place several days later, on 27 January. Stalin had, of course, his reasons for keeping Trotsky away during the elaborate ceremonies in the course of which the triumvirs presented themselves to the world as Lenin's successors. From Tiflis, Trotsky, his head spinning with fever, proceeded to the sea-side resort of Sukhum. There, in semi-tropical sunshine, amid palms, flowering mimosas, and camellias, he lay many a long day on the veranda of a sanatorium and recollected in solitude the strange fortunes of his association with Lenin, the friendship with which Lenin first received him in London in 1902, their subsequent sharp disagreements, their eventual reunion, and the stormy and triumphant years during which they stood together at the helm of the revolution. It was as if the triumphant part of himself had together with Lenin gone down to the grave.

More recollections, fever, darkness, loneliness. A warm message from Lenin's weak and disconsolate widow now brought a crumb of comfort to the man whose prowess and power had only so recently amazed the world: she wrote that, just before he

[1] Trotsky, *O Lenine*, pp. 166–8. [2] *Moya Zhizn*, vol. ii, p. 250.

died, Lenin had reread the character sketch of him Trotsky had written and was visibly moved by it, especially by the comparison Trotsky had drawn between him and Marx; and she wished Trotsky to know that Lenin had preserved to the end the friendly sentiment which he had shown him at their first meeting in London.[1]

Then gloom returned; and the sick man's imagination fed again on recollections, until a letter from his son Lyova jerked him back to the troubles of the day. Lyova described the great theatrically staged funeral in Moscow and the procession of huge crowds to Lenin's bier; and he expressed anguished astonishment at his father's absence.

Only now, it seems, as he read the disheartened letter from his adolescent son did it occur to Trotsky that he might have made a mistake when he did not return to Moscow. The multitudes which marched past Lenin's bier watched tensely the members of the Politbureau who stood guard over it and noted Trotsky's absence. Their imagination had been fired by the symbolism of the ceremonies; and in this mood they wondered why he was not there. Was it perhaps because of the difference which, according to the triumvirs, had separated him from the deceased man and because of his 'petty bourgeois deviation from Leninism'?

Trotsky's absence did not merely breed rumour and gossip in Moscow. It left the field free to his adversaries. This was a time of intense activity in the Kremlin and of important decisions. The succession to Lenin in government as well as party was being settled in the most formal manner. Rykov took Lenin's place as *Predsovnarkom*, chairman of the Council of the People's Commissars; and Rykov's place at the Supreme Council of the National Economy was taken by Dzerzhinsky. (Rykov was appointed *Predsovnarkom* because he had been Lenin's deputy—had Trotsky accepted the deputy's post, it would have been difficult to promote Rykov over his head.) Then the triumvirs made a new and more determined attempt to obtain control of the Commissariat of War. They dismissed from

[1] Many years later, after Trotsky had been exiled, Krupskaya told Count M. Károlyi and his wife: 'He [Trotsky] loved Vladimir Ilyich very deeply; on learning of his death he fainted and did not recover for two hours'. *Memoirs of Michael Károlyi*, p. 265.

the Commissariat Sklyansky, Trotsky's devoted assistant, and sent a special delegation to Sukhum to inform Trotsky that Frunze, who was Zinoviev's adherent, would take Sklyansky's place—a year later Frunze was to succeed Trotsky himself as Commissar of War. The Politbureau and the Central Committee were also giving effect to the decisions of the thirteenth conference directed against the opposition: still more adherents of the opposition were dismissed, demoted, or reprimanded. The propaganda department worked full blast to establish that cult of Lenin under which Lenin's writings were to be quoted as Gospel against all dissent and criticism, the cult which was designed primarily as an 'ideological weapon' against Trotskyism.

And, last but not least, the triumvirs stole yet more of Trotsky's thunder. He had dwelt on the weakness of the 'proletarian cells' as the chief cause of the party's bureaucratic deformation and had urged the party to recruit more members from the working class. This demand had undoubtedly gained him sympathy among workers. The triumvirs resolved to open at once a spectacular recruiting drive in the factories. But while Trotsky had advised a careful selection, they decided to recruit *en masse*, to accept any worker who cared to join, and to waive all customary tests and conditions. At the thirteenth conference they recommended the recruitment at a stroke of 100,000 workers. After Lenin's death they threw open the doors of the party even wider: between February and May 1924, 240,000 workers were inscribed.[1] This was a mockery of the Bolshevik principle of organization which required that, as the *élite* and vanguard of the proletariat, the party should accept only the politically advanced and the battle hardened. Among the mass of new entrants, the politically immature, the backward, the dullminded and the docile, the climbers, and the nest-featherers, formed a considerable proportion. The triumvirs hectically motioned the newcomers on to the band wagon, patted them on the back, flattered them, and exalted the keen and infallible class instinct and class consciousness that had brought them into the party.

This recruitment—the 'Lenin levy'—was presented as the spontaneous homage of the working class to Lenin and as the party's rejuvenation. The triumvirs were in effect saying to

[1] See Molotov's report on the 'Lenin Levy' in *13 Syezd RKP* (*b*), pp. 516 ff.

Trotsky: 'you thought you would endear yourself to the workers by playing them against the bureaucrats and by arguing that the proletarian element in the party should be strengthened. We have strengthened it, and we have done so without any of your scruples—we have wheedled a quarter of a million workers into the party. And what is the result? Has the party thereby become ennobled, more democratic, or more proletarian-socialist in outlook? Has bureaucracy been weakened?' The 'Lenin levy' in fact supplied the triumvirs with a devoted *clientèle* to which they presently appealed in the struggle against the opposition. Trotsky was aware what this demagogic exploitation of his idea meant; but he could not utter a word against the 'Lenin levy'. Had he done so, he would have been howled down as the enemy of the workers and as the hypocrite who at first pretended that he longed to see more proletarians in the party, but who now betrayed his fear of them and his true petty bourgeois nature. He made *une bonne mine au mauvais jeu*; and he even chimed in with the official eulogies for the Lenin levy.[1]

Trotsky's moody aloofness at a moment so critical for his and the party's fortunes was, of course, in some measure caused by his illness. Yet even more debilitating was his feeling that the tide was running against him. This was an unfathomed tide, and he tried to gauge and assess it in Marxist terms. He concluded that the revolution was on the ebb and that he and his friends were being hit by a groundswell of reaction. The nature of the reaction was confused and confusing: it looked like, and up to a point it was, a prolongation of the revolution. He was convinced that it was his duty to resist; but he did not see clearly by what means to resist and what were the prospects. It was a turbid and a slimy tide that threw him back. None of the great

---

[1] In a speech at Tiflis (on 11 April 1924) Trotsky said: 'The most important political fact of the last few months . . . has been the influx of factory workers into the ranks of our party. This is the best form in which [the working class] . . . demonstrates its will . . . and votes confidence in the Russian Communist Party. . . . This is a true, reliable, and infallible test . . . far more genuine than any parliamentary election.' (Quoted from Trotsky, *Zapad i Vostok*, p. 27.) Looking back on this 'test' twelve years later, Trotsky wrote: 'Availing itself of the death of Lenin, the ruling group announced a "Leninist levy". . . . The political aim of this manoeuvre was to dissolve the revolutionary vanguard in raw [inexperienced and meek] human material. . . . The scheme was successful . . . the "Leninist levy" was a death blow to the party of Lenin.' *The Revolution Betrayed*, pp. 97–98.

PLATE VII

(*a*) Lenin's body brought to Moscow for the funeral

Carrying the coffin is Dzherzhinsky, head of the G.P.U. (in front) and at the right
Sapronov (bare-headed), a leader of the Workers' Opposition

(*b*) Molotov and Bukharin in 1921 or 1922

PLATE VIII

Otto Schmidt, Berlin

Trotsky resting in the Caucasus in 1924

issues over which they wrestled at the Politbureau could be made to stand out in clear outline. Everything was blurred. The greatest issues were dragged down to the level of sordid intrigue. Had he, as his adversaries claimed, coveted personal power, he would have behaved quite differently, of course. But in his whole being he shrunk from the scramble; and half-consciously perhaps he was glad to escape from it into his melancholy lonesomeness in the Caucasus.

In the spring his health was improved and he was back in Moscow. The party was just getting ready for the thirteenth congress convened in May. The Central Committee and senior delegates met on 22 May to acquaint themselves with Lenin's will which had hitherto been in Krupskaya's keeping. The reading of the will had the effect of a bolt from the blue. Those present listened in utter perplexity to the passage in which Lenin castigated Stalin's rudeness and disloyalty and urged the party to remove him from the General Secretariat. Stalin seemed crushed. Once again his fortunes trembled in the balance. Amid all the worshipping of Lenin's memory, amid the endless genuflexions and vows to 'hold Lenin's word sacred', it seemed inconceivable that the party should disregard Lenin's advice.

But once again Stalin was saved by the trustfulness of his future victims. Zinoviev and Kamenev, who held his fate in their hands, rushed to his rescue. They implored their comrades to leave him in his post. They used all their zeal and histrionic talents to persuade them that whatever Lenin held Stalin guilty of, the offence was not grave and that Stalin had made ample amends. Lenin's word was sacred, Zinoviev exclaimed, but Lenin himself, if he could have witnessed, as they all had, Stalin's sincere efforts to mend his ways, would not have urged the party to remove him. (In fact, Stalin's embarrassment suited Zinoviev who was already afraid of him, but did not dare to break the partnership. Zinoviev hoped to earn Stalin's gratitude and to put himself back into the position of the senior triumvir.)

All eyes were now fixed on Trotsky: would he rise, expose the farce, and demand that Lenin's will be respected? He did not utter a word. He conveyed his contempt and disgust at the spectacle only through expressive grimaces and shoulder shrugging. He could not bring himself to speak out on a matter in which his own standing was so obviously involved. It was resolved to disregard Lenin's advice on Stalin. But if so, then Lenin's will could not be published; for it would show up and render ridiculous all the mummeries of the Lenin cult. Against Krupskaya's protest the Central Committee voted by an overwhelming majority for the suppression of the will. To the end Trotsky, as though numb and frozen with detestation, kept his silence.[1]

In the last week of May the thirteenth congress assembled. The triumvirs asked it to repeat with bell, book, and candle the anathema on Trotsky which the less authoritative conference had pronounced in January. The congress turned into an orgy of denunciation. Zinoviev fumed and fulminated: 'It was now a thousand times more necessary than ever that the party should be monolithic.'[2] Months before he had urged his partners to order Trotsky's expulsion from the party and even arrest; but Stalin cool-headedly refused to comply and hastened to declare in *Pravda* that no action was contemplated against Trotsky, and that a party leadership without Trotsky was 'unthinkable'.[3] At the congress Zinoviev struck out again; and in a moment of fatal recklessness he demanded that Trotsky should not merely 'lay down arms' but appear before the congress and recant. Not before Trotsky had done so, Zinoviev stated, would there be peace in the party.[4] This was the first time in the party's experience that a member had been confronted with the demand for recantation. Even this congress, zealous as it was to pronounce anathema on Trotsky, was shocked. The mass of delegates rose to give an ovation to Krupskaya when she, without supporting Trotsky, made a strong and dignified protest against Zinoviev's 'psychologically impossible demand'.[5]

Once only did Trotsky defend himself.[6] He spoke calmly and persuasively, with an undertone of resigned acceptance of defeat; but he refused adamantly to retract a single one of his criticisms. He was anxious not to pour oil on the flames and not

---

[1] Bajanov, who acted as secretary at this meeting, gives an eye-witness description of the scene (op. cit., pp. 43–47). Trotsky acknowledges implicitly the authenticity of Bajanov's account. (Trotsky, *Stalin*, p. 376.) In *The Suppressed Testament of Lenin* Trotsky adds this detail: 'Radek . . . sat beside me during the reading . . . and leaned to me with the words: "Now they won't dare go against you." I answered: "On the contrary, they will have to go the limit, and moreover as quickly as possible." ' p. 17.

[2] *13 Syezd RKP (b)*, p. 112.

[3] *Pravda*, 18 December 1923.

[4] *13 Syezd RKP (b)*, p. 113.

[5] Ibid., pp. 235–7.

[6] Ibid., pp. 153–68.

to burn his boats. He pleaded that he had framed all his criticisms in the terms of the Politbureau's resolution on the New Course, and that there was nothing in what he had said and written which had not been said or written in one form or another by his adversaries also. He even dissociated himself from some of the Forty Six who had demanded freedom for inner-party groupings. 'The allegation that I am in favour of permitting groupings is incorrect', he said. 'True, I did make the mistake of falling ill at the critical moment and did not have the opportunity . . . of denying this and many other allegations. . . . It is impossible to make any distinction between a faction and a grouping.' He repeated, however, that it was because of wrong policies and of the faulty inner-party régime that differences of opinion, which should have been only transient, became fixed and hardened and led to 'factionalism'. To Zinoviev's call for a recantation he replied:

Nothing could be simpler or easier, morally and politically, than to admit before one's own party that one had erred. . . . No great moral heroism is needed for that. . . . Comrades, none of us wishes to be or can be right against the party. In the last instance the party is always right, because it is *the only historic instrument which the working class possesses for the solution of its fundamental tasks*. I have said already that nothing would be easier than to say before the party that all these criticisms and all these declarations, warnings, and protests were mistaken from beginning to end. I cannot say so, however, because, comrades, I do not think so. I know that one ought not to be right against the party. One can be right only with the party and through the party because history has not created any other way for the realization of one's rightness. The English have the saying 'My country, right or wrong'. With much greater justification we can say: My party, right or wrong—wrong on certain partial, specific issues or at certain moments. . . . It would be ridiculous perhaps, almost indecent, to make any personal statements here, but I do hope that in case of need I shall not prove the meanest soldier on the meanest of Bolshevik barricades.[1]

He ended his plea by saying that he would accept the party's verdict even if it were unjust. But acceptance meant for him submission to discipline in action, not in thought. 'I cannot say so, comrades, because I do not think so', these words stood out in their stark simplicity and unyieldingness amid all the subtle

[1] Ibid., pp. 165–6. (My italics. I.D.)

reasonings, incisive arguments, and imaginative appeals in which his speech abounded. His calm and restraint infuriated the party secretaries. Bent but unbroken, disciplined but unrepentant, he seemed to them all the more defiant. His voice sounded in their ears like the cry of their own uneasy conscience; and they tried to blast it with insult. They drew from him no rejoinder. Only at the end of the congress he went out to the Red Square to address a meeting of Moscow's 'communist' children, the 'pioneers'. He greeted them as the 'new shift' who would one day come into the workshop of the revolution to replace those who had grown old, weary, and corrupt.[1]

* * * * * * * * *

By this time the whole Communist International had been drawn into the controversy. The triumvirs had to explain and justify their attitude to the foreign communists, from whom they were anxious to obtain a clear endorsement of Trotsky's condemnation in order to produce it to the Russian party. Yet the European communists—and in these years the influence of the International was still virtually confined to Europe—were alarmed by what was going on in Moscow and shocked by the violence of the attacks on Trotsky. To them Trotsky had been the embodiment of the Russian revolution, of its heroic legend, and of international communism. Because of his European manner of expression, he had appealed to them more than any other Russian leader. He had been the author of the International's stirring manifestoes, which in ideas, language and *éclat*, recalled the *Communist Manifesto* of Marx and Engels. He had been the International's strategist and tactician as well as inspirer. European communists could not see what it was that set Zinoviev, the International's President, and the other Russian leaders, against Trotsky; and they feared the consequences of the conflict for Russia and international communism. Their first impulse was therefore to defend Trotsky.

Before the end of the year 1923 the Central Committees of two important Communist parties, the French and the Polish,

---

[1] The speech is appended to the record of the congress. Ibid. Max Eastman, who was present at the congress, relates that he urged Trotsky to take a more militant attitude and to read from the platform Lenin's testament, but Trotsky would not listen. Eastman's account is confirmed by Trotsky himself, in a letter to Muralov, written from his exile at Alma Ata in 1928 (*The Archives*).

protested to Moscow against the defamation of Trotsky and appealed to the antagonists to compose their differences in a comradely spirit.[1] This happened shortly after Brandler had, on behalf of his party, asked that Trotsky should assume the leadership of the planned communist insurrection in Germany. The triumvirs resented the protests and feared that Trotsky, defeated in the Russian party, might yet turn the International against them. Zinoviev saw in the action of the three parties a challenge to his presidential authority.

At this time the International was agitated by the defeat it had just suffered in Germany. The questions connected with the defeat, the crisis which led up to it, and the policy of the German party, questions which in themselves provided enough ground for controversy, at once became entangled with the contention in the Russian party.[2]

The German crisis began when the French occupied the Ruhr early in 1923. The Ruhr was aflame with German resistance; and soon the whole of the Reich was embraced by a strong nationalist movement of protest against the Versailles Treaty and its consequences. At first the bourgeois parties led the movement; and the communists were swept aside. But then these parties, uncertain of the outcome, began to vacillate and withdraw, especially when social strife threatened to deepen the political turmoil. Germany's economy was thrown out of balance. Depreciation of money developed with catastrophic speed. The workers, whom inflation cheated of wages, were furious and impatient for action. The communists, who had lain low since the March rising of 1921, felt a strong wind in their sails. In July their Central Committee called upon the working class to prepare for a revolutionary decision. Its confidence in

[1] Souvarine spoke of the French protest at the thirteenth congress of the Russian party (*13 Syezd RKP (b)*), pp. 371–3). The Polish protest is in the archives of the Polish Communist party. (Deutscher, 'La Tragedie du communisme polonais entre deux guerres' in *Les Temps modernes*, March 1958.)

[2] The sources used for this account of the German crisis are: Trotsky's numerous essays, Brandler's reminiscences and explanations given to the author, Ruth Fischer, *Stalin and German Communism*, Thalheimer, *1923, Eine verpasste Revolution?*, Radek's, Zinoviev's, and Bukharin's analyses, Kuusinen's essay in *Za Leninizm, The Lessons of the German Events* (a record of the January 1924 session of the Executive of the Comintern devoted to the debate over Germany); the records of the congresses and conferences of the Comintern, and the Soviet and German Communist parties, at which the issue was thrashed out; and, finally, the extensive discussion that went on in the international communist press for over ten years after 1924.

its strength and capacity for revolutionary action, however, did not go deep; nor was it shared by all those concerned with its policy. Radek, who was in Germany as representative of the International's Executive, warned Moscow that the German party took too hopeful a view and that it might be heading for another abortive insurrection. Zinoviev and Bukharin spurred the Germans on, yet without proposing any definite course of action. At this stage, in July, Trotsky said that he was not sufficiently informed about conditions in Germany to express an opinion.

Presently, Trotsky arrived at the conclusion that Germany was indeed about to enter an acutely revolutionary situation and that the German party should not merely be encouraged to take a bold line but assisted in the working out of a clear plan of revolutionary action culminating in armed insurrection. The date of the insurrection should be fixed in advance so that the German party could conduct the struggle through the preliminary phases, prepare the working class, and deploy its forces with a view to the *dénouement*. The Executive hesitated. Not only Radek—Stalin, too, doubted the reality of the 'revolutionary situation' and held that the Germans should be restrained.[1] Zinoviev went on prodding them but balked at the plan for insurrection. The Politbureau, absorbed in its domestic preoccupations, discussed the matter casually; and Zinoviev conveyed its broad view to the leaders of the International. Somewhat half-heartedly it was decided to give the German party the cue for revolution, to assist it in military preparations, and, in the end, even to fix a date for the rising. The date was to be as close as possible to the anniversary of the Bolshevik insurrection—it was to be 'the German October'.

In September Heinrich Brandler, the leader of the German party, arrived in Moscow to consult the Executive. A bricklayer in his earlier years and a disciple of Rosa Luxemburg, a shrewd and cautious tactician and able organizer, he was not convinced that circumstances favoured revolution. When he expressed his doubts to Zinoviev—doubts very similar to those Zinoviev himself had entertained on the eve of the Russian October—the latter, torn between hesitation and the desire to act resolutely, sought to overrule Brandler's objections with heated argument

---

[1] See my *Stalin*, pp. 393–4.

and table-thumping. Brandler yielded. In his own party, especially in its Berlin branch which was led by Ruth Fischer and Arkadi Maslov, impatience for action and confidence had mounted high. He thought that he had found the same confidence in Moscow, for he assumed that Zinoviev spoke for the whole Politbureau. He diffidently concluded that if the leaders of the only victorious Communist party thought, as the Berliners did, that the hour had struck, then he ought to waive his objections.

It was at this point, feeling, as he himself put it, that he was not 'a German Lenin', that Brandler asked the Politbureau to assign Trotsky to lead the insurrection. Instead of Trotsky the Politbureau delegated Radek and Pyatakov. A plan of action was laid down, which centred on Saxony, Brandler's homeland where communist influence was strong, the Social Democrats headed the provincial government, and where they and the Communists already acted in a united front. Brandler and some of his comrades were to join the government of Saxony and use their influence in order to arm the workers. From Saxony the rising was to spread to Berlin, Hamburg, central Germany, and the Ruhr. According to Brandler—and his testimony on this point is confirmed by other sources—both Zinoviev and Trotsky pressed this plan on him.[1] Moreover, Zinoviev through his agents in Germany forced the pace of events so much that the coalition government in Saxony was formed on orders sent by telegram from Moscow; it was en route, while he was returning to Germany, that Brandler learned from a newspaper bought at a railway station in Warsaw that he was a Minister.[2]

Even if conditions in Germany had favoured revolution, the artificiality and the clumsiness of the plan and the remoteness of its direction and control would have been enough to produce a failure. The conditions were probably less favourable than they were assumed to be, and the social crisis in Germany less deep. Since the summer the economy had begun to recover, the Mark was stabilized, and the political atmosphere had become calmer. The Central Committee failed to arouse the mass of workers and to prepare them for insurrection. The scheme for arming the

---

[1] Ruth Fischer, op. cit., pp. 311–18; Zinoviev's speech in *The Lessons of the German Events*, pp. 36–37 ff. and in *13 Konferentsya RKP (b)*, pp. 158–78; and Trotsky, *Uroki Oktyabrya*.   [2] This has been related to the writer by Brandler himself.

workers miscarried: the Communists found the arsenals in
Saxony empty. From Berlin the central government sent a military expedition against the Red province. And so when the
moment of rising arrived, Brandler, supported by Radek and
Pyatakov, cancelled the battle orders. Only through a fault in
liaison did insurgents move into action at Hamburg. They
fought alone and, after a hopeless combat lasting several days,
were routed.

These events were to have a powerful impact on the Soviet
Union. They destroyed the chances of revolution in Germany
and Europe for many years ahead. They demoralized and
divided the German party and, coinciding with similar setbacks in Poland and Bulgaria, they had this effect on the International as a whole. They imparted to Russian communism a
deep and definite sense of isolation, a disbelief in the revolutionary capacity of the European working classes—even a disdain
for them. Out of this mood there developed gradually an attitude of Russian revolutionary self-sufficiency and self-centredness which was to find its expression in the doctrine of Socialism
in One Country. Immediately, the German débâcle became an
issue in the Russian contest for power. Communists both in
Russia and Germany delved into the causes of the defeat and
were eager to fix the responsibilities. In the Politbureau the
triumvirs and Trotsky laid the blame on each other.

On the face of it, there existed no connexion between the
German fiasco and the Russian controversy. The lines of division
were different and they even cut across one another. Radek and
Pyatakov, the two 'Trotskyists', had been from the outset at
least as sceptical as Stalin was about the chances in Germany; it
was they who advised Brandler to cancel the orders for insurrection. On the other hand, Zinoviev had, after hesitation,
sanctioned the plan for the rising, of which Trotsky was the
initiator; but he also sanctioned the cancellation of the marching
orders. Trotsky was convinced that the German party and the
International had missed a unique opportunity; and he held
that Zinoviev and Stalin were at least as much responsible for
that as Brandler. The triumvirs replied that the rising had been
bungled on the spot by the two Trotskyists; and they insisted
on Brandler's 'opportunism' and on the necessity to depose him
as leader of the German party.

Vis-à-vis Brandler the triumvirs were actuated by mixed motives. The rank and file of the German party had turned bitterly against him; and the organization of Berlin clamoured for his dismissal. Zinoviev was eager to appease the clamour and to save his own and the International's prestige by making a scapegoat of Brandler. In deposing him and installing Fischer and Maslov in the leadership of the German party, Zinoviev made of that party his fief. He had yet another reason for insisting on Brandler's exemplary punishment: he suspected Brandler and his friends in the German Central Committee of sympathy with Trotsky. In denouncing Brandler as Trotsky's follower Zinoviev also sought to burden Trotsky with the blame for Brandler's 'capitulation'. At last Brandler, unable to make head or tail of the rivalries, anxious to disentangle the German question from the Russian issues, and eager to save his position, declared his support to the official Russian leadership, that is to the triumvirs. This, however, did not save him.

Such was the situation in January 1924, when the Executive of the International met to hold a formal inquest on the German defeat. The meeting was preceded by much wire-pulling and many shifts in the Central Committees of foreign parties, designed to secure in advance the Executive's support for Zinoviev. When the Executive met, Trotsky was ill in a village not far from Moscow. He did not state his views, but asked Radek to convey their joint protest against Brandler's demotion and the changes in the German Central Committee. Radek conveyed the protest, but being mainly interested in defending his own and Brandler's policy, he gave the Executive the impression that Trotsky associated himself with that policy; and this enabled the triumvirs to link Trotsky once again with the 'right wing' in the German party.[1] In truth, Trotsky never ceased to be critical of Brandler's conduct; and the fact that Brandler had now declared his support for the triumvirs could not have commended him to Trotsky. Nevertheless, Trotsky objected on principle to the installation in Moscow of a 'guillotine' for foreign communist leaders. Foreign parties, he held, must be allowed to learn from their own experience and mistakes, to

---

[1] *The Lessons of the German Events*, p. 14. See also Trotsky's letters about this to A. Treint and A. Neurath, written in 1931 and 1932, and published in *The New International* (February 1938).

manage their own affairs, and to elect their own leaders. Brandler's demotion established a pernicious precedent.

Thus Trotsky demanded for the International the same inner freedom which he claimed for the Russian party; and he did it with the same result. Zinoviev had by now complete mastery over the International. He had deposed some of those foreign leaders who had appealed to the Politbureau to restrain its vehemence against Trotsky. Others allowed themselves to be browbeaten and apologized for their *faux pas*. Consequently, the Executive, although it failed to carry its inquest on Germany to a clear conclusion, left Zinoviev with his reputation untarnished; and it endorsed the demotions and promotions he had ordered. This allowed him presently to obtain from the International an endorsement of the triumvirs' action against Trotsky and the Forty Six.

In May, at the thirteenth congress of the Russian party, the leaders, old and new, of all the European parties appeared on the platform to echo the anathema on Trotsky. Only one foreign delegate, Boris Souvarine, Editor of *L'Humanité*, himself half-Russian and half-French, raised his voice against it, declaring that the French Central Committee had decided by twenty-two votes against only two to protest against the attacks on Trotsky, without necessarily declaring thereby its solidarity with the opposition; but that he, personally, shared Trotsky's views and would not abjure them. Souvarine's lonely voice only stressed Trotsky's defeat.[1]

A month later, the fifth congress of the International—the so-called 'congress of Bolshevization'—met in Moscow to put its seal under the excommunication of Trotsky, to which a denunciation of Radek and Brandler was added. Characteristic of the mood of the congress was a speech by Ruth Fischer, the new leader of the German party. A young, trumpet-tongued woman, without any revolutionary experience or merit, yet idolized by the communists of Berlin, she railed against Trotsky, Radek, and Brandler, those Mensheviks, opportunists, and 'liquidators of revolutionary principle' who had 'lost faith in the German and European revolution'. She called for a monolithic International, modelled on the Russian party, from which dissent and contest of opinion would be banished. 'This world

[1] *13 Syezd RKP (b)*, pp. 371–3.

congress should not allow the International to be transformed into an agglomeration of all sorts of trends; it should forge ahead and embark upon the road which leads to a single Bolshevik world party.'[1] Spokesmen of the French, English, and American delegations followed suit; and, shrinking from no abuse or insult, they challenged Trotsky to appear before the congress and state his views.[2] Trotsky refused to enter into any disputation. For one thing, he felt that all disputation was now useless. For another, having already been threatened with expulsion from the party if he indulged in any further controversy, he may have suspected that the challenge was a trap. And so he declared that he accepted the verdict of the Russian party and had no intention of appealing against it to the International. Even his silence, however, was received as proof of his malfeasance: echoing Zinoviev, delegates demanded from him nothing less than recantation.[3] He turned a deaf ear; and in the course of the full three weeks the congress heard nothing but foul-mouthed vituperation against the man to whom the previous four congresses had listened with deep respect and adoration. This time not a single voice rose to vindicate him. (Souvarine had by now been expelled from the French party for having translated and published Trotsky's *New Course*.)[4] Yet Trotsky still wrote the last of his great Comintern manifestoes for this congress. But he was not re-elected as full member of the Executive; Stalin took his place.

What accounted for the change that had come over the International? Only a few months earlier its three greatest parties had enough courage and dignity to rebuke the triumvirs. Now all gave a spectacle of submission and self-abasement. Zinoviev, we know, had in the meantime shuffled, displaced, or broken up at will the German, French, and Polish Central Committees. But why did these Committees and the parties behind them accept his dictates? Most of the deposed leaders had guided their parties from the day of foundation and had enjoyed high moral authority; yet nowhere did the rank and file stand up for them and refuse to accept the Executive's orders and to acknowledge Zinoviev's nominees as leaders. It took Zinoviev only a few weeks or at the most a few months to bring

[1] *5 Vsemirnyi Kongress Kominterna*, vol. i, pp. 175–92.
[2] Ibid., pp. 550–9.   [3] Ibid., vol. ii, pp. 156–7.   [4] Ibid., p. 181.

about what appeared to be a complete upheaval in the entire communist movement. But the ease with which he brought it about indicated a deep-seated weakness in the International. Only a diseased body could be thus subverted at a stroke.

Lenin and Trotsky had founded the International in the expectation that it would soon rally to its banners the majority of at least the European labour movement.[1] They expected it to become what its name said it was: a world party, transcending national boundaries and interests, not a decorous and platonic association of national parties in the style of the Second International. They believed in the basic unity of the revolutionary processes in the world; and this unity made it essential in their view that the new organization should possess a strong international leadership and discipline. The Twenty One Conditions of membership, which the second congress adopted in 1920, were designed to give the International a constitution appropriate to this purpose, and to establish, among other things, a centralized and strong leadership in the Executive. Trotsky had supported that constitution wholeheartedly.[2] By itself it was not calculated to assure the preponderance of the Russian party in the International. All parties were represented on the Executive in a democratic manner. Its few Russian members enjoyed in principle no privilege. Internationalism implied the subordination of national viewpoints to the broader interest of the entire movement but certainly not to any national-Russian viewpoint. Had revolution won in any of the important European countries or had at least the Communist parties there grown in strength and confidence, such international leadership and discipline might have become real. But the ebb of revolution in Europe tended to transform the International into an adjunct to the Russian party. The self-assurance of its European sections was weak; and it dwindled from year to year. The defeated parties developed a sense of inferiority; and they came to look to the Bolsheviks, the only successful practitioners of revolution, to tackle their problems, to solve their dilemmas, and to make their decisions for them. The Bolsheviks responded, first from a sense of solidarity, then from habit, and finally from self-interest, until they were only too eager to handle the leading strings in which the foreign parties had so willingly put themselves. International

---

[1] *The Prophet Armed*, p. 452.   [2] Ibid., p. 467.

leadership and discipline became in fact Russian leadership and discipline; and all the wide prerogatives which the Twenty One Points had vested in the international Executive of Lenin's and Trotsky's expectations passed almost imperceptibly to the Russian members of the Executive.

Lenin was disturbed by this state of affairs. He recalled Engels's forebodings about the preponderance of the German party in the Second International and pointed out that the supremacy of the Russian party might be not less harmful.[1] He tried to give foreign communists more self-reliance and even suggested that the Executive should be transferred from Moscow to Berlin or another European capital in order to remove it from the constant pressure of Russian interests and preoccupations. However, most foreign communists preferred to see the hub of their International placed in the safety of Red Moscow rather than expose it to persecution and police raids in bourgeois capitals.

Lenin's misgivings proved all too justified. As the years passed the intervention of the Russian members of the Executive in the affairs of foreign communism grew ever more meddlesome. Zinoviev ruled the International with relish, flamboyance, lack of tact and scruple. But even Trotsky found himself, as member of the Executive, involved in the exercise of a tutelage which was inherent in the situation. As chairman of the French commission of the Comintern, he supervised with plenary power the day-to-day work of the French communists. The German, the Italian, the Spanish, and the British parties eagerly sought his advice on every major issue and even on the detail of their activity; and he gave his advice freely.

This led him to make pronouncements and to engage in a voluminous correspondence which in themselves form a running commentary on the history of these crucial years, a commentary rich in thought, sparkling with wit, and often astonishingly far-sighted.[2] But parts of the correspondence also reflect the tutelage. Here, for instance, he summons peremptorily Frossard, the French leader, to face grave but not unjustified charges at the

---

[1] Lenin, *Sochinenya*, vol. xxxiii, pp. 392–4. More explicit remarks to this effect are in Lenin's, still unpublished, statements made at the Executive of the International.

[2] See his *Pyat Let Kominterna*, published also in English under the title *The First Five Years of the Communist International*, vols. i and ii.

International's assizes in Moscow. There he censures communist editors and prescribes the tactical line and even the topics and the style for their newspapers. Here he chides *L'Humanité* for publishing the writings of dubious contributors. There again he lays down a date by which the French party must expel, as it had undertaken to do, all Free Masons and 'all careerists'. On several occasions he acts as umpire to rival groups and lays down the law for them.[1] These, it is true, are extreme and exceptional instances. He never hectored or cajoled his subordinates in the Comintern, as Zinoviev and then Stalin did; and he always expected them to speak their mind on the affairs of the Russian party as frankly as he expressed himself on the conduct of their parties. It was not his fault if foreign communists rarely felt self-confident enough to speak their mind. He still treated the Executive as a truly international body and acted on its behalf from the general principles of communism and not from any peculiarly Russian angle. It was in this spirit that he used the wide powers which the Twenty One Points had vested in the Executive.

The actual preponderance of the Russian party, however, made it all too easy to use the Twenty One Points as the constitutional framework for the establishment of a Russian *de facto* dictatorship. This was what Zinoviev did even before 1923, when he was still curbed by Lenin and Trotsky. Later all restraints had gone. Moreover, inner democracy could not survive in the International after it had withered in the Russian party. The habits of 'substitutism' spread to the entire movement; and the chiefs of the Bolshevik Old Guard came to look upon themselves as the trustees not merely of the Russian working class but of the working classes of the world.

In 1923-4 Zinoviev and Stalin indeed set out to refashion the European movement after the new Russian image. They could not tolerate in the International the opposition which they were bent on suppressing in their own party. Just as they had used the Russian 1921 ban on inner-party factions to destroy Trotsky's influence at home, so they used the wide powers they wielded under the Twenty One Points to destroy his influence abroad. Trotsky had endorsed both the 1921 ban and the Twenty

---

[1] Ibid., vol. ii, pp. 124-84; Rosmer, *Moscou sous Lénine*, pp. 236-60; Frossard, *De Jaurès à Lénine.*

One Points. His adversaries planned their moves so that every step they made appeared as a plain application of the principles and precedents laid down with Trotsky's consent, if not on his initiative. They struck him down with his own weapons— only that he had never used these weapons for any comparable purpose or with comparable brutality. He had occasionally threatened foreign communists with disciplinary sanctions; they demoted, dismissed, and denounced them wholesale. He had demanded that the Comintern should, in accordance with its programme, tolerate no bourgeois pacifism, no Free Masonry, and no 'social-patriotism'. They purged it of 'Trotskyism' which had hitherto been almost synonymous with communism.

.   .   .   .

In May the thirteenth congress of the Russian party closed the debate which had started with the proclamation of the New Course. Trotsky could not reopen the controversy without incurring the charge of a breach of discipline; and he made no attempt to reopen it. He once described admiringly the self-discipline which had induced Jaurès to put, when necessary, 'his bovine neck under the yoke of party discipline'. He now put his own neck under a much harder yoke and refrained from discussing in public the party's economic policy and inner régime which had been declared taboo. Yet he could not reconcile himself to being branded as a semi-Menshevik guilty of a 'petty bourgeois deviation from Leninism'. Debarred from discussing the crucial and topical issues of policy, he fell back on history to vindicate himself. The opportunity offered itself when the State Publishers, carrying out an earlier decision of the Central Committee to produce a many-volumed edition of Trotsky's *Works*, prepared for the presses the book which contained his speeches and writings of 1917. He prefaced it with a long essay entitled 'The Lessons of October'. The volume appeared in the autumn of 1924; and at once it stirred up a storm.

Trotsky's speeches and writings of 1917 provided a strong reply to the obloquy about him as the unreformed Menshevik, for they reminded the party of his role in the revolution and of the unswerving militancy with which he then confronted the Mensheviks. Such a reminder was needed. The historical memory of nations, social classes, and parties is short, especially

in times of great upheaval when the breathtaking events of one
year crowd out of peoples' minds the events of preceding years,
when in political life generations or age groups succeed one
another at a furious pace, when the veterans of early struggles
rapidly dwindle in numbers, disperse, or grow exhausted and
weary, and when the young plunge into new struggles more or
less unaware of what has gone before. In 1924 those who had
belonged to the Bolshevik party since the early days of 1917
already formed less than 1 per cent. of the membership. To the
mass of young members the revolution was already a myth as
vague as it was heroic. The earlier political struggles with all
their tangled alignments appeared even more remote and unreal. The young communist took it for granted, for instance,
that Bolsheviks and Mensheviks had always opposed one
another in irreducible enmity as they had done within his
memory. It was almost inconceivable to him that they should
have formed, in the course of many years, two factions of the
same party, evoking common principles, quarrelling and breaking with one another, but also repeatedly trying to heal the
breach. It was even more inconceivable that many Bolshevik
leaders should have sought to make peace with the Mensheviks
as late as 1917.

The young were therefore shocked when they learned that
the Commissar of War had once been a Menshevik or semi-
Menshevik; and many were inclined to believe the triumvirs
when they maintained that once a Menshevik always a Menshevik. Nothing could shake that belief more severely than the
perusal of Trotsky's speeches and writings of 1917, which
showed up the recent anti-Trotskyist campaign as mendacious.
Thus Trotsky, merely by republishing his old texts, called out
his adversaries; but he challenged them directly in the 'Lessons
of October'.

Trotsky advanced in this essay his own interpretation of the
party's history and tradition, an interpretation which not merely
vindicated him but also impugned the records of most of his
assailants. The party's history, he wrote, fell into three distinct
periods: the years of preparation for 1917; the decisive trial of
1917; and the post-revolutionary era. Each of these periods had
problems, peculiarities, and a significance of its own. But it was
in the second period that Bolshevism rose to its climax. A revolu-

tionary party is tested in actual revolution just as an army is tested in actual battle. Its leaders and members are ultimately judged according to their conduct under this trial; compared with this their behaviour during the preparatory period is of little importance. A Bolshevik should not be judged by what he said or did before 1917, in the course of the confused and in part 'irrelevant manœuvres of émigré politics', but by what he said and did in 1917. The argument, although Trotsky gave it the impersonal form of historical narrative, was *pro domo sua*: his own pre-revolutionary associations with Menshevism belonged to the 'irrelevant manœuvres of émigré politics', but his position as leader of the October insurrection was unassailable. By the same criterion the record of his adversaries was against them: they may have been good 'Leninists' during the years of preparation, but they were found wanting in 1917.

He related the two major crises the party had gone through in 1917: in April, when Lenin had to overcome the resistance of the party's right wing, the 'old Bolsheviks', as Lenin himself called them, before he could persuade the party to set its course for socialist revolution; and on the eve of the October Revolution, when the same right wing balked at insurrection. The hesitancy and the errors of some of the leaders, Trotsky argued, did not detract from the Bolshevik achievement. The party was a living organism with its frictions and divergencies of opinion. However, Bolsheviks should be aware of the facts: even a revolutionary party of necessity includes conservative elements which hamper its progress, especially when the party faces a sharp turn and must take bold decisions. The edge of this argument was turned in the first instance against Zinoviev and Kamenev, the 'strike-breakers of revolution', but also against Rykov, Kalinin, and other leaders of the Old Guard who had opposed Lenin's policy in 1917. In effect, Trotsky called into question the triumvirs' right to speak as the only authentic interpreters of Bolshevik doctrine and, more broadly, the Old Guard's pretension to represent the Leninist tradition in its purity. The implied but obvious moral of his story was that this tradition was by no means as simple and constant as people were made to believe: the Old Guard represented that 'old Bolshevism' which Lenin had abjured because it clung to outdated slogans and irrelevant recollections, while Trotsky's attitude was in full harmony with the Bolshevism of 1917 under the sign of which the party had won.

From history and topical allusion Trotsky then passed to the latest critical event, the failure of communism in Germany. His main themes in the 'Lessons of October' were the role of leadership in a revolutionary situation and the strategy and tactics of insurrection. No Communist party, he argued, can create at will revolutionary opportunities, for these arise only as the result of a relatively slow decay of a social order; but a party can miss its opportunity through lack of determined leadership. In the affairs of revolution, too, there is a tide which must be 'taken at the flood'; missed, it may not return for decades. No society can live long in the tension of acute social crisis. If it finds no relief from that tension in revolution, it finds it in counter-revolution. It may take only a few weeks or even days for the scales to turn one way or the other. If during these weeks or days communists shrink from insurrection and delay action, believing that the revolutionary situation will drag on and offer them new chances, then indeed 'all the voyage of their life is bound in shallows and in miseries'. Such would have been the voyage of the Bolsheviks if the opponents of insurrection had had their way; and so was German communism bound in shallows and in miseries in 1923. Russia had offered the positive proof for the decisive role of revolutionary leadership; Germany offered the negative evidence. The same conservative frame of mind which the Bolshevik right wing showed in 1917 was responsible for the defeat in Germany. It was obvious at whom this sting of Trotsky's conclusion was aimed: the man who had spoken for the Bolshevik right wing in October 1917 was now President of the Communist International.

The triumvirs returned a massive riposte; and they summoned hosts of propagandists and historians and even foreign communist writers for the counter-attack.[1] Throughout the autumn and the winter the country's political life was entirely overshadowed by this controversy which has entered Bolshevik annals under the odd name of 'the literary debate'. Since it was

[1] The most important replies to Trotsky were collected in a large volume *Za Leninizm*—the contributors were Stalin, Zinoviev, Kamenev, Bukharin, Rykov, Sokolnikov, Krupskaya, Molotov, Bubnov, Andreev, Kviring, Stepanov, Kuusinen, Kolarov, Gusev, and Melnichansky.

impossible to deny roundly Trotsky's assertions about Zinoviev's and Kamenev's attitude in 1917, their defenders replied that he had fantastically magnified their errors, that there had been only fortuitous and superficial dissensions between them and Lenin, and that no special right wing or conservative trend of opinion had ever existed in the party. Trotsky, they said, had invented this in order to discredit not merely the Old Guard but the whole body of the Leninist tradition, and in order to ascribe to himself and to Trotskyism wholly imaginary merits.

To prove the point, the triumvirs and their historians had to oppose to Trotsky's account their own versions of the events of 1917, versions designed to enhance their own prestige and to belittle the part Trotsky had played. This was done timidly at first, but then with growing boldness and disregard for truth. Thus, it was not denied at first that Trotsky had acted an outstanding part; but this, it was said, was not superior to that acted by his present adversaries. Then Stalin himself intervened with a version of his own. He declared that the Military Revolutionary Committee of the Petrograd Soviet, over which Trotsky presided, had not at all been the headquarters of the October insurrection, as all historical accounts, without a single exception, had maintained hitherto. He asserted that a more or less fictitious 'Centre', of which Trotsky was not even a member but on which Stalin sat, directed the rising.[1] This version was so crudely concocted that even the Stalinists received it at first with embarrassed irony. But once put out, the story began to crop up stubbornly in the new historical accounts until it found its way to the textbooks, where it was to remain as the only authorized version for about thirty years. Thus that prodigious falsification of history was started which was presently to descend like a destructive avalanche upon Russia's intellectual horizons: it began as a mere attempt to bolster the reputations of Zinoviev and Kamenev whom it was eventually to depict, as it was also to depict Bukharin, Rykov, Tomsky, and so many other Bolshevik leaders, as the saboteurs and traitors of the October Revolution and as foreign spies. In 1924 most of the future victims of the falsification were united in a frantic effort to cast Trotsky into the shade.

Yet, as long as Trotsky stood on the ground of the events of

[1] Stalin, *Sochinenya*, vol. vi, pp. 324–31.

1917, his position was formidable. The triumvirs did therefore their utmost to shift him from that ground back to the pre-revolutionary era, the era of his opposition to Bolshevism. They established a canon of rigid continuity in the party's policies and a canon of its virtual infallibility. Whoever, they said, had, like Trotsky, opposed Bolshevism consistently throughout a long period was fundamentally in the wrong; and this was bound to show itself even in his later attitudes. Making a parody of determinism, the canon-makers instilled into the party's mind the idea that no political error or deviation, whether collective or individual, could be treated as a casual occurrence. (The rule did not apply, of course, to the triumvirs' own errors.) Each error had its deep causes or 'roots' in the peculiar make-up, petty bourgeois or otherwise, of any given group or individual. A major error weighed on him who committed it with the fateful gravity of original sin. Trotsky's fall dated back to his early Menshevik days, not merely to the 'manœuvres of émigré politics' but to his fundamental attitude towards the major problems of the time. In the October interval his petty bourgeois soul struggled to achieve grace. The party hoped to help him and to 'assimilate' him. But again and again his stubborn Menshevik nature reasserted itself.

In this light the disagreements which Trotsky had had with Lenin since the revolution also acquired a hitherto unsuspected sinister meaning. He had had two such major disagreements: over the peace of Brest Litovsk and over policy towards the trade unions. (The other disagreements in which Lenin acknowledged his own mistakes were ignored.) Innumerable pamphlets and articles were published which dwelt on these two cases and gave new accounts of them to prove that in both Trotsky's ineradicable anti-Leninism had revealed itself, and to establish a straight connexion between his opposition to Lenin and his attack on Lenin's successors. The contexts of the old controversies, the real alignments, the motives, the hesitations, the self-contradictions, and the human virtues and failings of the actors, were all omitted from the new accounts. The party was shown a picture of itself and of its leaders which resembled those early medieval frescoes of the Last Judgement, where the virtuous, whose faces express nothing but piety, climb straight to heaven while the sinners, concentrated symbols of vice, rush to damnation.

As the controversy switched backward and forward and back again to the years 1905–6, the fount of all of Trotsky's errors and deviations was at last discovered in his theory of permanent revolution. This was declared to be his master-heresy. Yet ever since 1917 the party had had no quarrel with that theory; Trotsky's early essays on it had been republished in the original and in many translations as an authoritative statement of communist doctrine. Even now its two chief tenets—that the Russian Revolution had to pass from the bourgeois to the socialist phase; and that it would be the prologue to world revolution—were still the party's household ideas; and they could not be openly refuted. The polemicists dug up a few barbed remarks which Lenin made in 1906: himself still holding that the Russian Revolution would be only bourgeois in character, Lenin then said that Trotsky spoke of a socialist consummation because he 'jumped' over the bourgeois phase and 'underestimated' the importance of the peasantry. In view of what had happened in 1917 these remarks had lost all relevance. This did not now prevent the polemicists repeating in circles that Trotsky's characteristic propensity was to 'jump over necessary intermediate stages' and to 'underestimate the peasantry'. True enough, it was not easy to square this charge with the other accusation that he was an unreformed Menshevik—the Mensheviks, far from 'jumping' over the bourgeois phase of the revolution, refused to go beyond it—and it took a great deal of purely scholastic argument to cope with this logical difficulty. However, as in all disputations of this sort, it was not the logic or the historical truth of the argument that mattered, but its undertone, its bearing on current policy, and the impression it made on the uninitiated.

That the insistence on Trotsky's inclination to 'underestimate the peasantry' had a bearing on current policy is obvious: the triumvirs and Rykov had begun to label Trotsky as the muzhik's enemy even the year before. Now they gave that label retrospective validity and historical colouring. More significant still was the broader undertone. To the popular understanding Permanent Revolution suggested a prospect of continuous upheaval and of endless struggle, and the impossibility for the Russian Revolution to settle down and achieve a measure of stabilization. In denouncing Permanent Revolution, the triumvirs appealed to the popular longing for peace and stability.

In truth, Trotsky's theory did claim that the fortunes of Bolshevik Russia depended *ultimately* on the spread of revolution abroad. Yet the hopes for its spread had been dashed many times and had just suffered the most severe setback in Germany. The Bolsheviks felt more isolated than ever. They found a psychological defence in their complacent sense of Russia's revolutionary self-sufficiency. Trotsky's theory offended and mocked that sense. Hence the intense irritation which the mere mention of Permanent Revolution began to arouse among Bolshevik cadres. They felt a vehement emotional urge to deprive Trotsky's theory of all ideological respectability. It was no matter of chance that in the autumn of 1924 Stalin, revising his own earlier views, formulated the doctrine of Socialism in One Country, which became the counterpart to Permanent Revolution. Stalin extolled the self-sufficiency of the Russian Revolution and thereby he offered the party an ideological consolation for its frustrated internationalist hopes.[1]

It is easy to see why and how the 'literary debate' weakened Trotsky's position even further. It fixed in the public mind a contradictory image of Trotsky as, on the one hand, an inveterate semi-Menshevik and, on the other, an equally inveterate 'ultra radical' and extremist seeking to involve the party in dangerous ventures at home and abroad. At home, it was said, he strove to embroil the Bolsheviks with the peasants whom he had never understood. Abroad, he always saw revolutionary opportunities where none existed. The same aberration had led him to oppose the Brest Litovsk Peace and to blame Zinoviev for the defeat of revolution in Germany. That Trotsky had also criticized Zinoviev for encouraging abortive risings abroad, that he had been opposed to the march on Warsaw in 1920, that he had consistently striven to normalize relations with the capitalist countries, and that he had been the first to advocate the N.E.P. policy in order to pacify the peasants—these and similar facts which contradicted the image of the ultra radical adventurer did not matter. Fact, fiction, and scholastic quibble were so jumbled together that Trotsky became the Quixote of communism, pathetic perhaps, but also dangerous, whom only the wisdom and the statesmanship of the triumvirs could restrain and render harmless.

[1] See *Stalin, A Political Biography*, pp. 281–93.

Many a party member, even some of Trotsky's own adherents, held that in the 'Lessons of October' he had chosen his ground wrongly.[1] He should have concentrated, they said, on issues that mattered instead of digging up Zinoviev's and Kamenev's errors of 1917. True, he had done this in self-defence after the triumvirs had raked up all the long-forgotten incidents of his controversies with Lenin and after they had prevented him from discussing current affairs. But most people quickly forgot 'who had started it all'; and they reproached him with not letting bygones be bygones. Official writers quoted against him the excerpts from Lenin's suppressed will in which Lenin begged the party not to hold against Zinoviev and Kamenev their 'historic errors'. Even Krupskaya, mindful of that advice, was prevailed upon to rebuke Trotsky and to say that he had made too much of the disagreements between Lenin and his disciples, because the fate of the revolution depended on the attitude of the party and of the working class as a whole, not on dissensions within a narrow circle of leaders.[2] This was a telling criticism, directed as it was against the advocate of inner-party democracy. Bolshevik self-esteem had anyhow been wounded by Trotsky in whose recollections the party's leadership appeared as a sluggish, hesitant body of men who would never have done their duty if they had not been prodded and pushed into action by Lenin.

The debate had a further consequence which greatly embarrassed Trotsky. Some elements of the dispersed anti-Bolshevik opposition, who had hitherto hated him like death, began to pin their hopes on him.[3] This was inevitable. In a single-party system some of the suppressed enemies of the government, no longer able to fight under their own banners, will applaud any important dissenter even if he belongs to the ruling party and no matter what the reasons for his dissent. They tend to view as their hero any one whom the ruling group itself stigmatizes as its dangerous adversary. The circumstance that Trotsky demanded freedom of expression, if only within the party, commended him at least to some anti-Bolsheviks who saw

[1] Trotsky, *The Stalin School of Falsification*, p. 90.
[2] Krupskaya, 'K Voprosu ob Urokakh Oktyabrya' in *Za Leninizm*, pp. 152–6.
[3] M. Eastman, *Since Lenin Died*, pp. 128–9; Bajanov, *Avec Staline dans le Kremlin*, p. 86.

no future for themselves without any freedom of expression. This was by no means the prevalent attitude among the anti-Bolsheviks. Many or perhaps most of them viewed with glee the fall of the man on whom they placed the chief blame for their defeat in the civil war. But the triumvirs made the most of any sign of real or spurious sympathy for Trotsky which could be detected outside the party, while he was all the more anxious to say and do nothing that might encourage such sympathy. This accounted largely for his restraint and long silences, and for his constant and emphatic reiteration of his solidarity with the triumvirs in the face of common enemies.

Finally, the 'literary debate' had an important effect on the triumvirs themselves. Its result was to discredit all the chief controversialists with the sole exception of Stalin, whose prestige was, on the contrary, enhanced. Trotsky had concentrated his attack on Zinoviev and Kamenev who had clearly expressed and placed on record their objections to the October insurrection. Stalin, having been less articulate and much more elusive in 1917, was now far less vulnerable. Indeed, Zinoviev and Kamenev stood at present in need of his moral support; and they were glad to receive from him testimonials of good Bolshevism.[1] This helped Stalin to establish himself definitely as the senior triumvir. Thus, unwittingly, Trotsky helped to defeat his future allies and to promote his chief and most dangerous adversary.

.   .   .   .

The storm raised by the 'Lessons of October' made Trotsky's position as Commissar of War untenable. The triumvirs had denounced him in such terms that they could not leave him in charge of the country's military affairs, although only a year earlier they were still afraid to accept his resignation. They now openly worked to remove him from the Commissariat.

In no phase of the struggle did Trotsky make the slightest attempt to appeal against them to the army. He restrained those of his followers who, like Antonov-Ovseenko, had been tempted to draw into the controversy the military cells which were, under

[1] Stalin, *Sochinenya*, vol. vi, pp. 326–7. (The text of Stalin's statement in defence of Zinoviev and Kamenev as 'good old Bolsheviks' is somewhat toned down in his *Works* (compared with the original published in *Za Leninizm*, pp. 88–89) but clear enough.)

the party's rules and regulations, entitled to have their say. Official spokesmen, it should be added, never reproached Antonov-Ovseenko with any offence graver than this—there was no question of any plot or preparations for a *coup*; and they repeatedly acknowledged Trotsky's restraining influence.[1] When hints were dropped about his Bonapartist ambition, this was done in private gossip only. Trotsky was not accused of making any single move designed to use his position as Commissar of War to his political advantage. He acknowledged as a matter of course, the Politbureau's jurisdiction over the army. Consequently, he accepted, though not without protest, the dismissal and demotion of his followers from the most influential posts in his Commissariat and the appointment to them of his adversaries.[2]

It would be futile to speculate whether Trotsky would have succeeded if he had attempted a military *coup*. Early in the conflict, before the General Secretariat had begun to shift and shuffle the party personnel in the army, his chances of success might have been high; they dwindled later. He never tried to test the chances. He was convinced that a military *pronunciamento* would be an irreparable setback for the revolution, even if he were to be associated with it. He had declared at the thirteenth congress that he saw in the party 'the only historic instrument which the working class possessed for the solution of its fundamental tasks'; and he could not try and smash that instrument with the army's hands. In any conflict with the party, he held, the army would have to rely on the support of counter-revolutionary forces and this would have condemned it to play a reactionary part. True, he saw 'degeneration' in the party. But this consisted in the breach between the leaders and the rank and file and in the party's loss of its democratic base. The task, as he saw it, was to reconstitute that base and to reconcile the leaders and the rank and file. Ultimately the revolution's salvation lay in a political revival 'down below', in the depth of society. Military action 'from above' could only usher in a régime even further removed from a workers' democracy than was the present government. Such was the 'logic of things'; and

[1] At the thirteenth conference even official speakers referred to Trotsky's restraining influence, see, e.g., Lominadze's speech in *13 Konferentsya RKP (b)*, p. 113.   [2] *Moya Zhizn*, vol. ii, pp. 253–4.

he did not believe that he could stand against it. He placed his own person and action within the framework of the social forces which determined the course of events; he saw his own role as subordinate to those forces; and his aim, the revival of proletarian democracy, dictated to him the choice of his means.

In the course of the year 1924 the direction of the Commissariat of War slipped from his hands. Through Frunze and Unschlicht the triumvirs gradually extended their control over the whole body of the army's political commissars; and now they had no qualms about drawing the armed forces into the inner-party conflict. They submitted to the military cells resolutions condemning Trotsky for publishing the 'Lessons of October'; and they convened a national conference of the political commissars and placed before them a motion demanding Trotsky's dismissal from the Department of War. At this time Trotsky once again succumbed to an attack of malaria and he did not, it seems, even state his case to the commissars. The conference duly passed the motion demanding his dismissal. Then he suffered the same rebuff from the communist cell on the Military Revolutionary Council, the Council over which he had presided since the day it was formed. To crown it all, a plenary session of the Central Committee was called for 17 January 1925, and 'the Trotsky case' figured as the first item on the agenda.

On 15 January Trotsky addressed a letter to the Central Committee in which he excused, on grounds of illness, his absence from the forthcoming session; but he stated that he had delayed his planned departure from Moscow—he was to go to the Caucasus again—in order to answer questions and offer explanations which might be required of him. Concisely and with subdued anger, he replied to the main accusations levelled against him— this was his only reply to the critics of the 'Lessons of October'. Then he asked to be immediately relieved from his duties as President of the Revolutionary Military Council and declared: 'I am ready to carry out any work whatever assigned to me by the Central Committee, in whatever position or without any position and, it goes without saying, under any conditions whatever of party control.'[1]

At the Politbureau Zinoviev and Kamenev proposed to ask

[1] The full text of the letter is in Eastman, *Since Lenin Died*, pp. 155–8.

the Central Committee to expel Trotsky from the Politbureau and the Committee. Once again, to their irritation, Stalin refused to comply; and Zinoviev and Kamenev wondered whether he might not make peace with Trotsky at their expense. The Central Committee decided that Trotsky should continue to sit on the Committee and the Politbureau; but it threatened him once again with expulsion if he engaged in any new controversy.[1] The Central Committee then formally declared the 'literary debate' as closed; but in the same breath it instructed all propaganda departments to continue the campaign 'which would enlighten the whole party . . . about the anti-Bolshevik character of Trotskyism, beginning from the year 1903 and ending with the "Lessons of October"'. Another campaign was to make clear to the country at large, not only to party members, the danger to the 'alliance of workers and peasants' which Trotskyism carried with it. As Trotsky was not allowed to reply, this became a 'one-sided debate'. The Central Committee finally 'declared it impossible that he should continue to work on the Revolutionary Military Council'.

Thus, with the badges of infamy stuck over the badges of his fame, with cries of denunciation ringing in his ears, gagged and forbidden even to defend himself, he left the Commissariat and the army which he had led for seven long and fateful years.

---

[1] Popov, *Outline History of the CPSU*, vol. ii, p. 216; *KPSS v Rezolutsyakh*, vol. i, pp. 913–21.

CHAPTER III

'Not by Politics Alone . . .'

'Not by politics alone doth man live . . .' was the title Trotsky gave to a short essay of his which appeared in *Pravda* in the summer of 1923.[1] Least of all could he himself live by politics alone. Even at the most vital moments of the struggle for power his literary and cultural activities took up a great part of his energy; and he became still more deeply absorbed in them when he left the Commissariat of War and the inner-party controversy slackened for a time. Not that he sought to escape from politics. His interest in literature, art, and education remained political in a wider sense. But he refused to dwell on the surface of public affairs. He turned the struggle for power into a struggle for the 'soul' of the revolution; and he thereby gave new dimensions and new depth to the conflict in which he was involved.

How intensely indeed he was engaged in literary work during the most crucial clashes in the Politbureau can be seen even from the following few facts. In the summer of 1922, when he refused to accept the office of Vice-Premier under Lenin and, incurring the Politbureau's censure, went on leave, he devoted the better part of his holiday to literary criticism. The State Publishers had collected his pre-revolutionary essays on literature for republication in a special volume of his *Works*; and he intended to write a preface surveying the condition of Russian letters since the revolution. The 'preface' grew in size and became an independent work. He gave to it nearly all his leisure but failed to conclude it. He resumed writing during his next summer holiday, in 1923, when his conflict with the triumvirs, complicated by the expectation of revolution in Germany, was mounting to a climax; and this time he returned to Moscow with the manuscript of a new book, *Literature and Revolution*, ready for the printer.

In the course of the same summer he wrote a series of articles

[1] *Pravda*, 10 July 1923; *Sochinenya*, vol. xxi, pp. 3-12.

on the manners and morals of post-revolutionary Russia which were later collected in *Problems of Everyday Life*. The topics he discussed were: family life under the new régime; bureaucracy 'enlightened and unenlightened'; 'civility and politeness'; 'vodka, the Church, and the cinema'; 'swearing in the Russian language'; and so on. He addressed many meetings of educationists, librarians, agitators, journalists, and 'worker-correspondents'; and in his speeches he dwelt on the dullness, shabbiness, and lifelessness to which the press had sunk, and insisted on the need to restore purity and force to the Russian language, now littered with party jargon and cliché. In the same summer and in the following autumn he worked on such diverse subjects as a comparative analysis of trade cycles in the nineteenth and the twentieth centuries (about which he published a short but weighty treatise in the *Vestnik* of the Socialist Academy)[1] and the controversy between two schools in psychology, Pavlov's and Freud's. He had long been familiar with Freud's theory; and he studied Pavlov's works and prepared himself to intervene in the controversy with a plea for freedom of research and experimentation and for tolerance towards the Freudian school. In 1924 he also wrote, and published in book form, the biographical sketches of Lenin in which, by presenting the founder of Bolshevism in all his humanity, he implicitly made his critique of the official 'icon' of Lenin and of the incipient Lenin cult.

In these writings he sought to strike at the root and not merely at the symptoms of the evils which beset the revolution: at the spiritual backwardness of Mother Russia which was no less important than her economic poverty. He spoke of the need for 'primitive cultural accumulation' as being at least as urgent as the need for industrial accumulation. He exposed the soil in which Stalinism was beginning to grow, and he sought to change the climate in which it was to flourish. Hence the importance he attached to manners and morals and to 'small matters' of everyday life: he showed how these affected the affairs of state. His treatment of such topics is best illustrated by what he wrote about the peculiarly Russian habits of swearing:

Abusive language and swearing are a legacy of slavery, humiliation, and disrespect for the dignity of man, one's own dignity and that of other people. . . . I should like to hear from our philologists,

[1] *Sochinenya*, vol. xii, pp. 357–63.

linguists, and folklore experts whether they know in any other language such dissolute, sticky, and low terms of abuse as those we have in Russian. As far as I know, nothing or almost nothing of the kind exists outside our country. Swearing in our lower classes was the result of despair, embitterment, and above all, of slavery without hope and escape. The swearing of our upper classes, the swearing that came out of the throats of the gentry and of those in office, was the outcome of class rule, of slave-owners' pride, and of unshakeable power. . . . Two streams of Russian abuse—the swearing of masters, officials, and police, full and fat, and the hungry, desperate, and tormented swearing of the masses—have coloured the whole of Russian life with despicable patterns. . . .

The revolution, however, is primarily the awakening of the human personality in the masses, in those masses which were supposed to possess no personality. In spite of occasional cruelty and the sanguinary relentlessness of its methods, the revolution is . . . marked by growing respect for the dignity of the individual and by an ever increasing concern for the weak. A revolution does not deserve its name if it does not help with all its might and all the means at its disposal—if it does not help woman, twofold and threefold enslaved in the past, to get on to the road of individual and social progress. A revolution does not deserve its name if it does not take the greatest possible care of the children . . . for whose benefit it has been made. But how can one create . . . a new life based on mutual consideration, on self-respect, on the real equality of women . . ., on the efficient care for children—in an atmosphere poisoned with the roaring, rolling, ringing, and resounding swearing of masters and slaves, that swearing which spares no one and stops at nothing? The struggle against 'foul language' is an essential condition of mental hygiene just as the fight against filth and vermin is a condition of physical hygiene. . . .

Psychological habits, coming down from generation to generation and permeating the whole climate of life, are extremely tenacious. . . . How often do we in Russia make a violent rush forward, strain our forces, and then let things drift in the old way? . . . This is true not only of the uncivilized masses but also of the advanced and so-called responsible elements in our present social order. There is no denying that the old pre-revolutionary forms of foul language are still in use now, six years after October, and that they are even *en vogue* high up 'at the top'. . . . Our life is made up of the most striking contrasts.[1]

In this struggle against the persistent and resurgent traditions of a way of life which had been rooted in serfdom Trotsky was

[1] *Pravda*, 16 May 1923; *Sochinenya*, vol. xxi, pp. 26–31.

to suffer a defeat as cruel as that which he suffered in the political field. But he showed a deep historical insight into the nature of the forces by which he was to be overwhelmed. The 'two streams of Russian abuse' were to merge in Stalinism and to impose their 'despicable patterns' on the revolution itself. Fifteen years later, during the great purges, the two streams swelled into a flood: it was then possible for an Attorney-General to rail at men in the dock, who had occupied the most exalted positions in state and party, in such terms as 'you son of a bull and a pig'; and the highest magistrates wound up their obsessive perorations with the scream: 'shoot the mad dogs!' The cursing rolled on from the courtrooms to factories, farms, editorial offices, and university halls; and for several years its din deafened the whole of Russia. It was as if centuries of swearing had become condensed in a single moment, come to life in Stalinism, and burst upon the world.

. . . .

The October Revolution had given fresh impulses to cultural life; but it had also thoroughly upset it and created enormous difficulties. This would have been the effect of any revolution, even in the most favourable circumstances, and even with the educated elements of the nation on its side. The effect was immensely aggravated when the chief moving force of the revolution was an oppressed, property-less, and of necessity uneducated class. True, the Bolshevik leaders were men of the intelligentsia; and some of them possessed wide and profound education. But they were a mere handful. The 'cadres' consisted mostly of self-educated workers and half-educated people of petty bourgeois descent. The party had trained them in politics, organization, and sometimes in the broad philosophy of Marxism. But all too often their approach to cultural affairs only showed that a little knowledge could be worse than complete ignorance.

The majority of the intelligentsia had met the October Revolution with hostility. Some died in the civil war. Many emigrated. Of those who had survived and stayed in Russia many came to serve the new régime as 'specialists'. A few even became enthusiastically converted to the revolution and did their best to raise the nation culturally. But most of the men of the intelligentsia were either too rigid in conservative habits of mind or else too intimidated or too mediocre and servile to exercise a large and fruitful intellectual influence. They took it ill when they were placed under the orders of self-educated or half-educated commissars. On the other hand, the commissars often lacked confidence in themselves, and were suspicious and inclined to disguise inner uncertainty by bluff and bluster. They were also fanatically convinced of the justice of their cause and sure that they had found in Marxism, in which, of necessity, they were also only half-educated, the master key to all problems of society, including those of science and art. All the stronger were the intelligentsia confirmed in their characteristic pre-judices, and in the supercilious conviction that Marxism could teach them nothing, that its *Weltanschauung* was a mere 'farrago of half-baked half-truths'. Thus a gulf was fixed between them and the new ruling groups.

Trotsky, like Lenin, Bukharin, Lunacharsky, Krasin, and a few others, did his utmost to bridge the gulf. He pleaded with the commissars and the party secretaries to treat the intelli-gentsia with consideration and respect; and he urged the intel-ligentsia to show greater understanding for the needs of the time and for Marxism. These pleadings had their effect, but the gulf, though narrowed, was still there. Then it again began to widen. As the party hierarchy began to free itself from all forms of public control and get accustomed to arbitrary government, it became more and more inclined to impose its dictates upon the scientist, the man of letters, and the artist as well. It also began to develop its own ambitions and to encourage 'cultural' aspirations which flattered its parvenu vanity and yet appeared to have the merits of revolutionary innovation. The slogans of 'proletarian culture', 'proletarian art', and 'proletarian litera-ture' were coined, soon acquiring the same kind of popularity that the 'proletarian strategic doctrine' had enjoyed in the army somewhat earlier.[1]

Trotsky saw it his task to curb intolerance and to expose the futility of the slogans about proletarian culture and art. This was not easy. The idea of proletarian culture appealed to some Bol-shevik intellectuals, and to young workers in whom the revolu-tion had awakened a craving for education but in whom it had also released iconoclastic instincts. In the background there was

[1] See *The Prophet Armed*, pp. 482-5.

the peasants' anarchic hostility towards all that had been associated with the gentry's way of life, including its 'cultural values'. (When the muzhik set fire to his landlord's mansion he often let go up in flames the library and the paintings—he saw in them only part of the landlord's possessions.) Theorizing Bolsheviks rationalized this iconoclastic mood into a pseudo-Marxist rejection of the old 'class culture' which was to be swept away. *Proletkult* proclaimed the advent of proletarian science and art. The doctrinaires of this group of writers and artists argued with some plausibility that just as there had been feudal and bourgeois epochs in the history of civilization, so the proletarian dictatorship ought to inaugurate a culture of its own, permeated by Marxist class consciousness, militant internationalism, materialism, atheism, and so forth. Some maintained that Marxism by itself already constituted that new culture. The originators and adherents of such views strove to obtain for them the party's support and even to make of them the guiding principles of educational policy.

Both Lenin and Trotsky repudiated the *Proletkult* theory. Lenin, however, confined himself to a few brief and sharp statements and left the field to Trotsky, to whom it was more congenial. We shall presently see how Trotsky conducted the argument against *Proletkult*. The pretensions of *Proletkult*, however, were only the most extreme expression of an inclination which was widespread far beyond *Proletkult* circles, especially among party men in charge of educational and cultural affairs—an inclination to settle such affairs by word of command, to lay down the law, and to intimidate those who were too well educated, too intelligent, or too independent-minded to obey. It was this frame of mind, from which the cultural policy of Stalinism was to take its rise, that Trotsky untiringly sought to overcome: 'The state is an organization of compulsion', he said in an address to educationists, 'consequently Marxists in office may feel tempted to arrange even their cultural and educational work among the toiling masses on the principle: "here is truth revealed to you—go down on your knees before it". Of course, ours is a stern government. The workers' state has the right and the duty to apply compulsion. We turn ruthless force against the enemies of the working class. But in the education of the working class this "here-is-truth-down-on-

your-knees" method . . . contradicts the very essence of
Marxism.'¹

Such exhortations and warnings fill many pages in *The Culture of a Transition Period*, Volume XXI of Trotsky's *Works.*
Words of command addressed to scientists and bans on their
theories 'can bring us nothing except harm and shame', he insisted, anticipating the harm and the shame of Stalin's pronouncements on linguistic and biological, not to speak of
sociological heresies. It should be added that Trotsky did not
argue in this spirit only after he had been driven into opposition. As early as in January 1919 he wrote, for instance:

> Our party . . . never was and never can become a flatterer of the
> working class. . . . The conquest of power does not by itself transform
> the working class or invest it with all virtues: it only opens before it
> the opportunity of learning, developing its mind, and freeing itself
> of its inadequacies. By an intense effort the leading groups of the
> Russian working class have carried out work of gigantic historic
> significance. But even in these groups there is still too much half-
> knowledge and semi-competence.²

With this half-knowledge and semi-competence he came to
blows again and again. Lenin, when he introduced N.E.P.,
told the Bolsheviks that they had 'to learn to trade'. It was
not less important, Trotsky added, that they should 'learn to
learn'.³

It was baneful, he reiterated, to approach the 'cultural heritage' of the past with nihilistic contempt. The working class had
to take possession of that heritage and to guard it. The Marxist
should not accept it all indiscriminately; he ought to view the
cultural legacy dialectically and see its historically formed contradictions. The achievements of civilization had so far served
a double purpose: they had assisted man in gaining knowledge
and control of nature and in developing his own capacities; but
they had also served to perpetuate society's division into classes
and man's exploitation by man. Consequently, some elements of
the heritage were of universal significance and validity while
others were bound up with obsolete or obsolescent social sys-

---

¹ He delivered this address in June 1924, just after the thirteenth congress had
denounced his 'deviation from Leninism'. *Sochinenya*, vol. xxi, pp. 133–63.
² *Sochinenya*, vol. xxi, pp. 97–98.   ³ Ibid., p. 260.

tems.[1] The communist approach to the cultural legacy should therefore be selective. As a rule, the main body of the strictly scientific thought of the past was relatively little distorted through the fact that it grew up in class society. It was in ideological creation, especially in notions on society itself, that man's domination by man was mirrored most directly. But even there elements which reflected class oppression and served to perpetuate it were intricately combined with other elements through which man took cognizance of himself, sharpened his mind, enlarged his intelligence, gained insight into his emotions, learned to control himself, and therefore to some extent surmounted the limitations of his social circumstances. That was why works of art created hundreds and even thousands of years ago still fascinated modern man and continued to strike a chord in him even while he was engaged in proletarian revolution or in building socialism. To be sure, the builder of socialism should review critically, using the criteria of dialectical materialism, all inherited values; but this could have nothing to do with flat rejection or pseudo-Marxist humbug. Before the cultural values of the past could be subjected to criticism they must be thoroughly assimilated; and before the Marxist decided to revise from his angle any field of knowledge, he must first master it 'from the inside'.

.   .   .   .

Addressing the old intelligentsia, Trotsky argued from the opposite angle: he sought to persuade them that they could not live by the cultural heritage alone, and that they should re-educate themselves and find their place in Soviet society. He was concerned in particular with the outlook of the scientists and technologists, whom he repeatedly addressed on the relation between Marxism and science. His own interest in the subject was stimulated when, after his departure from the Commissariat of War, he became head of the Board for Electro-technical Development and of the Committee for Industry and Technology. A new field of study opened before him—one which

[1] Trotsky spoke of the dual role of the machine which has raised the worker's productive power but has, under capitalism, also served as an instrument of exploitation. Yet socialism cannot and does not renounce the use of the machine. This is obvious to everyone, but the same reasoning applies to most achievements of civilization.

had attracted him in his early youth and which he then abandoned for the sake of revolutionary activity. He now became 'half-administrator and half-student'. 'I was especially interested', he writes,[1] 'in the Committee for Industry and Technology, which, because of the centralized character of Soviet industry, had developed on quite a large scale. I diligently visited many laboratories, watched experiments . . . and listened to explanations given by outstanding scientists. In my spare time I studied textbooks on chemistry and hydrodynamics. . . .' These interests are strongly reflected in his writings of the years 1925-6. Sitting at the feet of the scientists, he also acted as their tutor in the sociology and the Marxist philosophy of science. He was probably influenced by Engels's *Dialektik der Natur*, the first German and Russian editions of which appeared in Moscow in 1925. He made no explicit reference to that work, but it is unlikely that he should not have read it; and on some points he follows closely Engels's line of thought.

At least three of his excursions into the philosophy of science deserve to be mentioned here: an address on Mendeleev, delivered at an All-Russian Congress of Scientists in September 1925, on the occasion of an anniversary of the great chemist; a lecture on 'Culture and Socialism' given at the Red Square Club in February 1926; and a speech on 'Radio, Science, Technology, and Society' at the Congress for the promotion of radio held in March of the same year.

There was nothing of the professional philosopher about Trotsky. He never plumbed the depths of gnosiology as Lenin did in *Empiriocriticism and Materialism*. He attempted no systematic exposition of the principles of dialectics; he preferred to apply them in political and historical analyses rather than to expound them in the abstract. Yet it is difficult to read his works without becoming aware of the well-formed philosophy behind them, of the deep thought he had given to questions of method, and of his wide if not very systematic erudition. He carried this erudition lightly, avoided the pundit's ponderous pronouncements, and as if deliberately spoke the language of the dilettante. For all this, or perhaps because of it, his few essays on the dialectics of science belong to the most illuminating and lucid Marxist statements on the subject.

[1] Trotsky, *Moya Zhizn*, vol. ii, p. 262.

Nothing was further from Trotsky's mind than any attempt to impose politics upon science. He asserted the scientist's right and even duty to remain politically disinterested in the course of research and study. This, however, should not prevent the scientist from seeing the place of science in society. There was no contradiction between the disinterestedness of the individual scientist and the deep involvement of science as a whole in the social conflicts of its age. Similarly, an individual soldier or revolutionary may fight and give his life disinterestedly, but an army and a party must have definite interests and aspirations to defend.

Detachment and rigorous objectivity in research are necessary, but not enough. It is a most vital interest of science itself that the scientist should have a broad and up-to-date philosophical outlook. This, as a rule, he does not possess. Hence a characteristic cleavage in the scientist's mind. In his special field or in his laboratory he is implicitly a materialist but outside it his thinking is most often confused, unscientific, inclined to idealism or even to plainly reactionary views. In no great thinker was this cleavage more in evidence than in Mendeleev. As a scientist he was one of the greatest materialists of all times; yet he was caught up in all the conservative beliefs and prejudices of his time and was devoted to the decaying Tsardom. When he formulated his Periodic Law, he testified to the truth of that principle of dialectics which occupies a central place in Marxist thought and asserts that quantitative changes, whether in natural or social processes, at certain points turn into changes of quality. According to the Periodic Law quantitative alterations in atom weights result in qualitative differences between chemical elements. Yet Mendeleev could not envisage the approach of the one great qualitative change—the revolution—in Russian society.

'Know in order to be able to predict and act' was the maxim of the great discoverer, who compared scientific creation to the throwing of an iron bridge across a precipice: it is not necessary, Mendeleev said, to descend and to look for a support for the bridge at the bottom of the precipice; it is enough to find support on one of the banks and then to throw across a precisely weighted arch which will rest securely on the other bank.

The same is true of all scientific thought. It must base itself on the granite foundation of experience; but the generalization, like the arch of the bridge, detaches itself from the world of facts in order to intersect it again at another precisely anticipated point. . . .

That moment of scientific creation . . . when generalization transforms itself into prognostication and prognostication tests itself successfully through experience invariably gives the human mind the proudest and the truest satisfaction.[1]

Mendeleev the citizen, however, shunned all sociological generalization and political prediction. He viewed with utter lack of comprehension the emergence in Russia of the Marxist school of thought which formed itself in the course of a controversy with the Narodniks precisely over a prognostication about the way Russian society would develop.

Mendeleev's case then illustrates the predicament of the modern scientist: his lack of an integrated vision of the world and even of science. Of necessity science works empirically; and specialization and fragmentation of knowledge accompanies its progress. Yet the greater the specialization and fragmentation, the more urgent is the need for a unifying conception of the world—otherwise the thinker's mind becomes constricted within his speciality and even within it his progress is impeded. Lack of philosophical insight and distrust of generalizing thought have been responsible for much avoidable scientific confusion and groping in the dark. Marxism offers the scientist an integrated vision of nature and human society, a vision which, far from being an arbitrary concoction or a figment of the metaphysical mind, accords intimately with the varied empirical experience of science.[2]

---

[1] *Sochinenya*, vol. xxi, p. 276.

[2] Engels in *Dialektik der Natur* points out that Descartes anticipated by about 200 years the findings of science about the conservation of energy, when he asserted that the mass of movement in the universe does not change. Had scientists grasped Descartes's thought they might have arrived at their findings much earlier. This was *a fortiori* true of Kant's 'nebular hypothesis'. 'Had the great majority of students of nature had less aversion for [philosophical] thought, the aversion which Newton expressed in his warning: Physics Beware of Metaphysics—they would have necessarily drawn from Kant's . . . discovery conclusions which would have saved them endless detours. . . . Kant's discovery was the starting point for all further progress [i.e. for overcoming the static and adopting the dynamic view of nature as a whole]. Had inquiry at once proceeded in this direction, the science of nature would have been much more advanced by now than it is. But what good could come from philosophy? Kant's work had no immediate impact and it was only years later that Laplace and Herschel . . . vindicated it.' *Dialektik der Natur*, pp. 14, 62.

The unity and the diversity of man's thought was Trotsky's grand theme. Taking once again Mendeleev's work as his starting point, he surveyed the structure of modern science. Mendeleev had discovered that chemistry has its basis in physics and that chemical reactions are caused by the physical and mechanical qualities of particles. Physiology, Trotsky proceeded, stands in the same relation to chemistry as that in which chemistry stands to physics—not for nothing is it described as 'the applied chemistry of living organisms'. 'Scientific, i.e. materialistic, physiology has no use for any special supra-chemical Life Force (as conceived by Vitalists and Neo-vitalists) in order to explain the processes with which it is concerned. Psychology in its turn rests on the foundations of physiology. As the physiologist in his strict research can make no use of the concept of Life Force, so can the psychologist not cope with a single one of his specific problems by reference to the "soul". He has to relate psychical experiences to the phenomena of physiological existence.' This is what Freud's school does when it reveals that man's sexual urges underlie so many of his mental states; and this is *a fortiori* what the Pavlov school does when it treats the human soul as a complicated system of physiologically conditioned reflexes. Finally, the modern science of society is inseparable from the insight man has gained into the laws governing nature; it sees society as a peculiar part of nature.

Thus on the foundations laid by mechanics and physics rises the vast structure of contemporary science, all its varied parts interrelated and forming a single whole. Yet unity is not uniformity. The laws governing one science cannot be substituted for those ruling another. Even though Mendeleev has proved that chemical processes are in the last instance physical or mechanical, chemistry cannot be reduced directly to physics. Still less can physiology be reduced to chemistry, or psychology and biology to physiology. Nor can the laws governing the development of human society be simply deduced from the laws which apply to nature. In a sense it may remain the ultimate objective of science to explain the infinite variety of natural and social phenomena by a few general and elementary laws.[1] But

---

[1] Engels, in the work already quoted, expresses the view that, at least 'in the present state of knowledge', these general and elementary laws can be formulated only in philosophical terms, that is in the terms of dialectics, but not in those of natural science.

scientific thought is progressing towards that objective in such a way that it appears more and more removed from it, namely, by way of the division and specialization of knowledge and of the formulation and elaboration of ever new, particular, and detailed laws. The view, for instance, that chemical reactions are ultimately determined by the physical qualities of particles was the beginning of all chemical knowledge; but by itself it offered not a single clue to any single chemical reaction. 'Chemistry works with its own keys; and it finds those keys only in its own laboratories, through empirical experience and generalization, hypothesis and theory.' Physiology, connected though it is through the solid channels of organic and physiological chemistry with chemistry at large, has methods and laws of its own. So has biology; and so has psychology. Every science seeks support in the rules of another only 'in the last instance'; and every science applies itself to so particular a sphere, in which elementary phenomena appear in such complex combinations, that every such sphere requires an approach, methods of inquiry, and hypotheses which are peculiar to it alone. It is through diversity that the unity of science asserts itself.

In the study of nature the autonomy of every sphere of research is taken for granted; no serious student permits himself to confound the laws prevailing in one sphere with those valid in another. Only in reasonings on society, in history, economics, and politics, is such confusion and arbitrariness of method still endemic. Here no law need be acknowledged; or else the laws of natural science are crudely projected into the study of society, as they are, for instance, by Darwinists who dabble in sociology and by the neo-Malthusians.[1]

[1] Trotsky illustrated this point by quoting J. M. Keynes who, on a visit to Moscow, in 1925, in the course of a lecture at the Supreme Council of the National Economy explained unemployment in Great Britain by the rate of the increase of the British population. Keynes (according to a report in *Ekonomicheskaya Zhizn* of 15 September 1925) went on to say: 'I suppose that Russia's poverty before the war was caused largely by an excessive growth of population. At present, too, a considerable excess of the birth rate over the death rate is noticeable. This is the greatest danger for Russia's economic future.' At that time there was still unemployment in Russia. But already three years later, when planned economy was established, and for decades afterwards, one of the 'greatest dangers' was shortage of manpower and too slow a growth of population, a fact which strikingly demonstrates the impropriety of applying the Malthusian or neo-Malthusian concept of the 'pressure of population upon means of subsistence' to the economics of an industrially expanding society.

Trotsky then surveyed broadly the advance of science and technology 'in the last few decades' and its philosophical implications. That advance, he asserted, constituted an almost uninterrupted triumph for dialectical materialism, a triumph which, paradoxically, the philosophers and even the scientists were reluctant to acknowledge. 'The successes of science in mastering matter are, on the contrary, accompanied by a philosophical struggle against materialism.' The discovery of radioactivity in particular had encouraged philosophers to draw anti-materialistic conclusions. Yet their arguments were effective only in criticism of the old physics and of the mechanistic variety of philosophical materialism connected with it. Dialectical materialism had never tied itself to the old physics—indeed, it had philosophically transcended it in the middle of the nineteenth century, well ahead of the scientists. Insisting only on the primacy of being—'matter'—in relation to thought, dialectical materialism identifies itself with no particular conception of the structure of matter and treats every such conception as being only of relative validity—a stage in the progress of empirical knowledge. The scientists, on the other hand, find it difficult to dissociate philosophical materialism from this or that phase of their inquiry into the nature of matter. If only they learned to approach the issues in a larger spirit, to combine inductive and deductive reasoning, and empirical and abstract thought—they would be able to see their own discoveries in better perspective, avoid attributing to them absolute philosophical significance, and even anticipate more clearly the transitions from one phase of science to another. Many scientists dwelling on the allegedly anti-materialistic implications of radio-activity could not even see whither the discovery of radio-activity led them; and they viewed sceptically the possibility of splitting the atom. Criticizing this attitude, Trotsky went on record with this prediction:

The phenomena of radio-activity lead us straight to the problem of releasing the inner energy of the atom. . . . The greatest task of contemporary physics is to extract from the atom its latent energy—to tear open a plug so that that energy should well up with all its might. Then it will become possible to replace coal and petrol by atomic energy which will become our basic fuel and motive power.

Countering the sceptics, he exclaimed:

This is by no means a hopeless task, and what vistas its solution will open up! . . . *scientific and technological thought is approaching the point of a great upheaval; and so the social revolution of our time coincides with a revolution in man's inquiry into the nature of matter and in his mastery of matter.*[1]

Trotsky made this prophecy on 1 March 1926. He was not to live to see it come true; he was to die almost on the eve of its fulfilment.

Of his excursions into the philosophy of science one deserves to be especially recalled—his plea in defence of Freudian psycho-analysis. Already in the early 1920's the Freudian school of thought found itself under a ferocious attack which was to banish it from the Soviet Union for many decades. To influential party men, who had hardly any first-hand knowledge of Freud's theory, the school with its over-emphasis on sex appeared suspect and incompatible with Marxism. However, intolerance of Freudism was not confined to Bolsheviks; it was, at least, just as strong in politically conservative academic circles, among Pavlov's followers who were bent on establishing a virtual monopoly for their own teachings. They had this advantage over the Freudians that their school had grown up on Russian soil, and that to Marxist intellectuals it appealed as being the more obviously materialistic of the two. Thus party men and Academicians formed a curious alliance against psycho-analysis.

Trotsky, we know, was perturbed by this as early as 1922. In that year he wrote a letter to Pavlov in which he sought to vindicate Freudism and tactfully entreated Pavlov to exercise influence in favour of tolerance and freedom of research. Whether he sent the letter is not known; but he included it in volume XXI of his *Works*. Pavlov, it seems, ignored the plea. In the heat of the subsequent political crisis Trotsky could not pursue the matter. But he took it up again in 1926; and this time he protested in public against the sycophancy by which the Pavlov school was already surrounded. He spoke with proper respect and admiration about the teaching of Pavlov himself as being 'completely in harmony with dialectical materialism' and as 'destroying the partition between physiology and psychology'. Pavlov sees 'the basic reflexes as physiological and the system of reflexes as resulting in consciousness'; he also views 'the accumulation of physiological quantity as producing a new "psycho-

---

[1] *Sochinenya*, vol. xxi, p. 415. (My italics.)

logical" quality'. But Trotsky spoke with irony about the exaggerated pretensions of the Pavlov school, especially about its boast that it could explain the subtlest play of the human mind and even poetic creation as the work of conditioned reflexes only. Indeed, Trotsky remarked, Pavlov's method is 'experimental and painstaking: it approaches its generalizations step by step: it starts from the dog's saliva and advances towards poetry'; but 'the road to poetry can hardly be seen yet'.

He protested against the disparagement of Freudism all the more strongly because he held that Freud's teaching, like Pavlov's, was inherently materialistic. The two theories, he argued, differ in their methods of inquiry, not in philosophy.[1] Pavlov adopts the strictly empirical method and actually proceeds from physiology to psychology. Freud postulates in advance the physiological urge behind psychical processes; and his approach is more speculative. It is arguable that the Freudians give too much weight to sex at the expense of other factors; but a controversy over this would still remain within the framework of philosophical materialism. The psycho-analyst 'does not ascend from the lowest [physiological] to the highest [psychological] phenomena and from the basic to the complicated reflexes. Instead, he attempts to take all the intermediate stages at one jump, a jump from above downwards, from the religious myth, the lyrical poem, or the dream straight down to the physiological basis of the human psyche.' In a striking image Trotsky clinched the comparison:

The idealists tell us . . . that the 'soul' is a bottomless well. Both Pavlov and Freud think that physiology forms its bottom. Pavlov, like the diver, plunges down to the lowest depth and painstakingly investigates the well from there upwards. Freud stands over it and with a penetrating gaze attempts to pierce its ever shifting and troubled waters and to explore or guess the shape of things down below.

Pavlov's experimental method had, of course, a certain advantage over Freud's partly speculative approach which sometimes led the psycho-analyst to fantastic surmises. Yet,

[1] In his letter to Pavlov, Trotsky argued about the affinity of the two schools as follows: 'Your teaching about conditioned reflexes embraces, so it seems to me, Freud's theory as a particular instance. The sublimation of sexual energy . . . is nothing but the formation on a sexual basis of the conditioned reflexes n plus one, n plus two, and of the reflexes of further degrees.' Ibid., p. 260.

it would be too simple and crude to declare psycho-analysis as incompatible with Marxism and to turn one's back on it. In any case we are not obliged to adopt Freudism either. Freudism is a working hypothesis. It can produce, and it does produce deductions and surmises which point to a materialist psychology. In due time experimentation will provide the tests. Meantime we have neither reason nor right to declare a ban on a method which, even though it may be less reliable, tries to anticipate results towards which the experimental method advances only very slowly.[1]

Trotsky's plea fell on deaf ears. The psycho-analytical theory was presently banished from the universities. Less specifically but even more categorically he defended Einstein's theory of relativity;[2] but to the ecclesiastical 'materialism' of the Stalin era that theory, too, became anathema; and only after Stalin's death was it to be 'rehabilitated'.

.   .   .   .

In his essays on the philosophy of science, Trotsky, well informed and at times inspired though he was, was nevertheless something of an amateur. There was nothing amateurish, however, about his literary criticism. He was Russia's leading critic in these years. His *Literature and Revolution* influenced strongly the writers of *Krasnaya Nov*, the leading intellectual journal of the time, and especially its editor A. Voronsky, who was an avowed Trotskyist and a distinguished essayist. Even now, nearly four decades after it was written, the book is still unsurpassed not merely as a survey of the revolutionary *Sturm und Drang* in Russian letters and as an advance denunciation of the stifling of artistic creation by Stalinism, but more generally as an essay in Marxist literary criticism. The book is written with an intimate feeling for art and literature, with original insight, captivating verve and wit, and—in the closing pages—a power of vision which rises to rare heights of poetic sublimity.

In literature, too, Trotsky declared war on the iconoclastic attitude and on pseudo-revolutionary conceit and arrogance. He demanded freedom of expression for all artistic and literary schools, at least as long as they did not abuse it for plainly and

---

[1] Ibid., pp. 430–1. Whether Trotsky was right in saying that Pavlov's method yielded results more slowly than Freud's it is for the experts to judge. He underlined that his defence of Freudism should not be mistaken for indulgence towards the 'vulgar pseudo-Freudism', *en vogue* among the bourgeois public.

[2] *Pod Znamyenem Marksizma*, Nr. 1.

unmistakably counter-revolutionary purposes. Again, the iconoclastic attitude and intolerance were in evidence not only,
and not even mainly, among party men. They were even more
characteristic of various groups of young writers and artists.
New rebellious schools proliferated in art and literature. In
normal circumstances these schools, with their innovations and
attacks on established artistic authorities, might have excited
curiosity and caused a flurry within relatively narrow circles, and
then they might have fought their way, as so many of their predecessors had done, from obscurity to recognition, without much
political flag-waving *en route*. But, circumstances being what they
were, the rivalries of the artistic coteries and their controversies transcended normal limits. The new schools claimed for
themselves momentous political significance, advertised themselves as pioneers of the revolution, and sought to discredit
the older schools as socially reactionary as well as artistically
outdated.

Proletkult, we know, clamoured for the official acceptance of
its 'school of thought' and even for a monopoly. Its writers,
Lebedinsky, Pletnev, Tretyakov, and others, found a forum in
two periodicals, *Kuznitsa* and *Oktyabr*, and later founded their
own militant *Na Postu*. Since Bukharin, as *Pravda*'s editor, and
Lunacharsky, as Commissar of Education, patronized *Proletkult*, it took Lenin's pronouncement to rebuff its pretensions.
When the *Proletkult* writers, upset by the rebuke, turned to Trotsky, begging his protection, he replied that he would in any case
defend their right to advocate their views freely, but that he was
in complete agreement with Lenin on the harmfulness and
inanity of all slogans about proletarian literature and art. Even
the more modest clichés about a 'new socialist epoch in art' or
a 'new revolutionary renaissance in literature' were worthless:
'The arts have revealed a terrible helplessness, as they always
do at the beginning of a great epoch. . . . Like the owl, the bird
of wisdom, so the singing bird of poetry makes itself heard only
after sunset. In day-time things are done, and only at dusk do
feeling and reason take in what has happened.'

It was wrong to blame the revolution for the artist's plight.
The 'singing bird of poetry' made itself heard still less in the camp
of the counter-revolution. In a scathing survey of the émigré
literature Trotsky pointed out that although most of the famous

Russian writers had gone abroad they had not produced there
a single noteworthy work. Nor had the 'internal émigrés'—
those writers in Russia who thought and felt as the émigrés did
—much to boast of, writers like Zinaida Gippius, Evgenii
Zamyatin,[1] and even Andrey Belyi. For all their indubitable
gifts, these writers, engrossed in a callous egotism, were incapable of responding to the drama of their time—at best they
escaped into mysticism. Thus even Belyi, the most outstanding
among them, 'is always preoccupied with his own self, tells stories
about his own self, walks around his own self, sniffs at his own
self, and licks his own self'.[2] Gippius cultivated a lofty, other-
worldly, mystical and erotical Christianity; yet 'it was enough
that the nailed boot of a Red Guardman should step on her
lyrical toe and at once she burst into a scream by which one
could recognize the witch obsessed with sacrosant property'.
(But as she did not lack talent, there was indeed a poetic quality
in her witch's scream!) In their attachment to the spurious
values of a lapsed social system and in their alienation from their
time, these writers were to Trotsky repulsive and grotesque. He
saw them as expressing all that was worthless in the old intelligentsia. He drew a thumb-nail sketch of one of the types of that
intelligentsia, an 'internal émigré' *par excellence*:

When a certain Constitutional Democratic aesthete, having made
a long journey in a stove heated goods wagon, tells you, muttering
between his teeth, how he, a most refined European, with a set of
superb false teeth, the best in the world, and with a minute knowledge of Egyptian ballet techniques, was reduced by this boorish
revolution to travelling with despicable lice-ridden bagmen, then
you feel rising up in your throat a physical nausea with his dentures,
ballet techniques, and generally with all his 'culture' pilfered from
Europe's market stalls; and the conviction grows upon you that the
very last louse of the most uncouth of our bagmen is more important
in the mechanics of history and more, so to speak, necessary than
this thoroughly 'cultured' and in every respect sterile egotist.[3]

---

[1] Some of these writers later became émigrés. On Zamyatin's novel '*We*',
written in emigration, George Orwell modelled his '*1984*'.

[2] *Literatura i Revolutsia*, p. 36.

[3] 'Bagmen'—people who, during the civil war and the famine, travelled with their
bags over the country in search of food. Sometimes petty black-marketeers were
also described as bagmen. Because of the destruction of rolling stock, people
travelled mostly in goods wagons. Ibid., pp. 26–27.

Having disposed somewhat summarily of the 'internal émigrés', Trotsky went on to discuss the more creative trends in literature. He criticized and defended the *paputchiki* or 'fellow-travellers'. He coined this term to describe those writers who, without embracing communism, 'travelled a stretch of the road with the revolution', but were liable to part company with it and go their own way.[1] Such were, for instance, the 'Imagists', a literary school of which Yessenin and Kluyev were the outstanding poets. They had brought the muzhik's personality and imagination into poetry—Trotsky showed how they composed their colourful and crowded poetic images in the manner in which the muzhik liked to adorn his *izba*. In their poems one could feel both the attraction and the repulsion which the revolution exercised on the peasantry. The ambiguity of their attitude gave artistic tension and social significance to their work. They were the 'poetic Narodniks of the October era'. That this frame of mind should find a stirring expression was only natural in a peasant country—and it was found not only among the Imagists. Boris Pilniak, whose talent Trotsky valued highly, shared with them the attachment to Russia's primordial primitivism which the revolution had sapped. Consequently he 'accepted' Bolshevism and 'rejected' communism, conceiving the former as the elemental 'peculiarly Russian', and in part Asiatic aspect of the revolution and the latter as the modern, urban, proletarian, and predominantly European element. More harshly Trotsky wrote about Marietta Shaginyan who had 'reconciled' herself to the revolution only from a sort of fatalistic Christianity and utter artistic indifference to anything that lay, metaphorically speaking, 'outside her private drawing room'. (Shaginyan was one of the very few writers of this group to survive the Stalinist purges and emerge as a Stalin Prizewinner.)

Trotsky described Alexander Blok also as a *paputchik*, but placed him in a class of his own. Blok's poetry had received a first and mighty stimulus from the revolution of 1905. It was his misfortune that his best creative years fell in the doldrums between two revolutions, between 1907 and 1917; he could never make peace with the emptiness of those years. His poetry was then

---

[1] The term is used throughout the chapter in this its original sense, and not in the meaning it has since acquired in English usage.

romantic, symbolic, shapeless, unreal; but underneath it there was the assumption of a very real way of life. . . . Romantic symbolism is an escape from reality only in so far as it evades its concrete quality . . .; essentially, however, symbolism is a way of transforming and elevating life. . . . Blok's starry, snow-drifty, and formless lyricism reflects an environment and an epoch . . . beyond which it would, like a cloudy patch, be suspended in a vacuum. It will not survive its time and author.

But 1917 once again shook Blok and gave him 'a sense of movement, purpose, and significance. He was not the poet of the revolution. But having withered in the dull impasse of pre-revolutionary life and art, he now grasped with his hand the wheel of revolution. From that contact came "The Twelve", the most significant of all his poems, the only one which will survive into the centuries.' Unlike most later critics, Trotsky did not treat 'The Twelve' as an apotheosis of the revolution but as the 'swan song of that individualistic art which sought to join the revolution'. 'Essentially, this was a cry of despair over a perishing past; but so great was the cry and so intense the despair that it rose to a cry of hope for the future.'

The Futurists were the most vigorous and vociferous literary grouping in these years. They clamoured for a break with all that was *passé*, insisted on the allegedly basic connexion between art and technology, introduced technical-industrial terms into their poetic idiom, and identified themselves with Bolshevism and internationalism.[1] Trotsky devoted a detailed and discerning study to this trend. He dismissed the technological raptures of the Futurists as reflexes of Russian backwardness:

Except for architecture, art bases itself on technology . . . only in so far as the latter forms the basis of civilized activity at large. In practice the dependence of art, especially of verbal art, on material technology is negligible. One can write a poem about skyscrapers, *dirigibles*, and submarines even when one lives in the backwoods of the Ryazan gubernia; one can write it with a pencil stump on rough wrapping paper. The fact that there are skyscrapers, *dirigibles*, and submarines in America is enough to fire the fresh imagination of Ryazan—the poet's word is the most portable of all materials.

[1] 'Only "futurist art" is built on collectivism. Only futurist art represents the art of the proletariat in our times', wrote N. Altman, the 'theorist' of the group in *Iskusstvo Kommuny* in 1918.

The identification of Futurism with proletarian revolution was also questionable. It was not by chance that in Italy the same poetic school was absorbed by fascism.[1] In both countries the Futurists, when they made their first appearance, were artistic rebels without definite political leanings. They might have gone the way of all literary flesh, fought and gained recognition, and settled in respectability, had they not been caught up by violent political upheavals before they had had the time to mellow. Then their literary rebelliousness took a political colour from the upheaval around them, the Fascist upheaval in Italy, the Bolshevik in Russia. This was all the more natural as both fascism and bolshevism attacked, from their opposed angles, the political *passéism* of the bourgeoisie. The Russian Futurists had, no doubt, been genuinely attracted by the dynamic force of the October Revolution; and so they mistook their Bohemian rebellion for the genuine artistic counterpart of the revolution. Because they themselves had broken with certain artistic traditions, they flaunted their contempt for the past and imagined that together with them, the revolution, the working class, and the party stood for a break with 'ages of tradition' in every field. They took, Trotsky remarked, 'too cheap a view of the ages'. The cry against tradition had its justification as long as it was directed at a literary public and against the inertia of established styles and forms. But it sounded hollow when 're-addressed to the working class, which does not need to break and which cannot break with any literary tradition because it is not at all in the grip of any such tradition'. The all-out crusade against *passéism* was a storm in the intelligentsia's tea-cup, an outburst of Bohemian nihilism. 'We, Marxists, have always lived in tradition, and we have not, because of this, ceased to be revolutionaries.'

The Futurists further claimed that their art was collectivist, aggressive, atheistic, and therefore proletarian. 'Attempts', Trotsky retorted, 'to derive by way of deduction an artistic style from the nature of the proletariat, from its collectivism,

---

[1] In an appendix to *Literature and Revolution* Trotsky published a memorandum on the origins of Italian Futurism and its relation to fascism, written at his request by Antonio Gramsci, the Italian communist theorist and founder of *Ordine Nuovo*. Shortly afterwards Gramsci returned to Italy and spent the rest of his life in Mussolini's prisons. During his stay in Moscow Gramsci enjoyed Trotsky's confidence.

dynamism, atheism, etc., are pure idealism and can produce only clever philosophical home-spun, arbitrary allegories, and . . . provincial dilettantism.'

We are told that art is not a mirror but a hammer: it does not reflect things but transforms them. But nowadays they teach one to handle even a hammer by means of a 'mirror', by means, that is, of a sensitive film which fixes all phases of the movement. . . . How can we transform ourselves and our lives without looking into the 'mirror' of literature?

His critical view of the Futurists did not prevent Trotsky from acknowledging their literary merits; and he acknowledged these all the more generously because influential party men looked askance at their experimental obscurity and eccentricities. He warned Communists to beware of that 'hasty intolerance' which treats experimental art as a fraud or as the whim of a decadent intelligentsia.

The struggle against the old poetical vocabulary and syntax was, despite all its . . . extravagances, a progressive rebellion against the closed vocabulary . . ., against an impressionism which sips life through a straw, and against a symbolism lost in . . . heavenly emptiness. . . . The Futurist's work has in this respect been vital and progressive . . . it has eliminated from poetry many words and idioms which had become hollow; it has made other words and idioms full-blooded once again; and in some cases it has successfully created new words and idioms. . . . This applies not only to individual words, but also to the place of each word among other words, to syntax.

True enough, the Futurists had over-reached themselves in innovation; but 'the same has happened even with our revolution: such is the "sin" of every living movement. The excesses are and will be discarded, but the essential cleansing and the indubitable revolutionizing of the poetic language will have lasting effects.' The same should be said in favour of new techniques in rhythm and rhyme. These must not be approached in a narrowly rationalistic spirit; man's need for rhythm and rhyme is irrational; and 'the sound of the word forms the acoustic accompaniment to its meaning'. 'Of course, the overwhelming majority of the working class cannot yet be bothered with these issues. Even its vanguard has not yet had the time for them—

there are more urgent tasks. But we also have a future before us. And this demands from us a more attentive, a precise, a crafts-man-like, an artistic attitude towards language, the essential tool of culture, not only in poetry but even more so in prose.' In handling and weighing words, their meanings and shadings and sounds, 'micrometrical instruments' are needed. Instead, uncouth banality and routine were rampant. 'In one of its aspects, the better aspect, Futurism is a protest against slap-dashness, that most powerful literary school which has its very influential representatives in every field.' From this point of view Trotsky found something to say even for the 'formalist' school and the chief expounder of its ideas, Victor Shklovsky, although he criticized their exclusive concentration on form: while the formalist believes that at the beginning was the word, the Marxist thinks that at the beginning was the deed—'the word follows the deed as its sound-shadow'.

A special essay in *Literature and Revolution* deals with Maya-kovsky, the most gifted Futurist who was later canonized as *the* bard of communism. Trotsky held that Mayakovsky was artistically at his worst precisely where as a Communist he was at his best. This was not surprising: Mayakovsky took pains to be a Communist; yet a poet's outlook depends not on his con-scious thought and exertion but on his semi-conscious percep-tion and sub-conscious feeling and on the stock of images and impressions the poet has absorbed in early childhood. The revolution was for Mayakovsky a 'genuine and profound ex-perience' because it turned with its thunder and lightning against the obtuseness and inertia of the old society which Mayakovsky hated in his own way and with which he had not had the time to make peace. He adhered enthusiastically to the revolution but did not and could not merge with it. To this Mayakovsky's poetic style testifies:

The dynamic élan of the revolution and its stern courage appeal to Mayakovsky much more closely than do the mass character of its heroism and the collectivism of its affairs and experiences. As the Greek anthropomorphist naïvely assimilated the forces of nature to himself, so our poet, the Mayakomorphist, crowds with his own self the squares and streets and fields of the revolution. . . . His dramatic pathos rises frequently to extraordinary tension, but behind the tension there is not always real strength. The poet is too conspicuous

—he allows too little autonomy to events and facts. It is not the revolution which wrestles with obstacles but Mayakovsky who displays his athletics in the arena of words, sometimes performing genuine miracles, but frequently lifting with heroic effort notoriously empty weights. . . . About himself Mayakovsky speaks all the time in the first and third person. . . . To lift up man he raises him up to Mayakovsky. He adopts a tone of familiarity towards the most majestic historic phenomena. . . . He stands with one foot on Mont Blanc and with the other on the Elbrus. His voice out-thunders thunder. What is the wonder that . . . the proportions of earthly things vanish and that no difference is left between the small and the great? He speaks about love, the most intimate of feelings, as if it were the migration of peoples. . . . No doubt, this hyperbolic style reflects in some measure the frenzy of our time. But this does not provide it with an over-all artistic justification. It is impossible to out-clamour war and revolution, but it is easy to get hoarse in the attempt. . . . Mayakovsky shouts too often where one should speak; and so his cry, where cry is needed, sounds inadequate.

Mayakovsky's overloaded images, often beautiful in themselves, just as often destroy the unity of the whole and paralyse movement.

The excess of dynamic imagery leads to stand-still . . . every phrase, every idiom, and every metaphor is intended to yield the maximum and to reach the upper limit, the peak. That is why the thing as a whole has no maximum . . . [and] the poem has no peak. . . .

.   .   .   .

The refutation of the idea of 'proletarian culture' forms the central and most controversial part of *Literature and Revolution*. In the Preface Trotsky gives this succinct summary of his argument:

It is fundamentally wrong to oppose proletarian to bourgeois culture and art. Proletarian culture and art will never exist. The proletarian régime is temporary and transitory. Our revolution derives its historic significance and moral greatness from the fact that it lays the foundations for a classless society and for the first truly universal culture.

One should not reason, therefore, from historical analogy and conclude that since the bourgeoisie has created its own culture and art the proletariat will also do so. It is not merely the 'purpose' of proletarian revolution—its striving for classless culture

—that invalidates the parallel.[1] What militates against it even more strongly is a basic difference in the historic destinies of the two classes. The bourgeois way of life developed organically in the course of several centuries, whereas the proletarian dictatorship may last years or decades, but not longer; and its life span is filled with savage class struggles which allow little or no room for the organic growth of new culture.

We are still soldiers on the march. We have a day of rest. We must wash our shirts, cut and brush our hair, and first of all clean and grease our rifles. All our present economic and cultural work is nothing but an attempt to bring ourselves into some sort of order between two battles and two marches. . . . Our epoch is not the epoch of a new culture. We can only force open the gate to it. In the first instance we must acquire the most important elements of the old civilization. . . .

The bourgeoisie could create its own culture because even under feudalism and absolutism, even before it had gained political domination, it possessed wealth, social power, and education, and was present in almost every field of spiritual activity. The working class can gain in capitalist society at the most the ability to overthrow that society; but being a propertyless, exploited, and uneducated class, it emerges from bourgeois rule in a condition of cultural pauperism; and so it cannot originate a new and significant phase in the development of the human mind.[2] It was in fact not the working class but small groups of party men and intellectuals (who in this field, too, 'substituted' themselves for the class) that aspired to bring proletarian culture into being. Yet no 'class culture can be created behind the back of a class'. Nor can it be manufactured in Communist laboratories. Those who maintain that they have already found the proletarian culture in Marxism argue from ignorance: Marxism has been the product as well as the negation of bourgeois thought; and it has so far applied its dialectics mainly to the study of economics and politics, whereas culture

[1] 'The proletariat has taken power precisely in order to put an end to class culture for ever and to pave the way for a universal human culture. Not rarely we seem to be forgetting this.'

[2] 'The bourgeoisie assumed power when it was fully armed with the culture of its time. The proletariat assumes power when it is fully armed only with its acute need to obtain access to culture.'

is 'the sum total of knowledge and skill which characterizes society as a whole, or at least its ruling class'.

The contribution of the working class to literature and art is negligible. It is preposterous to speak of proletarian poetry on the strength of the work of a few gifted worker-poets. Such artistic achievement as these poets can claim they owe to their apprenticeship with 'bourgeois' or even pre-bourgeois poets. Even if their writings are inferior, they are still valuable as human and social documents. But it is an insult to the proletariat—'a piece of populist demagogy'—to treat such writings as new and epoch-making art. 'Art for the proletariat cannot be second rate. The *Proletkult* writers declaim much about "the new, monumental, dynamic" literature and painting. But where, comrades, is that art "of the great canvas and great style", that "monumental" art? Where is it? Where?' So far it had all been big talk, boasting, and baiting the opponents of *Proletkult*, the Imagists, the Futurists, the Formalists, and the *paputchiki*, without whose works Soviet literature would be utterly impoverished and left only with *Proletkult*'s dubious 'promissory notes'.

As might have been expected, Trotsky was accused of eclecticism, kow-towing before bourgeois culture, encouraging bourgeois individualism, and denying the party the right and the duty of 'exercising leadership' in literature and art. He replied:

> Art must find its own road. . . . The methods of Marxism are not its methods. The party exercises leadership in the working class but not over the [entire] historical process. There are some fields in which it leads directly and imperiously. There are other fields in which it supervises . . . and still others where it can only offer its cooperation. There are finally fields where it can only orientate itself and keep abreast with what is going on. The field of art is not one in which the party is called on to command.

Exaggerated attacks against individualism were out of place: individualism has played a dual role: it has had its reactionary effects, but it has also had progressive and revolutionary ones. The working class has suffered not from the excess but from an atrophy of individualism. The worker's personality is not yet formed and differentiated strongly enough; and to form and develop it is just as important as it is to train him in industrial skills. It is absurd to fear that the art of bourgeois individualism may sap his sense of class solidarity. 'What the worker will absorb from Shakespeare, Pushkin, Goethe, and Dostoevsky is . . . a more complex idea about the human personality, its passions and its feelings.'[1]

In the closing chapter of the book Trotsky discussed 'certainties and hypotheses' about the prospects. The 'certainties' referred only to the 'art of the revolution'; about 'socialist art', which could come to life only in a classless society, it was possible to make only guesses. The art of revolution, throbbing with all the class conflicts and the political passions of the time, belongs to a transition era—to the 'realm of necessity', not to that of freedom. Only in a classless society can human solidarity come to full fruition; and only then 'will those feelings which we, revolutionaries, are shy of calling by their names because the hypocrite and the canaille have made the words threadbare—only in classless society will the feelings of disinterested friendship, love for our fellow being, and heartfelt compassion ring out powerfully in socialist poetry'.[2]

The literature of revolution was still only groping for expression. It was argued that it must be realistic. In the broad philosophical sense this was true: the art of our epoch could not achieve greatness unless it was deeply sensitive to social reality. But it was preposterous to try and foster realism in the narrower sense, as a literary school. It was not true that such a school would be inherently 'progressive': by itself realism is neither revolutionary nor reactionary. Its golden age in Russia fell in the epoch of aristocratic literature. As a reaction against it came the tendentious style of the Populist writers, which then gave way to pessimistic symbolism, against which the Futurists reacted in their turn. The mutation of styles occurred against a definite social background and reflected changes in the political climate; but it also followed its own artistic logic and its own laws. Any new style grows out of the old style, as its dialectical negation: it revives and develops some elements of the old and abandons others.

Every literary school is potentially contained in the past, but develops through a hostile break with it. The relationship between

---

[1] *Literatura i Revolutsia*, p. 166. [2] Ibid., p. 170.

form and content . . . is determined by the fact that the new form is discovered, proclaimed, and evolved, under the pressure of an inner need, of a collective psychological demand which, like everything else . . . has its social roots. Hence the duality of every literary trend: on the one hand, any trend contributes something new to the techniques of artistic creation . . ., and on the other it expresses definite social demands. . . . These include individual demands because the social class speaks through the individual; and national demands because the nation's outlook is determined by that of its dominant class which is dominant also in its literature.[1]

The indubitable fact that literature has served as a vehicle for social aspirations does not justify anyone in neglecting or falsifying its artistic logic and in trying either to canonize or to ban any style. Some critics reacted crudely against symbolism. Yet, 'it was not Russian symbolism that had invented the symbol. It had only absorbed it into the modernized Russian language. The art of the future will certainly not renounce the formal achievements of symbolism.' Nor will it renounce the traditional genres and forms, even though some critics rejected these as obsolete, saying that satire and comedy had outlived their time, and that tragedy was dead because it was incompatible with a materialistic and godless philosophy of life. The burial of the old genres was at least premature. There was still room for a 'Soviet Gogol' or 'Soviet Goncharov' who would mercilessly expose 'the old and the new filth', the old and new vices, and the dull-mindedness which could be found in Soviet society.[2]

Those who spoke of the extinction of tragedy argued that religion, fate, sin, and penance are at the centre of the tragic motif. Against this Trotsky pointed out that the essence of tragedy lies in the wider conflict between man's awakened mind and his constricting environment, a conflict which is inseparable from man's existence and manifests itself in different forms at different stages of history. The religious myth had not created tragedy but only expressed it 'in the imaginative language of mankind's childhood'. Fate, as conceived by the ancients, and the medieval Christian Passions were not to be found in Shakespeare's drama, the artistic product of the Reformation. Shake-

---

[1] *Literatura i Revolutsia*, pp. 172–3.

[2] The new satirist had to contend with the Soviet censorship. Trotsky promised to lend him a hand in this struggle as long as his satire attacked social evils in the interests of the revolution.

speare marks, therefore, a significant advance upon Greek tragedy: 'his art is more human': it shows man's earthly passions transcending man himself and transformed into a sort of Fate. The same is true of Goethe's drama. Yet, tragedy can rise even higher. Its hero may become man defeated not by *hubris*, the gods, or even his own passion, but by society:

As long as man is not yet master of his social organization, that organization towers above him like Fate itself. . . . The struggle for communism which Babeuf waged before his time, in an immature society, was like the struggle of the classical hero against Fate. . . . The tragedy of restricted personal passion is too flat for our time— we live in an epoch of social passion. The stuff of contemporary tragedy is found in the clash between the individual and a collective or between hostile collectives represented by individuals. Our time is once again a time of great purpose . . . man attempts to free himself from all mystical and ideological fog and to reconstruct society and himself. . . . This is larger than the childish play of the ancients . . . or the monastic ravings of the Middle Ages, or the presumption of an individualism which wrenches the human personality from its social environment, exhausts it utterly and then hurls it into a vacuum of pessimism. . . .[1]

[The new artist will] project the great purposes of our time into art. It is difficult to foresee whether the dramatist of the revolution will create 'high' tragedy. But socialist art will certainly give it a new birth . . . as it will also give fresh life to comedy, because the new man will want to laugh, to the novel and to lyrical poetry, because the new man's love will be more beautiful and larger . . . and he will brood anew over issues of birth and death. . . . The decline of the old forms is by no means absolute or final . . . they will all have their renascence. . . . What matters is that the poet of the coming epoch should muse man's musings anew and feel man's feelings anew.[2]

Hypothetical though all anticipations of Socialist art were, Trotsky thought that one could discern odd pointers towards it in the confused, sometimes even meaningless, innovations in which Soviet art abounded during these years. In the theatre Meyerhold searched for a new 'biomechanical' synthesis of drama, rhythm, sound, and colour; and Tairov tried to 'break down the barrier' between stage and audience, theatre and life. Painting and sculpture struggled to get out of the impasse in

---

[1] *Literatura i Revolutsia*, pp. 180–1.   [2] Ibid., pp. 181–2.

which they had found themselves after the exhaustion of the representational styles. In architecture Tatlin's 'constructivist' school rejected the ornamental forms, advocated 'functionalism', and drew up ambitious blueprints for garden cities and public buildings worthy of a socialist society. These plans unfortunately took no account of material possibilities; but they contained, in Trotsky's opinion, rational elements and valuable intuitive premonitions:

We could not yet afford to give thought to architecture, the most monumental of all arts. . . . Large scale construction must still be delayed. The authors of these gigantic projects . . . have a breathing space for fresh reflection. . . . Tatlin, however, is unconditionally right when he discards the nationally limited style, allegorical sculpture, stucco moulding, arabesques, frippery and finery, and seeks to subordinate the whole design to the correct constructive use of building materials. . . . Whether he is also right in what appears to be his personal whim, the use of the revolving cube, the pyramid, and the glass cylinder, he has still to prove. . . . In the future such monumental tasks as the planning of garden cities, model housing estates, railways, and harbours will touch to the quick not only architects . . . but the broadest mass of the people. The antheap-like, imperceptible accumulation of town districts and streets, brick by brick, from generation to generation, will give place to titanic building . . . with map and compass in hand.

The wall between art and industry will crumble. The grand style of the future will aim at form creation, not ornamentation. . . . But it would be mistaken to see this as the . . . self-effacement of art before technology. . . . The gulf between art and nature may be expected to disappear, but it will do so not because art will go back, in Rousseau's sense, to man in his natural condition, but because it will bring nature nearer to itself, to art. The present location of mountains and rivers, fields and greens, steppe, forest, and maritime coasts, should by no means be considered as final. Man has already carried out some far from negligible changes in nature's map. But these are only school-boyish essays in comparison with what is to come. If faith could only promise to move mountains, technology, which takes nothing on faith, will really pull them down and shift them. Hitherto it has done this only for industrial commercial purposes (mines and tunnels). In the future it will do it on an incomparably wider scale, in accordance with comprehensive productive-artistic plans. Man will make a new inventory of mountains and rivers. He will seriously and more than once amend nature. He will

eventually reshape the earth to his taste . . . and we have no reason
to fear that his taste will be poor.

Here, at last, Trotsky unfolds his vision of man in the realm
of freedom, an up-to-date, Marxist version of

> The loathsome mask has fallen, the man remains
> Sceptreless, free, uncircumscribed, but man
> Equal, unclassed, tribeless, and nationless,
> Exempt from awe, worship, degree, the king
> Over himself; just, gentle, wise; but man
> Passionless?—no, yet free from guilt or pain.

There were those who, with Nietzsche, argued that a
classless society, if it ever came into being, would suffer
from excess of solidarity and that it would lead a passive and
herd-like existence in which man, his competitive and fighting
instincts extinguished, would degenerate. Yet socialism, far from
suppressing the human instinct for emulation, would redeem it
by turning it towards higher purposes. In a society free from class
antagonisms there would be no competition for profit and no
struggle for political power; and man's energies and passions
would concentrate on creative emulation in the fields of technology, science, and art. New 'parties' would spring into being
and contest with one another over ideas, over the planning of
human settlements, trends in education, styles in the theatre,
in music, and in sport, over schemes for gigantic canals, over
the fertilization of deserts, the regulation of climate, new chemical hypotheses, and so on. The contests, 'exciting, dramatic,
passionate', would embrace society as a whole, and not merely
priest-like coteries. 'Art will therefore not be starved of those
varieties of nervous energy and collective psychological stimuli'
which produce new ideas and images. People will divide into
rival artistic 'parties' according to temperament and taste. The
human personality will grow, refine itself, and develop that
priceless quality inherent in it—'the quality of never contenting
itself with what it has achieved'.

To be sure, these were remote prospects. Immediately ahead
was an epoch of fierce class struggle and civil wars from which
mankind would emerge impoverished and destitute. Then the
conquest of poverty and penury in all their forms would take
decades—and during this time the nascent socialist society

would be gripped by a 'passion for what are today the better sides of Americanism', for industrial expansion, records of productivity, and material comfort. But this phase, too, would pass; and then vistas which the imagination could not yet even encompass would open:

The present dreams of some enthusiasts . . . about imparting a theatrical quality and a rhythmical harmony to man's existence fit in with this prospect well and coherently. . . . The drudgery of feeding and bringing up children . . . will be lifted from the individual family by social initiative. . . . Woman will at last emerge from semi-slavery. . . . Socio-educational experiments . . . will evolve with a now inconceivable élan. The communist way of life will not grow up blindly like coral reefs in the sea. It will be built consciously. It will be checked by critical thought. It will be directed and corrected. . . . Man will learn to shift rivers and mountains, to build people's palaces on the heights of Mont Blanc and at the bottom of the ocean; and he will impart to his existence not only wealth and colour and dramatic tension but also a highly dynamic character. No sooner will one crust begin to form itself on the human existence than it will burst under the pressure of new . . . inventions and achievements.

At last man will begin in earnest to harmonize his own being. He will aim at bringing higher precision, purposefulness, economy, and consequently beauty into the movements of his own body at work, on the march, and at play. He will desire to master the half-conscious and unconscious processes of his own organism: breathing, blood circulation, digestion, reproduction; and he will seek, within unavoidable limits, to subordinate them to control by reason and will. . . . *Homo sapiens*, now stagnating, . . . will treat himself as the object of the most complex methods of artificial selection and psycho-physical training.

These prospects follow from the whole of man's development. He begins with expelling darkness from production and ideology—with breaking, by means of technology, the barbarous routine of his work and defeating religion by means of science. . . . Then by means of socialist organization he eliminates blind, elemental spontaneity from economic relationships. . . . Lastly, in the deepest and dimmest recesses of the unconscious . . . there lurks the nature of man himself. On it, clearly, he will concentrate the supreme effort of his mind and of his creative initiative. Mankind will not have ceased to crawl before God, Tsar, and Capital only in order to surrender meekly to dark laws of heredity and blind sexual selection. . . . Man will strive to control his own feelings, to raise his instincts to the height of his conscious mind, and to bring clarity into them, to channel his willpower into his unconscious depths; and in this way he will lift himself to new eminence, grow into a superior biological and social type—into the superman, if you like.

It is as difficult to say beforehand what are the limits of self-mastery that man may be able to reach as it is to foresee how far he can develop his technical mastery of nature. Social constructiveness and psychophysical self-education will become the twin aspects of a single process. All the arts—literature, theatre, painting, sculpture, music, and architecture—will impart to that process a sublime form. ... Man will grow incomparably stronger, wiser, subtler; his body will become more harmonious; his movements more rhythmical; his voice more musical. The forms of his existence will acquire a dynamic theatrical quality. The average man will rise to the stature of Aristotle, Goethe, Marx. And above these heights new peaks will rise.

It is doubtful whether Trotsky knew that Jefferson had similarly anticipated 'progress . . . physical or intellectual—until every man is potentially an athlete in body and an Aristotle in mind'. He was influenced rather by the French Utopians, from Condorcet to Saint Simon. Like Condorcet he also found in the contemplation of the future 'an asylum in which the thought about his persecutors could not haunt him, and where he lived in his mind with man restored to his rights and dignity and forgot man tormented and corrupted by greed, fear, or envy'. His vision of the classless society had, of course, been implicit in all Marxist thought influenced as it was by French Utopian socialism. But no Marxist writer before or after Trotsky has viewed the great prospect with so realistic an eye and so flaming an imagination.

. . . .

The whole 'Trotskyist' conception of culture and art soon came under fire. It offended the half-educated party man by its very breadth and complexity. It outraged the bureaucrat to whom it denied the right to control and regiment intellectual life. It also antagonized the ultra-revolutionary literary sects whose pretensions it refused to accept. Thus, a fairly wide anti-Trotskyist 'front' formed itself in the cultural field; and it was kept in being, reinforced, and eventually absorbed by the political front. The struggle against Trotsky's influence as a literary critic became part of the endeavour to destroy his political authority; and so his opponents declared his views on art to be part and parcel of the wider Trotskyist heresy.[1] Their attack centred on his denial of the possibility of proletarian culture, for here he challenged most provocatively the vested interests that were forming themselves; and he was denounced for expounding a variety of bourgeois liberalism. Only very little of the great mass of the dogmatic argumentation produced in this connexion still retains interest. Most of it was virtually disavowed by its own inspirers, especially by Stalin himself when some time later he brutally disowned all the claims of the 'proletarian' writers and artists, disbanded their organizations and mercilessly persecuted them. In the middle 1920's, however, Stalin flattered every half-baked literary and cultural ambition in order to 'mobilize' on his side the intelligentsia and the semi-intelligentsia.

Of the arguments advanced against Trotsky one or two should be mentioned here, however. Thus Lunacharsky criticized Trotsky on the ground that, recognizing only the great feudal and bourgeois cultures of the past and the culture of socialism which was to emerge in the future, he treated the proletarian dictatorship as a cultural vacuum and viewed the present as a sterile hiatus between a creative past and a creative future. This was also the substance of a more specific criticism which Bukharin made at a conference on literary policy which the Central Committee convened in February 1925.[2] While agreeing that Trotsky had most impressively argued his case, that Lenin, too, had been extremely critical of 'proletarian culture', and that a revolutionary working class could exercise political

---

[1] Thirty-five years after the publication of *Literature and Revolution* the struggle against Trotsky's influence on Soviet literary criticism was still on. During the 'de-Stalinization' of the middle 1950's many of the writers who had been charged with Trotskyism and had perished during the great purges of the 1930's were rehabilitated; and soon the guardians of orthodoxy were confronted with a revival of 'Trotskyist' influence in literature. In May 1958 a writer in *Znamya* stated: 'A. Voronsky, critic and editor of *Krasnaya Nov*, well known in those years [the 1920's] was under the definite influence of Trotskyist views on literature. True, it has now been revealed that he was not connected with the Trotskyist underground. He has been rehabilitated in this respect, as have been other writers wrongly accused. All the same, his . . . theoretical principles were borrowed from bourgeois and idealistic aesthetics and merged with Trotskyist ideas.' The writer devoted several pages to the views on literature expounded by Trotsky himself in order to refute them anew, without, however, resorting to the extremes of Stalinist falsification and abuse.   [2] *Krasnaya Nov*, May 1925.

but not cultural leadership, Bukharin nevertheless held that the proletariat would in time achieve cultural preponderance as well and impart its own character to the spiritual creation of the last epoch of class society. Trotsky's mistake, Bukharin maintained, was that he imagined that the proletarian dictatorship and the transition to socialism would be of so short a duration as not to allow any distinctive proletarian class-culture to arise. He did not take into account the 'unequal tempo' of social and political development in different countries, the probability or even the certainty that this would break up the process of international revolution into many separate phases, prolonging greatly the proletarian dictatorship and consequently allowing time for the formation of a culture and art peculiar to it.

There was some truth in Bukharin's argument (which formed part of his and Stalin's case for socialism in a single country). When Trotsky stated: 'We are soldiers on the march. We have a day of rest. Our present . . . cultural work is but an attempt to bring ourselves into some sort of order between two battles and two marches', he did indeed suggest a rapid succession of the main 'battles' of international revolution which should have radically shortened the era of proletarian dictatorship and the transition to socialism. This expectation was ever present in his political forecasts and also in the accents in which he had expounded his conception of permanent revolution, although it was not essential to the conception itself. Yet the 'day of rest' between the Bolshevik onslaught of 1917–20 and the next great 'battle' of revolution was to last not less than a quarter of a century; and the Marxist may well wonder how long the 'day of rest' which has followed the Chinese Revolution may yet last. Trotsky undoubtedly underrated the duration of the proletarian dictatorship and, what goes with it, the extent to which that dictatorship was to acquire a bureaucratic character.

However, his all too evident mistake about this does not invalidate his argument against 'proletarian culture'. On the contrary, it gives to it even greater strength. The fact that the dictatorship and the transition to socialism was to last far longer than he anticipated did not make the era of transition more fruitful culturally and more creative. It made it less so. Stalinism did not beget any proletarian culture. It was instead engaged in 'primitive cultural accumulation', that is, in an exceptionally

rapid and extensive spread of mass education and in the assimilation of Western technology. That this took place within the
framework of the social relations created by the revolution
accounted for the tempo and the intensity of the process and
gave to it immense historic significance. All the same, the
accomplishment consisted almost entirely in the absorption by
the Soviet Union of the heritage of bourgeois and pre-bourgeois
civilization, not in the creation of a new culture. Even this
achievement was marred by the Stalinist cult with its dogmatic
despotism, fetishism, horror of any foreign influence, and fear of
independent initiative. The 'cultural accumulation' was 'primitive' in more than one sense: it was accompanied by the suppression or distortion of those finer and more complex cultural
values which Trotsky was anxious to preserve and develop under
a proletarian dictatorship. When he asserted: 'Our epoch is not
the epoch of a new culture—we can only force open the gate to
it', he unknowingly epitomized beforehand the cultural history
of the entire Stalin era and even of its sequel. Throughout that
era the Soviet Union, with bloody head and hands, could only
batter at the gate to a new culture—the gate it has now half-
forced.

CHAPTER IV

An Interval

AFTER Trotsky had left the Commissariat of War there followed a pause in his inner-party struggle; and it lasted throughout the year 1925 into the summer of 1926. During this time Trotsky did not express himself controversially in public on the issues that had been at the centre of the debates of 1923–4. Even behind the closed doors of the Central Committee and the Politbureau he did not attempt to keep up the discussion. He acknowledged his defeat and submitted to the restrictions which the Central Committee had imposed on him.

During this pause the '1923 Opposition' did not exist in any organized form. Trotsky had in effect disbanded it. 'We must not do anything at this moment', he advised his puzzled and bewildered followers, 'we must not come out into the open in any way. We should only maintain our contacts, preserve the cadres of the 1923 Opposition, and wait until Zinoviev has used himself up.'[1] Had he acted otherwise and initiated new protests or demonstrations of opposition, he and his adherents would at once have had to face the threat of expulsion from the party or at least from its leading bodies. He had every reason to assume that the triumvirs would not shrink from extreme reprisals.

How desperately Trotsky and his adherents were anxious at this time to avoid a renewal of the struggle can be seen from this incident: in 1925 Max Eastman, the American writer, published *Since Lenin Died*, a book in which he gave a true account, the first to see the light, of the struggle over the succession to Lenin and in which he quoted the substance of Lenin's testament. Eastman, who had also written a character sketch of Trotsky, *The Portrait of a Youth*, had been in Moscow, had become an adherent of the Opposition, had obtained from

---

[1] V. Serge, *Le Tournant obscure*, p. 97; *Mémoires d'un révolutionnaire*, p. 229. Serge attributes this 'directive' once to Trotsky himself and on a second occasion to Victor Elzin, Trotsky's assistant. Elzin, in any case, would have expressed Trotsky's view in this matter.

Trotsky himself the information about Lenin's last will and the contest over the succession; and had even begged Trotsky to act more aggressively and to read out the will at the thirteenth congress. He had submitted the manuscript of *Since Lenin Died* to Rakovsky in Paris and had received an indirect answer expressing full approval. He had therefore every reason to think that the work would meet with Trotsky's blessing as well.[1] Trotsky was indeed grateful to Eastman, with whom he remained in friendly relations until ten years later, when Eastman turned againt Communism. However, he found Eastman's friendly service embarrassing: the triumvirs charged Trotsky with having committed a gross indiscretion, pressed him to issue a denial of Eastman's disclosures, and threatened him with disciplinary proceedings if he refused. Trotsky's closest associates, whom he consulted, were so reluctant to be forced into a fight over the Eastman incident, that they urged Trotsky to disclaim all responsibility for it. The Politbureau was not content with this, however. It demanded a straight denial of Eastman's story about the testament; it even dictated the terms of the denial. Once again, 'the leading group of the Opposition', as Trotsky puts it, asked him to yield for the sake of peace.[2] And so on 1 September 1925 there appeared in the *Bolshevik* a statement signed by Trotsky that 'all talk about [Lenin's] "testament", allegedly suppressed or violated, is a malicious invention and is directed wholly against Lenin's real will and the interests of the party of which he was the founder'. The statement was reproduced by all foreign communist newspapers and was later eagerly quoted by Stalin.[3] Although such denials made for tactical considerations are not rare in politics, this was particularly galling for Trotsky. After he had watched almost passively the suppression of the testament, his virtual title-deed to the succession, he had now to come forward as a witness bearing false testimony against himself and for Stalin—all in order to postpone a fresh outbreak of inner-party hostilities.

[1] 'I showed the manuscript to Rakovsky . . .', writes Eastman in a letter to the author, 'and told him I would publish it or not according as he decided. Madame Rakovsky sent it back with enthusiastic praise and that was, I thought, as much "authorization" as could be obtained under the circumstances.'

[2] Trotsky explained these circumstances in a letter to Muralov, written from his exile at Alma Ata on 11 September 1928. *The Archives.*

[3] Stalin, *Sochinenya*, vol. x, p. 175.

In such circumstances it was not easy to 'maintain contacts and preserve the cadres of the 1923 Opposition'. For any political group inaction, no matter how well justified by tactical considerations, is a most trying experience. A small band of intellectuals and very advanced workers may fill the interval with studying and arguing within its own circle. But for any larger group, especially if it is composed of factory workers, inactivity most often amounts to political suicide. It saps their faith in their cause; it deadens their fervour; it breeds indifference or despair. Such were the effects of waiting in most groups of the Opposition: they shrank and fell apart. Thus, in Leningrad there were, at the beginning of 1926, not more than about thirty Trotskyists who, grouped around Alexandra Bronstein-Sokolovskaya, Trotsky's first wife, still kept in close touch with one another and met regularly. Many hundreds of previously organized oppositionists had vanished into a political no man's land. In Moscow the Trotskyist 'cadres' were much more numerous and alive; but in the great provincial cities and towns, in Kharkov, Kiev, Odessa, and elsewhere the Opposition's strength declined as much as it did in Leningrad.

The chiefs of the Opposition, bound by ties of political and personal friendship, formed a close circle around Trotsky which often met and deliberated. In it were some of the strongest intellects and characters that could be found in the Bolshevik party. As to political ability, experience, and revolutionary achievement, this circle was certainly superior to the team which led Stalin's faction and ruled the party. Rakovsky, Radek, Preobrazhensky, Yoffe, Antonov-Ovseenko, Pyatakov, Serebriakov, Krestinsky, Ivan Smirnov, Muralov, Mrachkovsky, and Sosnovsky, had been prominent in the early years of the revolution and the civil war and had held offices of the highest responsibility.[1] Marxists of large views, unconventional, resourceful, and full of verve, they represented the most advanced and internationally minded elements in the party.

Of all these men Radek was by far the most famous, though not the most important. He was, next to Trotsky, the most brilliant and witty Bolshevik pamphleteer. Of mercurial temper, a shrewd and realistic student of men and politics, uncannily

[1] Rakovsky, Yoffe, and Krestinsky now held ambassadorial posts in London, Paris, Tokyo, and Berlin; but they remained in close relations with Trotsky.

sensitive to the moods of the most diverse social milieus, Radek had prompted some of Lenin's most important initiatives in diplomacy and Comintern policy. Europe was his home. Like Dzerzhinsky, he had come to the Bolsheviks from the Social Democratic Party of the Kingdom of Poland and Lithuania, Rosa Luxemburg's party which had been strongly influenced by Trotsky's views.[1] He also had behind him many years of a stormy activity on the extreme left of German socialism; he had been a forerunner and one of the founding fathers of the Communist International. When shortly after the October Revolution he arrived in Russia, he at once gained admittance into the inner circle of the leaders; he accompanied Trotsky to Brest Litovsk; and he led, together with Bukharin and Dzerzhinsky, the Left Communists in their opposition to peace. After the collapse of the Hohenzollern monarchy, Lenin sent him on a clandestine mission to Germany, where he was to help to set afoot the newly formed Communist party. He made a perilous and adventurous journey across the 'cordon sanitaire' by which Russia was surrounded and arrived in Berlin incognito just before Rosa Luxemburg and Karl Liebknecht were assassinated. He was seized by the police and thrown into prison. There, while Berlin was swept by a White terror and his life hung by a thread, he contrived a feat of extraordinary versatility: he managed to make contact with leading German diplomats, industrialists, and generals; and in his prison cell he conducted with them, especially with Walter Rathenau, who was to be Foreign Minister in the Rapallo era, talks designed to tear open the first breach in the cordon sanitaire.[2] From his cell, too, he maintained clandestine contacts with the German Communist party and helped to shape its policy.

A pioneer of revolutionary socialism, Radek had in him also something of the gambler. He was as much in his element weaving a diplomatic intrigue as when, a mole of revolution, he tunnelled underground. Of observant eye and untrammelled mind, he diagnosed the ebb of revolution in Europe before other Bolshevik leaders saw it; and he advocated the united

[1] In that party, however, Radek and Dzerzhinsky had been Luxemburg's antagonists and had stood closer than the rest of the party to the Bolsheviks.

[2] See Radek's memoirs in *Krasnaya Nov*, no. 10, October 1926; R. Fischer, *Stalin and German Communism*, pp. 203–11.

front. When he returned to Germany in 1923, he still saw no flow and restrained Brandler from rushing into what he considered a hopeless attempt at revolution. His taste for the political gamble led him astray, however; and in his 'Schlagetter speech' he made an ambiguous appeal to the desperate extremists of German nationalism. On his return to Moscow he was made to bear the onus of the German defeat and of association with Trotsky. Barred from the European sections of the Comintern, he was, in 1925, appointed Principal of the Sun Yat-sen University in Moscow just at a time when the rumblings of the Chinese revolution made themselves heard—his job was to train propagandists and agitators for China's young Communist movement.[1] Restless, contemptuous of cant, Bohemian in appearance, sharp-tongued, and inclined to cynical postures, he was held by many to be an erratic and even a shady character. He was, however, the subject of much obloquy by adversaries who feared his disrespectful gaze, his banter and deadly pasquinade. The stuff of the man was certainly much more solid than it appeared, although it was to deteriorate dreadfully in later years, under the press of the Stalinist terror. His Bohemian exterior and cynical postures concealed a fervent faith which he was loath to exhibit; and even his snappy quips and jeers were hot with revolutionary passion.

Into the leading circle of the Opposition Radek discharged the electric shocks of his intellect and humour. He was greatly attached to Trotsky, with whom he had so much in common in range of international experience. Of that attachment he gave proof in his essay 'Trotsky, the Organizer of Victory', written in 1923.[2] Trotsky was somewhat wary of Radek's impulsive political improvizations but felt a warm affection for the man and admired his talent.[3] If he distrusted the gambler in Radek, he was nevertheless stimulated by his observations and ideas, and enjoyed the great jester and satirist.

Preobrazhensky's character stands out in stark contrast to Radek's. He was a theorist and probably the most original

---

[1] Before 1914 Radek had analysed revolutionary developments in the colonial and semi-colonial East in the Polish *Przegląd Socjal-Demokratyczny*, Rosa Luxemburg's theoretical paper.

[2] K. Radek, *Portrety i Pamflety*, pp. 29–34.

[3] See Trotsky's correspondence with Radek in *The Archives*, and 'Radek and the Opposition' in *Écrits*, vol. i, pp. 160–3.

Bolshevik economist. A Leninist since 1904, he had been co-author with Bukharin of the *ABC of Communism*, the once famous compendium on Bolshevik doctrine; and he had been secretary of the Leninist Central Committee. He left that office and made room for Molotov when the party's discipline had grown too rigid for him. As its critic he was Trotsky's forerunner—indeed he had criticized Trotsky's disciplinarian attitude at the eleventh congress, early in 1922. Later in the year, however, the two men drew together; Preobrazhensky was one of the few to whom Trotsky confided his plans and related his private talks with Lenin and their agreement to form the 'bloc' against Stalin. The author of important works on economic history, a man of rare erudition and analytical gifts, Preobrazhensky was primarily a scholar, pursuing his line of reasoning to no matter what unpopular conclusions it might lead him and no matter what damage it might do to his standing with the party. He thought in elaborate and massive theorems; and in his *New Economics* he made the first serious and still unequalled attempt to apply the 'categories' of Marx's *Das Kapital* to the Soviet economy. Only the introductory volume was allowed to appear and even that was suppressed soon thereafter and confined to oblivion. Yet the *New Economics* remains a landmark in Marxist thought. The anticipatory analysis it gave of the processes of primitive socialist accumulation will remain topical as long as there are under-developed countries in the world which strive to industrialize on a socialist basis. Many regarded Preobrazhensky rather than Trotsky as the author of the Opposition's economic programme—he created at any rate its theoretical groundwork. There were, however, implicit divergencies between his and Trotsky's views; but these did not become explicit and result in serious political conflict until 1928, the year when the two men were exiled from Moscow.

Pyatakov was the most outstanding industrial manager among the Bolsheviks. While Preobrazhensky supplied the Opposition with theorems, Pyatakov placed the theorems on the firm ground of practical experience. Lenin in his will describes Pyatakov as one of the two foremost leaders of the young generation—the other being Bukharin—and as an administrator of exceptional ability and drive but a man devoid of political judgement. This one-sidedness was characteristic of the Oppositionist as well: Pyatakov shared the Opposition's views on economic policy, but kept aloof from its 'battle of ideas' and quailed at its onslaughts on the party leadership. Yet he was far from being a timid character. Only a few years earlier he and his brother had led the Bolsheviks in the Ukraine when the Ukraine was occupied by Denikin; and there, behind enemy lines, he organized sabotage, set afoot partizan detachments, and directed the struggle. The White Guards seized the two brothers and put them, together with other Reds, before a firing squad. The execution was in progress and his brother had already been shot when the firing squad had to flee before the Reds who had captured the town and were converging on the spot where the massacre took place. Straight from the corpses of his brother and of his nearest comrades, Pyatakov went to assume command over the Red Guards. Such were the antecedents of the man who in and out of Opposition was to be the moving spirit and the chief organizer of the Soviet drive for industrialization for fifteen years, and who was to end in the dock, 'confessing' to having been a wrecker, a traitor, and a foreign spy.

Most of the other chiefs of the Opposition were men of heroic mould. Preobrazhensky had gone through fire and water when he led the Bolshevik underground movement in the Urals during the years of counter-revolution. Once, when caught by the Tsarist police and tried, he had Kerensky as his defence counsel. Kerensky, eager to save his client, declared in court that Preobrazhensky was not involved in any revolutionary movement. The defendant rose in the dock, disavowed his counsel, and proclaimed his revolutionary conviction. He led the Bolsheviks of the Urals in 1917 and during the early part of the civil war. Rakovsky, whose long and courageous struggle up to 1914 is related in *The Prophet Armed*,[1] directed the Communist forces during the civil war in Bessarabia, where the White Guards put a prize on his head. He returned to Russia and became Chairman of the Council of the People's Commissars in the Ukraine. Antonov-Ovseenko's part in the October insurrection and the civil war need not be recalled here.[2] Muralov had been, like Antonov, one of the legendary heroes of the 1905 revolution,

---

[1] *The Prophet Armed*, pp. 207–8.
[2] See op. cit., pp. 221, 298–301, 434–5.

and in October 1917 he led Moscow's Red Guards in their assault on the Kremlin. Afterwards he was commander of the military region of Moscow and Army Inspector. Trotsky describes him as a 'magnificent giant as fearless as kind'. An agronomist by education, he gave, in the intervals between battles, agricultural advice to peasants and 'medical treatment to men and cows'. Ivan Smirnov had led the army which defeated Kolchak in Siberia. Serebriakov was one of the most energetic political commissars on the fronts of the civil war. Sosnovsky had made his mark as an agitator in the fighting line and as vigilant observer and critic of morals and manners—his was one of the best pens in Bolshevik journalism.

For all their prowess and intelligence, these men did not for the time being see any clear road ahead. They were above all anxious to stay within the party; and they could stay in it only if they lay low. They watched events and the moves of their adversaries and waited for something to happen that would allow them to come to the fore.

. . . .

Though he lay low, Trotsky did not lay down his arms. By hint and allusion he kept up his criticism of the official régime and its policy. Everything he said, even when he said it in a deliberately inoffensive manner, was a reflection on what his adversaries did, and even more so, on what they thought—no matter whether he spoke about the uncouthness of the Russian bureaucrat, the debased style of the newspapers, or the false starts the party was making in cultural affairs. And he never turned his attention from those major issues of policy, foreign and domestic, in which stuff for future controversy was piling up.

In May 1925, nearly five months after he had left the Commissariat of War, he was appointed to serve on the Supreme Council of the National Economy, under Dzerzhinsky. There was heavy irony in the appointment: Dzerzhinsky was neither economist nor policy-maker; and only to slight Trotsky did the triumvirs assign to him a post under Dzerzhinsky. They did not even consult Trotsky; but he could not easily refuse. When he resigned from the Commissariat of War he had declared that he was 'ready to carry out any assignment, under any conditions of party control'; and he could not go back on this pledge. Far off were the days when he had been able to decline the office of Lenin's deputy.

Within the Council of the National Economy Trotsky became chairman of three commissions: the Concessions Committee, the Board of Electrotechnical Development, and the Industrial-Technological Commission. The Concessions Committee had been set up in the early days of N.E.P. when Lenin hoped to reattract former concessionaires and other foreign investors to assist in Russia's economic recovery. These hopes had come to nothing. The Bolsheviks were too frightened of foreign capital to be able to attract it; and foreign investors were too frightened of the Bolsheviks to co-operate with them. The Concessions Committee was at a loose end. In his office, in a tiny one-floor hotel outside the Kremlin, Trotsky occasionally received a foreign visitor who inquired about the chances of gold prospecting in Siberia, or of manufacturing pencils in Russia.

Presently, however, Trotsky made a stronghold of the cage to which he was confined. Assisted by the secretaries who had served in his military train during the civil war, he opened an inquiry into the state of Concessions and of Russia's foreign trade. This led him to investigate costs of industrial production at home and abroad, and to make a comparative study of the productivity of Russian and Western labour. The inquiry threw into sharp relief the nation's industrial backwardness—it showed that the productivity of Russian labour was only one-tenth that of American. With graphic diagrams he illustrated the poverty of Russia's industrial equipment. Thus, while the United States possessed 14 million and Great Britain one million telephones, the Soviet Union possessed only 190,000. The length of its railway lines was 69,000 kilometres against 405,000 in the United States. The consumption of electricity per head was only 20 kilowatts compared with 500 kilowatts in the U.S.A.[1]

Obvious though the facts were, their emphatic presentation came as a shock. Official spokesmen dwelt smugly on the advance of Russia's industry since the civil war, when output had been close to nil; or they compared current production with that of 1913; and congratulated themselves on the results. Trotsky argued that new scales of comparison were needed and that the progress of recent years should be measured by the standards of

[1] *Sochinenya*, vol. xxi, pp. 419–20.

the industrial west rather than by those of native backwardness.[1] The nation could not rise unless it had a ruthlessly clear
awareness of the low level from which it started. 'It is often said
that we work "almost" like the Germans, or like the French. I
am ready to declare a holy war on this word "almost". Almost
means nothing. . . . We must compare costs of production, we
must find out what a pair of shoes costs here and abroad, we
must compare the quality of the goods, and the time it takes to
produce them—only then can we make comparisons with
foreign countries.'[2] 'We must not lag behind others', he concluded. 'Our first and essential watchword . . . is not to lag behind! Yes, we are extraordinarily far behind the advanced
capitalist countries. . . .'

In launching this watchword—'We must not lag behind'
—Trotsky was several years ahead of Stalin; but unlike Stalin
he strove to open Russia's eyes to the full length of the distance
to be made up. He realized that this involved political risks—
people viewing Russia's poverty soberly and gauging her misery
to its depth might become cynical or despondent. Stalin, when
he embarked on industrialization, preferred to keep the masses
unaware of the prodigious climb and the inhuman effort required of them. Trotsky relied on the people's courage and
maturity. 'Let us, comrades, neither mock at ourselves, nor take
fright. But let us firmly remember these figures: we must make
these measurements and comparisons in order to catch up with
the West at any price, and to surpass it.'[3] Thus he re-emerged
from the petty administrative technicalities under which the
triumvirs had intended to bury him; he found his way back to
the central issue of policy; and he took up the call for industrialization which he had raised in 1922–3.

As Chairman of the Board of Electrotechnical Development
he became engrossed in electrification. He travelled up and
down the country, investigated resources, examined schemes for
power plants, planned their location, and produced reports.
From one such journey he returned to urge the Politbureau to
adopt a project for the utilization of the rapids of the Dnieper,
the project which became famous as the Dnieprostroy, one of

---

[1] *Sochinenya*, vol. xxi, pp. 44–45. See Trotsky's speech of 7 December 1925.
Throughout most of the Stalin era official propagandists avoided making com
parisons between Russia and the West.   [2] Ibid., pp. 397–405.   [3] Ibid., p. 419.

the feats of industrial construction in the next decade. When he first canvassed the idea, early in 1926, the Politbureau made little of it. Stalin remarked that the planned power-station would be of no more use to Russia than a gramophone was to a muzhik who did not possess even a cow.[1] Trotsky then appealed to the enthusiasm and the imagination of the young. In a speech to the Comsomol he said:

Recently we opened the power station of Shatura, one of our best industrial installations, established on a turf bog. The distance from Moscow to Shatura is only a little over a hundred kilometres. A stone's throw, it would seem; and yet what a difference in conditions! Moscow is the capital of the Communist International. You travel a few scores of kilometres and—there is wilderness, snow, and fir, and frozen mud, and wild beasts. Blacklog cabin villages, drowsy under the snow. From the train one's eyes catch the wolf's foot-prints in the snow. Where Shatura station stands, elks roamed a few years ago. Now metal pylons of exquisite construction ran the whole way down from Moscow . . . and under these pylons she-foxes and she-wolves will lead out their cubs this spring. Such is our entire civilization—extreme contradictions: supreme achievements of technology and generalizing thought and primordial Siberian wilderness.

Shatura stands on marshes; we have many marshes, many more than power stations. We have many other sources of fuel which only wait to be transformed into power. In the south the Dnieper runs its course through the wealthiest industrial land; and it is wasting the prodigious weight of its pressure, playing over age-old rapids and waiting until we harness its stream, curb it with dams, and compel it to give light to cities, to drive factories, and to enrich ploughland. We shall compel it![2]

Industrialization was no end in itself, of course; it was part of 'the struggle for socialism with which the whole future of our civilization is inseparably bound up'. Again, in contrast to the Stalin of later years, Trotsky insisted that while struggling to catch up with the West the U.S.S.R. must not seek to isolate itself from the West. He had been a staunch defender of the monopoly of foreign trade and he had originated the idea of 'socialist protectionism'; but the purpose of that protectionism,

[1] Trotsky quoted Stalin's statement verbatim from the record of the April 1926 session of the Central Committee. See Trotsky's 'Personal Statement' of 14 April 1927 in *The Archives*.   [2] *Sochinenya*, vol. xxi, p. 437.

he argued, was not to cut off socialist industry, but, on the contrary, to enable it to establish close and many-sided links with the world economy. True enough, the 'world market' would press on Russia's socialist economy and subject it to severe and even dangerous tests. But these tests could not be avoided; they should be faced boldly. The dangers to which Russia was exposed by contact with the more advanced capitalist economy would be compensated by decisive advantages to be derived from international division of labour and from the assimilation of superior Western technology. In isolation, Russia's economic development must be distorted and retarded. In arguing thus, Trotsky was again in implicit conflict with official economic thought which was already becoming fixed in conceptions of national self-sufficiency: socialism in a single country presupposed a closed Soviet economy. Trotsky argued in effect against the essential premisses of Stalin's doctrine even before the controversy over it had opened.

.   .   .   .

After the 1923 débâcle in Germany, Trotsky endeavoured to reassess the international situation and the prospects of communism. The Comintern, anxious to save its face, belittled the importance of its setback, forecast a new revolutionary situation in Germany, and encouraged 'ultra-left' policies.[1] When, early in 1924, the first British Labour government was formed under Ramsay MacDonald and when Edouard Herriot, heading the Cartel de Gauche, became French Prime Minister, some of the communist leaders viewed these governments as 'Kerensky régimes' destined to pave the way for revolution. Against this Trotsky pointed out that it was necessary 'to distinguish the ebb of revolution from its flow', that it would take time for the German working class to recover from defeat, and that no rapid revolutionary developments should be expected in Britain and France.

Yet he still held that the capitalist world was unable to regain any enduring balance. He saw the greatest single factor of its instability, and the central issue of world politics at large, in the ascendancy of the United States. In the years 1924 and 1925 he analysed again and again the economic rise of the United States

---

[1] See Zinoviev's speech at the fifth Comintern congress (*Pyatyi Vsemirnyi Kongress Kom. Internatsionala*), pp. 64 ff.; also R. Fischer's statements, ibid., pp. 175–92.

and its impact upon the world. He predicted emphatically the emergence of the United States as the leading world power bound to involve itself in the affairs of all continents and to spread out its networks of military and naval bases over all oceans. He couched his conclusions in such forceful terms that most of what he said sounded far-fetched in the 1920's. This was the time of the 'Dawes Plan', of America's relatively timid and only tentative intervention in European affairs which, after 1929, was to be followed by a relapse into isolationism lasting over a decade. The world-wide expansion of American power, which Trotsky foreshadowed, could still be seen, if at all, only in embryo. He saw, as he so often did, the full-grown being in that embryo. The economic basis for expansion was there: the national income of the United States was already two and a half times as large as the combined incomes of Britain, France, Germany, and Japan. The United States' ascendancy was accompanied by Europe's impoverishment, 'Balkanization', and decline. He concluded, therefore, that 'the superiority which Britain in her heyday held *vis-à-vis* Europe is insignificant in comparison with the superiority which the U.S.A. has gained over the entire world, including Britain'.[1]

It was true that the ruling classes of both America and Europe were slow to grasp the full weight of this shift—they lagged mentally behind the events. 'The American is only beginning to grow aware of his international importance. . . . America has not yet learned to make its domination real. But it will presently learn to make it so, and it will learn on Europe's body and bones.'[2] The traditions of American isolationism and pacifism, rooted in geography and history, were brakes on expansion; but they were bound to give way to the dynamic force of the new facts. The United States would find itself compelled to assume the leadership of the capitalist world. The urge for expansion was inherent in its own economy; and it was intensified by the fact that European capitalism depended for its survival on American assistance. Here Trotsky made his celebrated and hotly contested forecast that the United States would 'put Europe on American rations' and then dictate to Europe its will. Having taken Britain's place as the world's industrial workshop and bank, the United States was also taking Britain's place

---

[1] *Europa und Amerika*, p. 22.　　　　　[2] Ibid., p. 36.

as the world's greatest maritime power and empire.[1] For this it need not burden itself with colonial possessions which had so often been a drain on the strength of British imperialism as well as a source of wealth. 'America will always find enough allies and helpers all over the world—the strongest power always finds them—and with the allies will be found also the necessary naval bases.'[2] Consequently 'we are entering an epoch of the aggressive unfolding of American militarism'.[3]

To those who, over-impressed by the strength of American isolationism and pacifism, doubted this prospect, Trotsky replied that the United States was following in Germany's footsteps. Like Germany, but incomparably more powerful, it was a latecomer among the great industrial nations. 'How long is it since the Germans were looked upon as starry eyed dreamers, as a "nation of poets and thinkers"? Yet a few decades of capitalist development were enough to transform the German bourgeoisie' into an exponent of the most brutal imperialism. Far less time was needed for a similar transformation in the United States. In vain did Britain's rulers console themselves that they would act as political and diplomatic tutors to the inexperienced Americans. They might do so, but only during a brief spell, until the Americans had learned the arts of imperialism and gained self-confidence. In the end the weight of American power would tell. Even now the 'inexperienced Yankee' enjoyed definite advantages over the sophisticated and subtle British imperialist: he could afford to pose as the liberator of the colonial peoples of Asia and Africa, helping to free from the British oppression the Indians, the Egyptians, and the Arabs; and the world believed in his pacifism and generosity.

It was beyond American power to arrest the decay of bourgeois Europe, however. American predominance was itself a source of instability for Germany, France, and Britain, for it was primarily at their expense that American power expanded. The economic disequilibrium between Europe and America would again and again be reflected in their trade and balances of payment, in financial crises, and in convulsions of the whole

---

[1] At the Washington Naval Conference of 1922 Great Britain had, in fact, given up the traditional forms of British naval supremacy.

[2] *Europa und Amerika*, p. 42.

[3] See Trotsky's speech of 25 October 1925 published in *Pravda* on 5 November 1925.

capitalist system. Nor was the United States immune: the more the world was dependent on it, the more did the transatlantic Republic become dependent on the world and involved in the world's menacing chaos.

The conclusion? 'Bolshevism has no enemy more fundamental and irreconcilable than American capitalism.'[1] These were 'the two basic and antagonistic forces of our age'. Wherever communism might advance, it would run into barriers set up by American capitalism; and in whatever part of the world the United States might seek to expand, it would be confronted by the threat of proletarian revolution: '. . . if and when American capital penetrates into China, . . . it will find there, among the masses of the Chinese people, not the religion of Americanism but the political programme of Bolshevism translated into Chinese.'

In this duel of giants American capitalism had all the material advantages. But Bolshevism would learn from America and assimilate its superior technology. It would be easier for the Bolsheviks to achieve this than for American capitalists to put the world on American rations. 'Americanized Bolshevism will defeat and crush imperialist Americanism.'[2] The United States might pose as 'liberator' of the colonial peoples and thereby contribute to the decomposition of the British empire; but it would not succeed in establishing its own supremacy over the coloured races. Nor would it in the long run succeed in banishing communism from Europe.

We do not in any way underrate the power of the United States. In evaluating the prospects of revolution we start from a clear realization of the facts. . . . However, we are of the opinion that American power itself . . . is the greatest lever of European revolution. We do not overlook the fact that this lever will turn, politically as well as militarily, with a terrible momentum against European revolution. . . . We know that American capital, once its existence is at stake, will unfold incalculable fighting energy. All that we know from history and our own experience about the struggle of privileged classes for their domination may pale into insignificance compared with the violence that American capital will let loose on revolutionary Europe.[3]

---

[1] *Europa und Amerika*, p. 47. Trotsky relates that shortly after the October Revolution he said half-jokingly to Lenin that two names, those of Lenin and Wilson, were the 'apocalyptic antipodes of our time'.

[2] Ibid., p. 49.   [3] Ibid., p. 91.

How then, Trotsky asked, would communism be able to hold
its ground? He did not expect the clash between the two 'basic
antagonistic forces' to develop while communism was entrenched only on Europe's eastern fringe and in parts of Asia.
As always, he looked forward to revolution in Western Europe;
and he was convinced that, to withstand American onslaught
and blockades, the peoples of the Continent would have to
form 'the United States of Socialist Europe'.

> We, the peoples of Tsarist Russia, have held out through years of
> blockade and civil war. We have had to endure misery, privation,
> poverty, and epidemics. . . . Our very backwardness turned out to
> be our advantage. The revolution has survived because it could rely
> on its gigantic rural hinterland. . . . The outlook for industrialized
> Europe . . . would be different. A *disunited* Europe would not be able
> to hold out. . . . Proletarian revolution implies its integration.
> Bourgeois economists, pacifists, profiteers, cranks, and windbags like
> to chatter about the United States of Europe. But the bourgeoisie,
> divided against itself, cannot create it. Only the victorious working
> class will be able to unify Europe. . . . We shall serve socialist Europe
> as a bridge to Asia. . . . The United States of Socialist Europe
> together with our Soviet Union will exercise a tremendous magnetic
> attraction on the peoples of Asia. . . . And the gigantic bloc of the
> nations of Europe and Asia will then be unshakeably established,
> and it will stand up to the United States.[1]

The prospect of an Armageddon of global class struggle
presently came under severe criticism as sheer fantasy.[2] No
doubt, Trotsky threw into exaggerated relief what was at the
time only one of the tendencies at work in world politics. In the
following two decades other tendencies came to the fore: both
the United States and Russia relapsed into relative isolation;
Europe, with the Third Reich risen in its midst, became once
again the world's storm centre; and Hitler's conquests and
threats of domination made the U.S.A. and the U.S.S.R. temporary allies. However, Trotsky made his forecasts in the first

---

[1] *Europa und Amerika*, pp. 90–91.
[2] It will be remembered that both Trotsky and Lenin spoke for the United
States of Socialist Europe as early as at the beginning of the First World War.
(See *The Prophet Armed*, pp. 236–7.) The watchword was still included in the manifesto of the fifth congress of the Comintern which Trotsky wrote in 1924. Soon
thereafter, however, the slogan and the idea of the United States of Socialist
Europe were renounced by the Comintern as a Trotskyist day-dream.

years of the Versailles Peace, when Germany was still prostrate, when Hitler was merely an obscure provincial adventurer, and Germany's military power was incapable of asserting itself. Not more than a faint prelude had been enacted to the conflict of the two blocs, which was to unfold only after the Second World War. From the prelude Trotsky guessed the outline, the plot, and the ·eitmotif of the real drama. He ran so far ahead of his time that more than thirty years later much of his prediction still remains unconfirmed by events; but the truth of so much of it has since been demonstrated that few would venture to dismiss as chimerical the prophecy as a whole.

Against the general background of the changed relation between Europe and America Trotsky gave a more detailed prospect of a single country's future in *Where is Britain Going?* He wrote this book early in 1925, just when Moscow was beginning to attach great importance to a new link established between the Soviet and the British trade unions. In the previous November a delegation headed by A. A. Purcell, chairman of the British Trades Union Congress, had visited the Soviet capital and made a solemn pledge of friendship and solidarity with the Russian Revolution. The Soviet leaders eagerly responded, hoping that they had found solid allies in Purcell, Cooke, and other newly elected, leftish chiefs of the British trade unions; and they were all the more willing to cultivate the new 'friendship' because the Communist party of Great Britain was weak and insignificant. The Comintern's ultra-left policy was reaching a dead end; it was to be replaced by more moderate tactics. The question was mooted whether the revolution might not 'enter Britain through the broad gateway of the trade unions' rather than through the 'narrow path of the Communist party'. In May—Trotsky had just completed his book—Tomsky led a Soviet delegation to the annual Congress of the British unions and he formed, with the Politbureau's blessing, the Anglo-Soviet Trade Union Council, which was to occupy a large place in the inner party controversy of the following year.

In his book Trotsky spoke of the approach in Britain of a social crisis of the first magnitude. American predominance, the obsolescence of Britain's industrial equipment, and strains and stresses in the empire, all were combining to prepare it. Britain had emerged from the First World War victorious but battered and worn. Victory concealed her weakness, but not for long. British governments kept up the pretence of smooth and friendly co-operation with the United States, underneath which there was irreconcilable conflict. 'Peacefully' the British were surrendering their financial dominance, commercial privileges, and naval supremacy; but they could not go on doing this indefinitely, according to Trotsky,—at the end of the road was a clash of arms. Nor could the dissolution of the British empire, made inevitable by both the lapse of British rule on the seas and the rise of the colonial peoples, remain latent for any length of time. Lost to Britain were the strategic advantages of insularity. Finally, since 1918 the Versailles system and the disruption of the German economy had veiled Britain's industrial inferiority to Germany. But Germany, aided by the United States, was rapidly recovering strength and had already reappeared as Britain's most direct and dangerous competitor in the world market, upsetting her trade and payment balances and aggravating all the elements of British weakness. All this, Trotsky concluded, pointed to dangerous Anglo-American tensions, fraught with war, and to a violent flare-up of class struggle, indeed, to a revolutionary situation in the British Isles.

In retrospect both the realism of this analysis and the errors of perspective stand out clearly. Trotsky did not imagine that the British could escape an armed conflict with the United States, although he himself had shown convincingly that such a conflict would have been suicidal folly for bourgeois Britain. Although he was perhaps the first analyst to grasp all the implications of America's new superiority, his idea of the British empire had still an almost Victorian or Edwardian touch about it: he could not envisage that the British would 'peacefully' and 'to the end' surrender their supremacy to the United States. And he saw the decline of British power as a cataclysmic collapse, not as the chronic and long-drawn-out process it was to become.

Despite its errors of prognostication *Where is Britain Going?* is the most, or rather the only effective statement of the case for proletarian revolution and communism in Britain that has ever been made. This was Trotsky's encounter with Fabian socialism and its doctrine of the 'inevitability of gradualness'; and for a long time thereafter Fabianism could not recover intellectually from the assault.[1] With quick and sharp thrusts Trotsky stripped it of its socialist pretensions and showed up its dependence on Conservative and Liberal traditions, its staleness, its insularity, its parochial quaintness and empirical narrow-mindedness, its pacifist hypocrisy and national arrogance, its snobbery and meekness towards established opinion, its fetishistic attitude towards religion, monarchy and empire—in a word all the qualities which made MacDonald, Thomas, the Snowdens, and the other labour leaders of the time unfit to take the head of a militant socialist movement and which turned them into opponents of revolution glad to consume the fruit of past struggles, but shrinking in panic from new conflict and upheaval. Trotsky had no doubt that in the approaching crisis they would see their main task in keeping the working class mentally enthralled, morally disarmed and unable to act.

The ruthlessness of his argument was greatly enlivened but scarcely softened by the humour with which he conducted it:

British pigeon-fanciers, by means of artificial selection, achieve special varieties, with a continually shortening beak. But there comes a moment when the beak of a new stock is so short that the poor creature is unequal to breaking the egg shell and the young pigeon perishes, a sacrifice to compulsory restraint from revolutionary activities, and a stop is put to the further progress of varieties of short bills. If our memory is not at fault, MacDonald can read about this in Darwin. Having entered upon MacDonald's favourite course of analogies with the organic world, one can say that the political art of the British bourgeois consists in shortening the revolutionary beak of the proletariat, and so not allowing him to pierce the shell of the capitalist state. The beak of the proletariat is its party. If we look at MacDonald, Thomas, Mr. and Mrs. Snowden, we have to confess that the work of the bourgeoisie in selecting short-billed and soft-billed has been crowned with astonishing success. . . .[2]

The Fabian school prided itself on its peculiarly British tradition, which it refused to adulterate with alien Marxism. Trotsky retorted that the Fabians cultivated only the conservative patterns of their national tradition and neglected or suppressed its progressive strands.

---

[1] An American critic, writing in the *Baltimore Sun* (21 November 1925), remarked that the world had not heard anything like Trotsky's fiery invective since Luther's days.   [2] *Where is Britain Going?* p. 67.

The MacDonalds inherited from Puritanism not its revolutionary strength, but its religious prejudices. From the Owenites they received not communistic fervour but Utopian hostility to the class struggle. From the past political history of Britain the Fabians borrowed only the mental dependence of the proletariat on the bourgeoisie. History turned its nether parts to these gentlemen; and the writings that they there read became their programme.[1]

For the benefit of young Marxists Trotsky recapitulated the two major British revolutionary traditions, the Cromwellian and the Chartist. He saw the Puritans as being beneath their biblical cloaks essentially political innovators, fighters, and promoters of definite class interests, who stood half-way between the German Reformation with its religious philosophy and the French Revolution with its secular ideology. Luther and Robespierre met in Cromwell's personality.[2] Obsolescent though much of Cromwell was, especially his bigotry, he was still a great master of revolution with whom British Communists might usefully serve an apprenticeship. A note of affinity crept into Trotsky's appreciation of the Commander of the Ironsides: '. . . it is impossible not to be struck by certain features which bring the existence and character of Cromwell's army into close association with the character of the Red Army . . . Cromwell's warriors regarded themselves as Puritans in the first instance and only in the second as soldiers just as our warriors recognize themselves to be revolutionaries and communists first and soldiers afterwards.'[3] For all his lack of reverence for Parliament, Cromwell set the stage for British parliamentarianism and democracy. This 'dead lion of the seventeenth century', this builder of a new society was still more alive politically than were the many living dogs of the Fabian kennel. So were the militant Chartists to whose heritage British Labour would turn afresh, once it had lost faith in the magic of gradualness. Chartist watchwords and methods of action were still greatly to be preferred to 'the saccharine eclecticism of MacDonald and the economist stupidity of the Webbs'. The Chartist movement was defeated because it was ahead of its time—'an historic overture'; but it would be 'resurrected on a new and immeasurably broader historic basis.'[4]

Trotsky saw in the Communist party, weak though it was, the

---

[1] *Where is Britain Going?* p. 47.   [2] Ibid., p. 127.
[3] Op. cit., p. 126.   [4] Op. cit., pp. 130-1.

sole legitimate successor to these traditions. He dismissed as a 'monstrous illusion' the hope that any leftish Fabians or trade union chiefs could give a revolutionary lead to the British workers. That the Communist party in Britain was of negligible size and that Fabianism appeared to be formidable and unshakeable was true. But had not British Liberalism also appeared to be powerful and invincible just before it collapsed as a party? When the Labour party came to occupy the place vacated by Liberalism, it was led by the men of the Independent Labour party which had been a small group. The shock of great events makes old and seemingly solid political structures crumble and brings about the emergence of new ones. This had happened after the shock of the First World War and it would happen again. The rise of Fabianism was 'only a brief stage in the revolutionary development of the working class'; and 'Macdonald has a still shakier seat than had Lloyd George'.

It was with subdued misgivings that Trotsky asked whether British communism would prove equal to its task. But once again revolutionary optimism led him astray as it had sometimes led Marx. 'We do not intend to prophesy', Trotsky wrote, 'what will be the tempo of this process [of revolution in Britain], but in any case it will be measured in terms of years, or at the most in terms of five years, not at all by decades.'[1] In later years Trotsky argued that at the decisive moment, in 1926, Stalin's and Bukharin's tactical prescriptions, the policy of the Anglo-Soviet Council, crippled British communism. The historian must doubt whether these prescriptions, inept though they were, were the basic cause of the prolonged impotence of British communism which thirty years later still vegetated as a sect on the outer fringe of British politics. However, the great social crisis which Trotsky forecast was indeed about to open with the strike of the British coal-miners, the longest and the most stubbornly fought in industrial history; and during the general strike Britain moved towards the brink of revolution.

Trotsky's book aroused much controversy in Britain. H. N. Brailsford initiated it in a preface to the English edition. Acknowledging Trotsky's exceptional merits as an analyst and writer and his familiarity with English history and politics,

---

[1] Op. cit., p.14. 'The hive of revolution swarms too well this time!' Trotsky added, p. 52.

Brailsford wrote that Trotsky nevertheless failed to understand the democratic and nonconformist religious traditions of the British Labour movement and 'the instinct of obedience to the majority graven on the English mind'. Ramsay MacDonald,[1] George Lansbury,[2] and others, dismissed Trotsky's views as a foreigner's misconceptions. Bertrand Russell, on the other hand, held that 'Trotsky was perfectly familiar with the political peculiarities of the English Labour movement'; and he also agreed that socialism is incompatible with Church and Throne. Yet Russell could not see how anyone not an enemy of the British people could incite them to revolution, the sequel of which would be an American blockade or even a war in which Britain would be doomed to defeat.[3] Other writers resented the disrespect and derision with which Trotsky turned on MacDonald, although a few years later when MacDonald broke with the Labour party most of these critics tore the 'traitor' to shreds.

Trotsky answered his critics several times.[4] In a reply to Russell he denied any intention of inciting British workers to revolution in the interest of Soviet Russia. In no country, he wrote, should the workers undertake any steps in the interest of the Soviet Union which do not follow from their own interests. But he remained unconvinced by Russell's rationalistic pacifism:

Revolutions are as a rule not made arbitrarily. If it were possible to map out the revolutionary road beforehand and in a rationalistic manner, then it would probably also be possible to avoid revolution altogether. Revolution is an expression of the impossibility of reconstructing class society by rationalist methods. Logical arguments, even if Russell turns them into mathematical formulae, are impotent against material interests. The ruling classes will let civilization perish together with mathematics rather than give up their privileges. . . . You cannot get away from these irrational factors. Just as in mathematics we use irrational magnitudes in order to arrive at altogether realistic conclusions, so in revolutionary policy . . . one can bring a social system into rational order only when one makes frank allowance for the contradictions inherent in society so as to be able to overcome them by means of revolution. . . .[5]

---

[1] *The Nation*, 10 March 1926.
[2] Lansbury's *Labour Weekly*, 27 February 1926.
[3] *New Leader*, 26 February 1926.
[4] *Pravda*, 11 February and 14 March 1926.
[5] *Kuda Idet Angliya? (Vtoroi Vypusk)*, p. 59.

The British Communists at first received Trotsky's work with delight and enthusiasm—the giant had come to reinforce their puny ranks.[1] Later in the year, however, under the wing of the Anglo-Soviet Council, they had second thoughts, and began to feel embarrassed by Trotsky's attack on the leftish trade-union leaders. (Even earlier, in November 1925, he had already been criticized on this ground by the Russo-American Communist M. Olgin, until recently Trotsky's fervent admirer.[2]) In the spring of 1926 the British Communist party was already lodging a complaint with the Russian Politbureau about Trotsky's 'hostility' towards it; and Trotsky had to rebut the charge.[3]

. . . .

It was during this interval in the struggle between Trotsky and his adversaries that a great regrouping of men and ideas occurred within the Bolshevik party and that a new and funda-mental division appeared among its leaders and in its ranks—a division which forms the background to the political history of the following fifteen years.

The middle 1920's are often described as the halcyon time of N.E.P., as the only period between 1917 and the middle of the century when the Soviet people relaxed, enjoyed peace, and had a taste of well-being. This picture cannot be accepted at its face value. What gives the period a quasi-idyllic appearance is its contrast with the one that preceded it and the one that was to follow. The middle 1920's knew none of the bloody struggles and upheavals, and none of the famines, of the early 1920's and the early 1930's. The passage of time was healing the wounds the nation had suffered. Economic recovery was under way. The farmers tilled their land and reaped their harvest. The wheels of industry no longer stood still. Blown-up bridges and railways, burnt-out houses and bombshelled schools were rebuilt. Flooded coal-mines were restored. Links between town and country were re-established. Private trade flourished. Shoppers no longer car-ried sackfuls of depreciated bank-notes: the rouble, still some-what shaky, reacquired the mysterious respectability of money.

---

[1] See, for instance, R. Palme Dutt's review in *Labour Monthly*, April 1926.

[2] *Die Freiheit*, 15 November 1925.

[3] *The Trotsky Archives*, excerpts concerning Politbureau sessions of the first days of June 1926.

There was even a bustle of prosperity about the central squares and thoroughfares of the cities.

Yet this bustle was largely deceptive. The great and now unified Soviet republic, extending from the Polish and Baltic frontiers over the whole area of the former empire, remained engulfed in cruel poverty and riddled with social tensions. Only one-sixth of the nation lived in the towns; and not even one-tenth of its manpower was employed in industry. Recovery was painfully slow. Mines and factories still turned out less than three-quarters of their pre-war output; they produced no engines, no machine tools, no motor-cars, no chemicals, no fertilizers, and no modern agricultural machinery. The Soviet Union did not yet possess most of the industries essential to modern society. The flourishing private trade, much of it barbarously primitive and fraudulent, covered the national misery as with bubbling froth.

It is true that the peasants consumed the produce of their enlarged fields and for the first time since ages ate their bread to the full. But this was 'prosperity' at the rock bottom of civilization. It was enjoyed in the absence of any higher needs and amenities, in squalor, darkness, and primeval rural idiocy. About a third of the rural population, not growing its own food, was excluded even from that kind of well-being. Because the peasants ate more than before, the town dwellers had to eat less: they consumed only two-thirds of the food and only half the meat they used to consume under Tsarist rule. Less produce was also left for export: Russia now sold abroad only about a quarter of the amount of grain she used to export. As of old most of her people were in rags and barefoot. Only in two significant respects, it seems, had there been a marked advance: in hygiene and education. The Russians used more soap and had more schools than ever before.

Of the social tensions the chronic antagonism between town and country was the most dangerous. The town dweller had the sense of being ill used by the farmer, who was indubitably the chief beneficiary of the revolution. The muzhik, on the other hand, felt that he was skinned by the town people. There was some ground for such feelings on both sides. The urban workers earned far less than before the revolution; and there were two million unemployed, almost as many as were employed in large-scale industry. The workers contrasted their own want with the ampleness of food in the country. The peasants resented the fact that they had to pay for industrial goods more than twice the prices they paid before 1914, while for their own produce they did not obtain much more than the pre-war price. Each of these two classes imagined that it was exploited by the other. In truth both were 'exploited' by the nation's poverty.

Neither town nor country represented, however, any uniform interest. Each was torn by its own contradictions. The urban worker knew that the N.E.P.-man, the middleman, and the bureaucrat cheated him of the fruit of his labour. He paid high prices for the food for which the peasant received so little—the middleman controlling nine-tenths of the retail trade cashed in on the difference. In the factory the worker was confronted by the manager who, acting on behalf of the employer-state, deprived him of his share in running the factory, kept down wages, and demanded more work and harder work.[1] By the manager's side stood the trade-union official and the secretary of the party cell, who were less and less inclined to side with the worker and often acted as arbitrators in industrial disputes. The employer-state could in fact rarely afford to meet the workers' claims. The national income was small, productivity low, and the need for capital investment desperately urgent. When the manager, the party secretary, and the trade-union official urged the worker to produce more, the latter cursed his new 'bosses'; but he did not dare to press his claims or down tools. Outside the factory gates there waited long queues of men anxious to get jobs. Once again, as under capitalism, the 'reserve army of the unemployed' helped to depress the wages and the conditions of the employed.

The cleavages in the peasantry were less marked but not less real. The muzhiks had benefited from the agrarian upheavals and from N.E.P. in unequal degrees. The middle layer of the peasantry was strengthened. There were many more smallholders now, more *serednyaks*, who lived on the yield of their land, without having to work on the land of wealthier farmers and without employing labour on their own farms. Of every ten peasants three or four belonged to this category. One or perhaps two were kulaks employing hired labour, enlarging their farms, and trading with the town. Five out of the ten were poor

---

[1] Only one in five or six workers was employed in privately owned industry.

peasants, *bednyaks*, who had carved out for themselves a few acres from the landlords' estates but only rarely possessed a horse or farm tools. They hired the horse and the tools from the kulak, from whom they also bought seed or food and borrowed money. To pay the debt, the *bednyak* worked on the kulak's field or let out to him part of his own tiny plot.

At every step the realities of rural life came into conflict with Bolshevik policy. Lenin's government had decreed the national-ization of the land together with the expropriation of the land-lord. In theory and in law the peasants were in possession of the land without owning it. They were forbidden to sell and to rent it. The Bolsheviks had hoped to curb inequality in this way and to prevent the growth of rural capitalism. Slowly but surely life overlapped these barriers. In innumerable daily transactions, which no administration could trace, land passed from hand to hand; and capitalist relationships evolved: the rich grew richer and the poor poorer. True, this was only a rudimentary and ex-tremely crude form of rural capitalism: by the standards of any advanced bourgeois society even the Russian kulak was a poor farmer. But such standards were irrelevant. That the new stratification of the peasantry developed on an extremely low economic level did not soften its impact; it sharpened it. The possession of a few horses and ploughs, of a stock of grain, and of a little cash gave to one man more direct power over another than the ownership of much more capital may give to anyone in a wealthy bourgeois society. Ten years after the revolution the wages of the landless farm labourers (who should not be confused with the poor peasants) were nearly 40 per cent. less than the wages the landed gentry had paid them. Their working day was much longer; and their conditions were little better than those of slave labour. The old landlord employed many hands on his estate whereas the kulak employed only a few; and so the labourers could not organize against him and defend themselves as effectively as they used to organize against the landlord. The *bednyak* was sometimes even more exploited and helpless than was the labourer.

In these relations there were the makings of a violent social conflict; but the conflict could not unfold itself and find ex-pression. Much as the village poor may have resented the kulak's rapacity, they were utterly dependent on him and could rarely afford to stand up to him. More often than not the wealthy peasant led a submissive village community, diverted its resentment from himself, and turned it against the town, the workers, the party agitators and the commissars.

All these tensions within town and country and between them underlay the friction between the many nationalities of the Soviet Union. We have seen that friction at work in the transition from war communism to N.E.P. and have heard Lenin castigating the *dzerzhymorda*, the vile Russian bureaucrat, as the chief culprit. With the years matters grew worse. The ever stricter centralization of government automatically favoured the Russian against the Ukrainian, the Byelorussian, and the Georgian, not to speak of the more primitive nationalities and tribes of Soviet Asia. Great Russian chauvinism emanating from Moscow excited and exacerbated local nationalisms in the outlying republics. The kulak and the N.E.P.-men were nationalists by instinct. In Russia proper they were Great Russian chauvinists. In the other republics they were anti-Russian nationalists. The intelligentsia were extremely susceptible to the prevalent moods. Among industrial workers internationalism was on the wane. The working class was reconstituting itself and growing in size by absorbing fresh elements from the country, elements who brought with them into the factories all of the peasants' political inclinations, a distrust of things foreign and intense regional loyalties.

Every now and then the tensions snapped. In the autumn of 1924 a peasant rising swept Georgia and was quelled in blood. Less violent but more persistent signs of the peasantry's antagonism to the government showed themselves everywhere. In the elections to the Soviets which took place in March 1925 over two-thirds of the electorate abstained from voting in many rural districts; and the government had to order new elections. There was a sporadic agitation for independent peasant Soviets. Here and there energetic and politically minded kulaks furthered their interests and ambitions through the existing Soviets and even through the rural party cells. There were many scattered acts of terrorism in villages. Party agitators sent from the town were clubbed to death. 'Worker correspondents' reporting to newspapers on the exploitation of farm labourers were lynched. The strong farmer had used to the full the opportunities N.E.P.

had offered him; and now he felt constricted by its limitations and sought openly or surreptitiously to remove them. He pressed for higher food prices, for the licence to sell and rent land, for unrestricted freedom to hire labour, in a word for a 'neo-N.E.P.'.

All this foreshadowed a national crisis which might be delayed for a couple of years only to become more dangerous later on. The ruling party had to seek a solution. Yet the party itself was increasingly affected by the cleavages that rent the nation. Three major currents of Bolshevik opinion formed in 1925. The party and its Old Guard split into a right and a left wing and a centre. The division was in many respects new. In none of the many earlier factional struggles had there been anything like it. Never before had the dividing lines been so clear cut and stable. Factions and groups had sprung into being and vanished together with the issues which had given rise to their differences. Alignments had changed with controversies. Opponents in one dispute joined hands as friends in the next dispute, and vice versa. The factions and groups had not sought to perpetuate themselves and had had no rigid organization or discipline of their own. This state of affairs had begun to change since the Kronstadt rising; but it was only now that the change became complete and universal. From the Politbureau and the Central Committee down to the rank and file the party was torn, although lower down the differences remained unexpressed. Not only were the issues which caused the division largely new; new and fateful was above all its finality.

What was sometimes startling was the manner in which men regrouped themselves and took up new positions. As in any political movement, so among the Bolsheviks some people had always been inclined towards moderation; others had shown a propensity to radicalism, and still others had been habitual trimmers. In the present regroupment many remained true to character. Rykov and Tomsky, for instance, who had always been far from the Left Communists, quite naturally found their place at the head of the new right. Most of the trimmers, especially the professional managers of the party machine, took up positions in the centre. Of the persistent radicals some had already joined the Workers' Opposition, the Decemists, or the Trotskyists; others had still to decide where they stood. But strange and unexpected conversions also occurred. Under the pressure of new circumstances and difficulties and after much heart-searching, some Bolsheviks, among them the most eminent leaders, abandoned accustomed attitudes or postures and assumed new ones which appeared to negate all that they had hitherto stood for. Men burnt the things they had worshipped and worshipped the things they had burnt.

In part the new differences resulted from the fact that some of the groups and individuals exercised power while others did not. Many a Left Communist who had been in office for seven or eight years, had wielded great influence, and enjoyed the privileges of power, came to approach public affairs from the ruler's viewpoint, not from that of the ruled. On the other hand, a 'moderate' Bolshevik, who had lived all these years among the masses and shared their experiences, willy-nilly voiced their disillusionment and spoke like an 'ultra-left'. There were also other causes for realignment. Under the single-party system the broader class antagonisms which we have just surveyed could find no legitimate political expression; and so they found an illegitimate and indirect expression within the single party. Wealthy farmers could not send their representatives to Moscow to state claims and demands before any national assembly or to act as pressure groups. Workers could not hope that their nominal deputies would voice their grievances freely and fully. Yet every social class and group exercised its pressure in non-political forms. The wealthy peasants controlled the stocks of grain on which the provisioning of the urban population depended: 6-10 per cent. of the farmers produced more than half of the marketable grain surpluses. This gave them a potent weapon: by withholding supplies they periodically created acute food shortages in the towns. Or else they refused to buy over-priced industrial goods; and stocks of unsold goods piled up in factory yards and warehouses. Symptoms of overproduction thus appeared in a country which really suffered from underproduction. The workers were sullen and inefficient and sought to quell their despair with vodka. Wild and widespread drunkenness made frightful ravages in popular health and morale. Hard as the party tried to neutralize the conflicting social pressures, and to isolate itself from them, it was not immune. Food shortages and stocks of unsold industrial goods rudely awakened its members to realities. Some Bolsheviks were more sensitive to the workers' demands; others were more susceptible to the pressure from the peasants. The great cleavage between town and country tended to reproduce itself within the party and within its ruling circle.

It was several years since Zinoviev had spoken of the 'unconscious Mensheviks' who could be found side by side with the 'genuine' Leninists within the Bolshevik party and who formed in its ranks a potential party of their own. Even more important, it now turned out, was the potential party of 'unconscious Social Revolutionaries'. The authentic Social Revolutionaries, like the Narodniks, their political forebears, had been distinguished by their bias in favour of the muzhiks, among whom they refused to make any class distinctions, whom they treated neither as kulaks nor as *bednyaks*, whom they glorified as land labourers at large, whose interests they refused to subordinate to those of the industrial workers, and in whose striving for private property they saw nothing incompatible with socialism. Woolly in their theories and addicted to sentimental generalities, the Social Revolutionaries had represented an agrarian antithesis to the collectivism of the urban proletariat, a quasi-physiocratic variety of socialism. It was only natural that such an ideology should exercise a powerful influence in a nation four-fifths of which lived on the land and by it. The Bolsheviks had suppressed the party that had expounded this ideology, but they had not destroyed the interests, the emotion, and the mood that had animated it. That emotion and that mood now invaded their own ranks. There, in an environment traditionally hostile to Narodnik ideas, the mood could not be expressed in customary terms. It refracted itself through the prism of the Marxist tradition and came to be voiced in Bolshevik terms. This trend had received a strong impulse from the anti-Trotskyist campaign, in the course of which the triumvirs sought to discredit Trotsky as the muzhik's enemy. The accusation was partly a cold-blooded invention; but it also summed up a real feeling. Subsequently the neo-Narodnik trend gained in strength until, during the present pause in the struggle against Trotskyism, it led to the emergence of the new right wing in the party.

The man who came forward as the inspirer, theorist, and ideologue of the right was Bukharin. His appearance in this role was something of a puzzle. Ever since the peace of Brest Litovsk he had been the chief spokesman of Left Communism, rigidly committed to a 'strictly proletarian' viewpoint. He had denounced aggressively Lenin's 'opportunism', opposed Trotsky's army discipline, and defended the non-Russian nationalities against Stalin. Then, early in 1923, he had sympathized with Trotsky's radical ideas. In the years 1924–5, however, his name became the symbol of moderation, 'opportunism', and of the penchant for the well-to-do peasant. The conversion was by no means fortuitous. Bukharin's Left Communism had been based on his expectation of early revolution in Europe, the prospect on which all Bolshevik leaders had staked much, but perhaps none as much as Bukharin. All had seen in European revolution Russia's escape from her poverty and backwardness. None had believed that with a small working class surrounded by many millions of property-loving peasants they could advance far towards the socialist goal. Least of all had Bukharin believed it. With eager enthusiasm he had looked to the Western workers to rise, overthrow their bourgeoisie, and stretch out helping hands to Russia. He had surrounded those Western workers with a halo of revolutionary idealization and exaggerated beyond all measure their class consciousness and militancy. He had rejected the Brest Litovsk peace with the utmost indignation, because he was afraid that the sight of Bolshevik Russia bowing to the Hohenzollerns might discourage and demoralize the European working classes, and that Bolshevism cut off from the latter and left alone with the Russian peasantry would find itself in an impasse.

Bukharin now found that Bolshevism was indeed left alone with the Russian peasantry. He ceased to count on revolution in the West. Together with Stalin he proclaimed 'socialism in a single country'. With the same assurance with which he had hitherto spoken of the imminent collapse of world capitalism, he now diagnosed its 'stabilization'. From this new angle he took a fresh look at the domestic scene. He could not humanly accept the conclusion to which his whole earlier reasoning pointed: that the Russian Revolution was in a blind alley. He concluded instead that, as the Western workers had failed as allies, Bolshevism must acknowledge that the muzhiks were its only true friends. He turned towards them with the same fervour, the same hope, and the same capacity for idealization with which he had hitherto looked to the European proletariat. It is true that under Lenin's inspiration the party had always cultivated 'the alliance of workers and peasants'. But never since 1917 had the Bolsheviks offered friendship to the wealthy farmer; and Lenin had always treated the middle and even the poor peasants as 'vacillating allies', whom the lure of property might turn into enemies. So difficult and uncertain an alliance now failed to reassure Bukharin. He wished to base the alliance on what seemed a wider and firmer foundation. He hoped to persuade his comrades that they ought to appeal to the peasantry at large and cease playing the poor muzhik against the rich, and that they ought even to stake their hopes on the 'strong farmer'. This amounted to the abandonment of class struggle in rural Russia. Bukharin himself, inhibited by old habits of mind or by tactical motives, shrank from drawing all these conclusions; but they were drawn for him and made explicit by his disciples, Maretsky, Stetsky, and other young 'red professors', who expounded the neo-Narodnik or neo-Populist ideas in the universities, the propaganda departments, and the press.

Bukharin was guided by more practical considerations as well. Within the framework of N.E.P., the Bolshevik 'alliance' with the poor peasants against the rich had yielded few, if any, positive results. The poor peasants, and even the middle ones, could not feed the towns. They produced, at best, just enough to feed themselves. The well-being and even the survival of the urban workers depended on the small minority of wealthy farmers. These were, of course, eager to sell their goods; but they sold in order to grow wealthier, not just to survive. Their bargaining position was extremely strong. Indeed, never before had the dependence of the town on the country been so one-sided, so brutal, so naked. Government and party could not improve matters by vexing and incommoding the kulaks and inciting the poor against them. Pestered with requisitions and price controls, fretful at the restrictions on the sale and renting of land and on the employment of labour, the kulak ploughed less, reaped less, and sold less. The government had either to break his strength or to allow him to accumulate wealth. Not a single group within the party suggested that the kulaks be dispossessed—to all groups the expropriation of millions of farmers was still inconceivable and from a Marxist viewpoint impermissible.[1]

There was therefore a peculiar realism and consistency in Bukharin's conclusion that the party must allow the wealthy farmer to grow wealthier. The purpose of N.E.P., he argued, was to use private enterprise in Russia'a reconstruction; but private enterprise could not be expected to play its part unless it obtained its rewards. The overriding interest of socialism lay in increasing national wealth; and that interest would not be harmed if groups and individuals grew wealthier together with the nation—on the contrary, by filling their own coffers they would enrich society as a whole. This was the reasoning which induced Bukharin to address to the peasants his famous appeal: 'Enrich yourselves!'

What Bukharin overlooked was that the wealthy peasant sought to enrich himself at the expense of other classes: he paid low wages to the labourers, squeezed the poor farmers, bought up their land, and tried to charge them and the urban workers higher prices for food. He dodged taxation and sought to pass its burden on to the poor.[2] He strove to accumulate capital at the expense of the state and thereby slowed down accumulation within the socialist sector of the economy. Bukharin dwelt on that part of the social picture in which the interests of the different classes and groups and of the various 'sectors' were seen as complementary and as according with one another so that kulak, *bednyak*, worker, factory manager, and even N.E.P.-man, all appeared as a happy band of brothers. This aspect of the picture was real enough, but it formed one part of it only. He overlooked the other part, where all was discord and conflict and where the band of brothers turned into a pack of enemies seeking to cut one another's throats. A Bolshevik Bastiat, he extolled

---

[1] Since at least 10 per cent. of the twenty-odd million farmsteads belonged to the kulaks, dispossession would have at once affected between two and three million holdings, even if the middle peasants had been spared. The upper layer of the middle peasantry was often indistinguishable from the kulaks and so the number of those affected would have been much larger in any case.

[2] The single agricultural tax then in force favoured the kulak. The *bednyak* who let out to the kulak part of his own holding, in order to obtain the horse and the tools with which to cultivate the other part, paid as a rule the land tax on the plot yielded to the kulak. Indirect taxation was becoming more and more important in the Soviet budget and, as always, it fell heavier on the poor than on the well-to-do.

les harmonies économiques of Soviet society under N.E.P. and prayed that nothing should disturb those harmonies. He prayed from the heart because he had a strong premonition of the furies that would descend upon the land with the 'liquidation of the kulaks as a class'.

The first major controversy in which Bukharin developed his ideas was one where Preobrazhensky, the Trotskyist, was his opponent. Trotskyism, with its purely Marxist emphasis on class conflict and class antagonism and on the primacy of the socialist interest *vis-à-vis* the private, was the obvious antithesis to the neo-Populist attitude; and within their respective groups the two co-authors of the *A.B.C. of Communism* represented the opposed poles of Bolshevik thought. The controversy developed before the end of the year 1924, when Preobrazhensky published fragments of his *New Economics*.

Preobrazhensky based his whole argument on the imperative need for rapid industrialization—on that hung the whole future of Russia's socialist régime. Because of its backwardness, the U.S.S.R. could industrialize only by means of primitive socialist accumulation. Contrary to Bukharin's assumptions this was by definition antagonistic to private accumulation. Internationally, the contest between capitalism and socialism would be decided by the relative wealth, efficiency, and cultural strength of the two systems. Russia had entered the contest with an antiquated, essentially pre-industrial structure. She could not afford any 'free competition' with Western 'monopoly capitalism'. She had to adopt a 'socialist monopolism' and stick to it until her productive forces had reached the level already attained by the strongest capitalist nation, the United States.[1] (Preobrazhensky argued that even if Russia had not stood alone and if the whole of Europe had overthrown capitalist rule, the whole of Europe would still have to engage, albeit far less forcibly and for a shorter time, in primitive socialist accumulation, because its productive resources would be inferior to those of American capitalism.)

What is the essence, he asked, of primitive socialist accumulation? In an underdeveloped country, socialist industry by itself cannot produce the sinews of rapid industrialization. Its profits or surpluses can make up only a part, and a small one at that, of

---

[1] E. A. Preobrazhensky, *Novaya Ekonomika*, vol. i, part 1, pp. 101–40.

the required accumulation fund. The rest must be obtained from what would otherwise have gone into the wages fund and from the profits and incomes earned in the private sector of the economy. (To put it in Keynesian terms, the savings of the nationalized industry are far too small in relation to investment needs and so private savings must provide the nationalized industry with the major portion of its investment capital.) The needs of accumulation in the socialist sector set therefore rather narrow limits to private accumulation; and the government must impose the limits. The workers' state is compelled in a sense to 'exploit' the peasantry during this period of transition. It cannot pander to consumer interests; it must press on with the development of heavy industry in the first instance. The resulting relative shortage of consumer goods implies different levels of consumption for various social groups, material privileges for administrators, technicians, scientists, skilled workers, and others. Repugnant though this inequality may be, it does not produce new class antagonisms. The privileged bureaucracy does not form a new social class. Discrepancies in the earnings of bureaucrats and workers are not different in kind and social import from 'normal' differences in the wages of skilled and unskilled workers. They amount to inequality within one and the same class, not to an antagonism between hostile classes. Such inequality must and can disappear only with the growth of social wealth and universal education which should blur and eventually abolish the distinction between skilled and unskilled labour, and between manual work and brain work. In the meantime 'we should take the productionist and not the consumptionist point of view. . . . We do not live yet in a socialist society with its production for the consumer. We are only in the period of primitive socialist accumulation—we live under the iron heel of the law of that accumulation.'[1]

In this transition era the workers' state has already forfeited advantages peculiar to capitalism but does not yet benefit from the advantages of socialism. This is 'the most critical era in the life of the socialist state. . . it is a matter of life and death that we should rush through this transition as quickly as possible and reach the point at which the socialist system works out all its advantages. . . .'[2] Preobrazhensky did not suggest that during

[1] Ibid., p. 240.
[2] Ibid., p. 63.

the transition industrial wages and peasant incomes should actually be depressed (as they were in the Stalin era). What he meant and said was that as a result of intensive accumulation the national income would grow rapidly and that with it should rise the earnings of workers and peasants; but these would rise less rapidly so that a high proportion of the national income could be earmarked for investment.

He maintained that the 'law' of accumulation asserted itself as an 'objective force', comparable in some respects to the 'laws' of capitalism which determined the economic behaviour of men regardless of whether they were aware or unaware of those laws and regardless also of their own ideas and intentions. The law of primitive socialist accumulation would eventually compel the managers of the nationalized industry, i.e. the party leaders, to embark upon intensive industrialization, no matter how reluctant they were to do so. For the time being many of them received with apprehension and even aversion the proposition that the state-owned industry must, in order to expand, absorb resources from the private sector, and gradually socialize it and transform many millions of scattered, tiny, and unproductive farmsteads into large-scale and mechanized producers' co-operatives. However, the 'subjective views' of those responsible for the conduct of economic affairs need not be of decisive importance: 'the present structure of our state-owned economy often proves itself to be more progressive than our entire system of economic leadership'.[1] The new bureaucracy might resist the logic of the transition epoch; but it would have to act on it. Preobrazhensky still assumed that revolution would spread to western Europe in not too remote a future. Even so, the problem of primitive accumulation 'would stand in the centre of our attention for two decades, at the very least'.[2] It has stood there for nearly four decades, and it still does so.

Trotsky did not share Preobrazhensky's views fully, although the basic idea was common to them both. He refrained, however, from engaging in any public discussion of the differences. He did not wish to embarrass Preobrazhensky who soon came under severe attack. At the moment their differences were of no political consequence—only four years later, after Trotsky's and Preobrazhensky's banishment from Moscow, were

[1] E. A. Preobrazhensky, *Novaya Ekonomika*, vol. i, part 1, p. 184.   [2] Ibid., p. 254.

they to acquire significance and to contribute to a painful breach.

The very abstract manner in which Preobrazhensky presented his argument hardly appealed to Trotsky. He himself approached the same problem more empirically, though also less methodically. With the scholar's utter indifference to tactics, Preobrazhensky, when he dwelt on the necessity for the underdeveloped workers' state to 'exploit the peasantry', offered a handle to anti-Trotskyist propagandists. True, he spoke of exploitation only in the strictly theoretical sense in which the Marxist speaks of the exploitation by capitalism of even the best-paid workers on the ground that they produce more value than their wages represent. He argued that, in the exchange between the two sectors of the economy, the socialist sector would take out of the private one more value than it would put into it, although with the growth of the national income the mass of value would grow in the private sector too. Official critics, however, seized on the provocative phrase about exploitation, gave it the vulgar meaning, and so twisted it that Preobrazhensky was understood to say that the impoverishment and degradation of the peasantry were necessary concomitants of accumulation. He tried to correct himself and 'withdrew' the unfortunate phrase. The correction made matters worse: it suggested that the critics had not been quite wrong.

It will be remembered that at the twelfth congress, when Trotsky spoke about primitive socialist accumulation, Krasin asked whether this might not imply exploitation of the peasantry; and that Trotsky then jumped to his feet to deny it.[1] Preobrazhensky now posed the same question and answered it in the affirmative. On internal evidence, the answer was too blunt and too rigid for Trotsky. He, at any rate, refused to commit himself to the view that the peasantry would have, as a rule, to foot the bill of primitive accumulation from beginning to end.[2] Nor did Trotsky advocate a pace of industrialization as forced as that which Preobrazhensky expected. There were even deeper differences between them. Preobrazhensky, for all his

[1] See pp. 102–3.
[2] In the debate Bukharin underlined this difference between Trotsky and Preobrazhensky. Bukharin, *Kritika Ekonomicheskoi Platformy Oppozitsii*, p. 56.

references to international revolution, constructed his theorem in such a way that it implied that primitive socialist accumulation might be concluded by the Soviet Union alone or perhaps by the Soviet Union in association with other underdeveloped nations. This prospect appeared unreal to Trotsky, who did not see how the Soviet Union alone could raise itself to the industrial height attained by the West; and it was a prospect which created an opening for an intellectual reconciliation with 'socialism in a single country'. Nor could Trotsky agree with Preobrazhensky about the 'objective force' or logic of primitive accumulation which would impose itself on the party leaders and make them its agents, regardless of what they thought and intended. This was a view which must have appeared to Trotsky to be too rigidly deterministic, even fatalistic, and to rely too much on the automatic development of socialism and too little on the consciousness, the will, and the action of fighting men.

These were still platonic differences, however, containing only the seed of political disagreement. Even if Trotsky thought that Preobrazhensky had overstated the case for industrialization, it was still the same case that they were both defending. If he held that Preobrazhensky had shown too little political tact in dealing with the peasantry, he himself was just as critical as Preobrazhensky of the official pandering to the strong farmer. In abstraction, the theorem of the *New Economics* might have envisaged the transition to socialism within a single industrially underdeveloped nation-state. Yet, politically, Preobrazhensky held no brief for socialism in a single country. Finally, much though he trusted the laws of accumulation to prevail over the economic conservatism of the party leaders, he did not rely solely on the working of those laws—he was still a fighter calling on Bolsheviks to do their duty and not wait until necessity drove them to do it. Trotsky therefore watched Preobrazhensky's controversies sympathetically, if with reserve.

Bukharin attacked the whole of Preobrazhensky's conception as 'monstrous'.[1] He made the most of the dictum about the exploitation of the peasantry. If Bolsheviks were to act on Preobrazhensky's ideas, he stated, they would destroy the workers' alliance with the peasantry and would demonstrate that the proletariat (or those who ruled in its name) had become

[1] Bukharin, *Kritika Ekonomicheskoi Platformy Oppozitsii*, p. 21.

a new exploiting class, seeking to perpetuate its dictatorship. State-owned industry could not and must not expand by 'devouring' the private sector of the economy—on the contrary, only by leaning on it could it achieve any significant progress.[1] In Preobrazhensky's scheme the peasant market played a subordinate role: he saw the main outlet for the produce of the state-owned industry within that industry itself, in its ever-expanding demand for producer goods. Against this Bukharin argued that in a country like Russia the peasant market must form the basis of industrialization. It was primarily the rural demand for goods that ought to dictate the pace of industrial expansion. He was, as he said, afraid of, and alarmed by, the 'parasitically monopolistic tendencies' of a state-owned economy; and he saw in the peasantry's unfettered economic activity the main, if not the only, counterbalance to such tendencies.

Here, however, Bukharin was caught in a fundamental dilemma, for his argument turned against the very essence of socialism. Where, he asked, if not in the peasant market, would state-owned industry find 'the stimuli which would compel us to move ahead, which would guarantee our progress and replace the private economic stimulus, the stimulus of profit'?[2] As peasant property was, in the Marxist view, incompatible with fully fledged socialism, Bukharin in fact placed a question mark over Marxist socialism at large. He implied that the socialist sector could not find within itself any effective substitute for the profit motive, and so it had ultimately to take the impulse for its own progress from the profit motive which was active in the private sector.[3] In quasi-Narodnik fashion Bukharin looked to the peasant to save the nation from the monopolistic grip of the state-owned economy. He pleaded that the peasant should

[1] Ibid., p. 16.

[2] Preobrazhensky replied that the pressure of workers defending their consumer interests should provide the decisive counterbalance to the parasitic features of a bureaucratically managed economy. Such pressure could make itself felt only when the workers were free to defend their interests against the state, that is under the conditions of a workers' democracy.

[3] The party as a whole, and Bukharin with it, remained committed to Lenin's sketchy scheme for the development of co-operatives in farming. This commitment did not affect practical policy, however. Preobrazhensky argued that even Lenin's scheme was inadequate because its emphasis was not on producers' co-operatives, but on other less important forms of co-operation.

not merely be allowed to thrive on his farm, but that the peasant's needs should determine the pace of the nation's advance
towards socialism. Under such circumstances the advance would
be slow, even very slow; but that could not be helped: '. . . we
shall move ahead by tiny, tiny steps, pulling behind us our large
peasant cart.'[1] There was perhaps more of Tolstoy than of Marx
in this image of Russia's advance; and nothing could contrast
with it more than Preobrazhensky's: 'We must go through this
transition as quickly as possible. . . . We are under the iron heel
of the law of primitive accumulation.' Here were two irreconcilable programmes.

As long as two theorists conducted the argument in more or
less esoteric language, it did not generate much heat outside
narrow circles. But it was inevitable that the issues should be
taken up in more popular form and move into the centre of a
wider political debate. It was not the Trotskyist Opposition, reduced to silence and dispersed as it was, that took them up in
the first instance. The strongest reaction against Bukharin's neo-
Populism, his 'wooing' of the strong farmer and his virtual
reconciliation with Russia's industrial backwardness, came from
Leningrad. It was mainly in the party organization of that city,
led by Zinoviev, that a new left was forming itself as counterpart to the new right. Leningrad had remained the most proletarian of Soviet cities. It had the strongest Marxist and
Leninist traditions. Its workers felt more acutely than anyone
else the need for a bold industrial policy. The city's engineering
plants and shipyards, starved of iron and steel, were idle. Less
than anyone else could the Leningraders agree that the muzhiks
should dictate the tempo of industrial reconstruction. Less than
anyone could they reconcile themselves to the prospect that they
were to move ahead only slowly and drowsily drag along the
huge and heavy peasant cart. All the antagonism of urban
Russia to the inert conservatism of rural Russia was focused in
the old capital. The party organization, although it was managed in a bureaucratic manner and had long ceased to be
representative of the workers, could not help reflecting in some
measure the prevalent discontents. Its organizers and agitators
had to deal with vast numbers of unemployed and became influenced by their resentments and impatience. The popular

[1] Bukharin, *Kritika Ekonomicheskoi Platformy Oppozitsii*, p. 9.

mood infected various grades of the party hierarchy on the spot and impelled them to make a stand against the new right. Throughout most of 1925 Zinoviev led the attack against Bukharin's school. The whole Northern Commune was aroused. The Comsomol passionately threw itself into the struggle; and the press of Leningrad opened a barrage.

At the same time a new rift appeared in the Politbureau. Once the triumvirs had defeated Trotsky and removed him from the Commissariat of War, the bonds of their solidarity snapped. Molotov related afterwards that the discord began in January 1925 when Kamenev proposed that Stalin should take Trotsky's place at the Commissariat of War. According to Molotov, Kamenev and Zinoviev hoped in this way to oust Stalin from the General Secretariat.[1] (Much earlier, as early as October 1923, Zinoviev and Kamenev had toyed with this idea and had even sounded Trotsky. He, however, saw no advantage then in joining hands with Zinoviev, whom he regarded as the most vicious of his adversaries.[2]) Stalin himself traces the beginning of this conflict to the end of the year 1924, when Zinoviev proposed Trotsky's expulsion from the party and Stalin replied that he was against 'chopping off heads and blood letting'.[3] When Trotsky left the Commissariat, Zinoviev proposed that he should be assigned to a minor job in the management of the leather industry; and Stalin persuaded the Politbureau to make a less humiliating appointment. In a pique, Zinoviev appealed to the Leningrad organization, charging Stalin and other Politbureau members with a leaning for Trotsky and with being 'semi-Trotskyists' themselves.

In these petty manœuvres, however, no divergencies over policy had as yet shown themselves. Only in the last week of April 1925 did members of the Central Committee notice signs of a political breach between the triumvirs. In the text of a resolution prepared for the forthcoming party conference, Stalin intended to proclaim socialism in a single country. He had

---

[1] See *14 Syezd VKP (b)*, p. 484.

[2] Voroshilov's disclosures about this made in Trotsky's presence did not meet with Trotsky's denial. Ibid., pp. 388–9. Zinoviev confirmed them in substance. Ibid., pp. 454–6.

[3] 'To-day they chop off one head, to-morrow another, the day after to-morrow still another—who, in the end, will be left with us in the party?' Stalin, *Sochinenya*, vol. vii, pp. 379–80.

formulated the idea some months earlier, but now for the first time he sought to obtain official sanction for it and to incorporate it in party doctrine. Zinoviev and Kamenev objected. None of the triumvirs, however, wished to scandalize the party by revealing their disunion so soon after their show-down with Trotsky. They hushed up the matter and agreed on an ambiguous motion which in its opening passages reminded the party that Lenin had never been a believer in socialism in a single country, and in its conclusion upbraided Trotsky for not having been a believer in it either.[1] With this incongruous text in their hands, the triumvirs presented a common front to the conference. They still maintained it over decisions of immediate practical importance. The conference voted for an enlargement of the freedom of private farming and trade, for a reduction in agricultural taxation, for the abolition of restrictions on the lease of land and on the hiring of farm labour. In these decisions there showed itself a marked influence of Bukharin's school of thought. None of the leaders objected to them, however, in part because all had been alarmed by a bad harvest and all recognized the need to offer new incentives to the farmers; and in part because these resolutions too were framed ambiguously so that any interpreter could make out of them whatever he wished.

For another four or five months, throughout the summer, the dissension among the triumvirs did not come into the open. Zinoviev and the Leningraders campaigned only against Bukharin and Rykov, and against the neo-Populist 'red professors'. By doing so they helped Stalin to consolidate his position. The Politbureau still consisted of these seven members: Stalin, Trotsky, Zinoviev, Kamenev, Bukharin, Rykov, and Tomsky. The leaders of the new right, Bukharin, Rykov, and Tomsky, allied themselves with Stalin and with him formed the majority. The arithmetic of the Politbureau vote was so plain that had Zinoviev and Kamenev been eager only to oust Stalin, they would have sought to make common cause with Bukharin rather than attack him. They acted as they did because in this situation matters of conviction and fundamental differences were more important to them than calculations of personal advantage.

Meantime the crisis in the country deepened. The concessions

---

[1] *KPSS v Rezolutsyakh*, vol. ii, pp. 46–50; Popov, op. cit., vol. ii, p. 239.

made to the strong farmers failed to appease them. In the summer the deliveries of grain fell far below expectation. The government was suddenly compelled to stop the export of grain and to cancel orders placed abroad for machinery and raw materials which were to be paid with the proceeds. Industrial recovery suffered a severe if temporary set-back. Food became scarce in the towns and the price of bread went up. The party leaders had to consider anew what should be done to ease the tension between town and country. Bukharin urged the Polit-bureau to offer the farmers further concessions and new in-centives—it was at this time that he wound up one of his appeals to the peasants with the call: 'Enrich yourselves!' He insisted on the need to do away at last with the restrictions that ham-pered the accumulation of capital in farming. To those who were outraged by his demand and afraid of the kulak, he replied: 'As long as we are in tatters, . . . the kulak may defeat us econo-mically. But he will not do so if we enable him to deposit his savings in our banks. We shall assist him, but he will also assist us. Eventually the kulak's grandson will be grateful to us for our having treated his grandfather in this way.'[1] Bukharin's disciples again dotted the i's, spoke of the advent of the neo-N.E.P., and elaborated the view that it should be possible to integrate peacefully the well-to-do farmer into socialism. One of them, Bogushevsky, argued in the *Bolshevik*, the Central Committee's policy paper, that the kulak was no longer a social force to be reckoned with—he was a mere bogy, a 'phantom', or a 'de-crepit social type of which only a few specimens have survived'.[2]

Leningrad replied with an outcry of indignation. Its workers were daily finding fresh proof of the kulak's strength and striking power—at their bakers'. At the Moscow Committee Kamenev, showing with fresh statistics how dependent the towns had be-come for the bare necessities of life on a small minority of the peasantry, sounded an alarm at the Central Committee's in-clination to accept this state of affairs and to yield even further to the clamour for a neo-N.E.P. The Leningraders demanded that the party should make a new appeal to the poor peasantry against the rich. They pointed out that by its attempts to pro-pitiate the kulak, the party had antagonized the great mass of the poor and middle peasants and enabled the kulaks to become

---

[1] *Bolshevik*, nr. 8, 1925.   [2] *Bolshevik*, nrs. 9–10, 1925.

the virtual leaders of rural Russia. This was undoubtedly true.[1]
But the weak point in the critics' argument was precisely that
the poor and even the middle peasants did not produce the food
surpluses the town needed. More than ever was the party hierarchy therefore afraid of 'fanning class struggle in the country-
side' and incurring the kulaks' hostility. Rural committees
became wary of organizing farm labourers and supporting their
claims. There was much talk about an impending return of the
nationalized land into private hands. In Georgia the Commissar
of Agriculture published 'theses', i.e. the draft of a decree, to
this effect; and similar decrees were expected to be promulgated
in the rest of the Caucasus and in Siberia. Stalin himself saw no
reason why the title-deeds to the land should not be handed
over to the peasants 'even for the duration of forty years'. He, too,
firmly discouraged 'incitement to class struggle in the country'.[2]

The controversy now turned from current policy to the larger
underlying issues. Did we or did we not, the Leningraders asked,
carry out a proletarian revolution? Are we going to sacrifice the vital interests of the workers to those of the strong
farmers? What is happening to our party that makes it abandon
class struggle in the country and turns it into a promoter
of rural capitalism? What is it that impels our chief theoreti-
cian to cry out 'Enrich yourselves!'? Why are so many of
our leaders resigned and ready to reconcile themselves to
Russia's backwardness? Where is our revolutionary fervour of
earlier years? The Leningraders concluded that all they had
fought for was in jeopardy, that the party's ideals were being
falsified and the Leninist principles abandoned. They wondered
whether the revolution had not reached a point of exhaustion as
other revolutions, especially the French, had done in their time.
It was not Zinoviev or Trotsky or another of the illustrious
intellectuals but Peter Zalutsky, a self-taught worker and secretary of the Leningrad organization, who first came out in a
public speech with a significant analogy between the present
state of Bolshevism and Jacobinism in decline, and who first

---

[1] Later in the year, at the fourteenth congress, the Stalinist spokesmen admitted
the facts. Mikoyan, for instance, declared: 'We are making great efforts to regain
the middle peasant who has become the kulak's political prisoner.' *14 Syezd VKP (b)*,
pp. 188–9. More euphemistically Molotov stated: 'At present we do not as yet
truly lead the middle peasant.' Ibid., p. 476.

[2] Stalin, *Sochinenya*, vol. vii, pp. 123, 173–81 and *passim*.

raised an alarm about the 'Thermidorian' danger that threatened the revolution—we shall presently find this idea in the very centre of all Trotsky's denunciations of Stalinism.[1]

Bolshevism, Zalutsky said, might decay through its own lassitude. Its destroyers might come from its own midst, from among those of its own leaders who succumbed to reactionary moods. A cry for the rehabilitation of the revolution came from Leningrad. Let our rulers remain loyal to the working class and the ideals of socialism! Let equality remain our ideal! The workers' state may be too poor to make our dream of equality come true, but let it not mock at the dream!

Of this mood Zinoviev made himself mouthpiece. Early in September he wrote an essay 'The Philosophy of the Epoch' which the Politbureau allowed him to publish only after he had deleted the most provocative parts. 'Do you want to know what the mass of the people is dreaming about in our days?' ran one of the censured passages,

It is dreaming about equality. . . . If we wish to be genuine mouthpieces for the people, we ought to place ourselves at the head of its struggle for equality. . . . In what name did the working class, and behind it the vast mass of the people, rise in the great days of October? In what name did they follow Lenin into the fire? In what name . . . did they follow his banner in the first difficult years? . . . In the name of equality. . . .[2]

About the same time Zinoviev also published his book *Leninism* which combined an interpretation of party doctrine with a critical survey of Soviet society. He exposed the conflicts and tensions between the private and the socialist sectors and pointed out that even in the socialist sector there were strong elements of 'state capitalism'. National ownership of industry represented the element of socialism there; but relations between the employer-state and the workers, bureaucratic management, and differential wages bore the marks of capitalism. For the first time Zinoviev here came out with an open critique of socialism in a single country. Even if the Soviet Union were to remain isolated for an indefinite time, he maintained, it could achieve much progress in building socialism; but, poor and

---

[1] *14 Syezd VKP (b)*, pp. 150–2.
[2] The censured passages were quoted by Uglanov at the fourteenth congress. Ibid., p. 195.

backward and exposed to dangers from without and within, it could not hope to achieve *full* socialism. It could not raise itself economically and culturally above the capitalist West, abolish class differences, and let the state wither away. The prospect of socialism in a single country was therefore unreal; and Bolsheviks had no need to place before the people such a *fata morgana*, especially as this would imply the abandonment of the hope for revolution abroad and a break with Leninist internationalism. Here was the crux of the new division. The new right framed its policies in strictly national and isolationist terms. The left adhered to the party's internationalist tradition, despite all the defeats that international communism had suffered.

At this stage, in the summer of 1925, Stalin and his followers defined their attitude as that of the centre. Partly from conviction and partly from opportunist calculation, because he depended on Bukharin's and Rykov's support, Stalin backed the pro-muzhik policy. But he curbed his right-wing allies and disavowed their most outspoken statements like Bukharin's 'Enrich yourselves!'[1] Cautious, cunning, and caring not a straw for logical and doctrinal niceties, he borrowed ideas and slogans from both right and left and combined them often quite incongruously. In this lay a great part of his strength. He managed to blur every issue and to confuse every debate. To critics who attacked him for any of his pronouncements he was always able to produce another statement of his which contained the exact opposite. His eclectic formulas were a boon to officialdom and to habitual sitters-on-the-fence; yet they also attracted many honest but timid or muddled minds. As in any 'centrist' faction, so among the Stalinists some leaned to the left and others to the right. Kalinin and Voroshilov were close to Bukharin and Rykov, while Molotov, Andreev, and Kaganovich were 'left Stalinists'. The differences among his own supporters also induced Stalin to keep his distance from the right. Only over one issue—socialism in a single country—was his solidarity with Bukharin complete.

Early in October the Central Committee considered arrangements for the fourteenth congress which was convened for the

[1] Stalin, op. cit., p. 159.

end of the year. Four members of the Committee, Zinoviev, Kamenev, Sokolnikov, and Krupskaya, came out with a joint statement demanding a free debate in which party members could speak their minds on all the controversial issues that had arisen. With this the two triumvirs gave notice of their intention to appeal against Stalin and Bukharin to the rank and file.

Sokolnikov did not share all of Zinoviev's and Kamenev's views. As Commissar of Finance he had in recent years done his utmost to encourage private enterprise; and many regarded him as a pillar of the right. But he too had become uneasy over the trend of policy and Stalin's growing power; and so he endorsed the demand for a debate. Krupskaya stood firmly behind Zinoviev and Kamenev and encouraged them to divulge to the entire party the differences in the Politbureau, without mincing matters. She had not yet reconciled herself to the fact that in defiance of her husband's will Stalin had remained General Secretary; and she viewed with hostility the growing influence of Bukharin's school of thought. She had tried to speak out against it, but the Politbureau had not allowed her to do so. Her voice carried weight with party members who knew how long and how closely she had been associated with Lenin, not only as wife but as secretary and co-thinker. She was now eager to testify in favour of Zinoviev's interpretation of Leninism and against socialism in one country.

In asking for an open debate, the four members acted in accordance with statute and custom: the party had never yet held a congress without a preliminary discussion. The Central Committee nevertheless refused to allow a debate; and it obliged Zinoviev and Kamenev to refrain from any public criticism of official policy. The two triumvirs were thus placed in the same quandary in which they had previously placed Trotsky. To speak up in public was to act against the principle of cabinet solidarity which bound them as members of the Central Committee and the Politbureau. But not to speak up was to act against their own political conscience and interest. While they were silent and their followers attacked only the Bukharinists, Stalin unremittingly worked to dislodge them from power. Kamenev had hitherto exercised the dominant influence over the Moscow committee. In the course of the summer the General Secretariat quietly removed his lieutenants from their posts, and filled the vacancies with reliable supporters of the new majority. In Leningrad, however, Zinoviev and his followers were firmly entrenched; and for the time being Stalin could do nothing against them. Zinoviev himself had to keep up the pretence of the Central Committee's unanimity; but his followers were free to speak. They were all anger and passion; and they were ready to carry their attack on official policy into full congress.

Between October and December Moscow and Leningrad were engaged in an intense, bitter, and barely concealed tug of war. In both capitals the elections of delegates to the congress were rigged; Moscow elected only Stalin's and Bukharin's nominees, while all of Leningrad's delegates turned out to be Zinoviev's followers. When three days before the opening of the congress the Central Committee met again, it was clear that nothing could avert an open conflict. Zinoviev and Kamenev had resolved to challenge publicly the official policy report and to present their own counter-report. On 18 December, the day the Congress assembled, Zinoviev opened the attack and in *Leningradskaya Pravda* thus branded his adversaries:

They bandy about loud phrases on international revolution; but they portray Lenin as the inspirer of a nationally limited socialist revolution. They fight against the kulak; but they offer the slogan 'Enrich yourselves!' They shout about socialism, but they proclaim the Russia of N.E.P. as a socialist country. They 'believe' in the working class; but they call the wealthy farmer to come to their aid.

.   .   .   .

The exchanges between Bukharinists and Zinovievists had gone on for many months now and the conflict between the triumvirs had been simmering for nearly a year. This, it might have seemed, was the realignment for which Trotsky had waited, the opportunity to act. Yet throughout all this time he was aloof, silent about the issues over which the party divided, and as if unaware of them. Thirteen years later, when he stood before the Dewey Commission in Mexico, he confessed that at the fourteenth congress he was astonished to see Zinoviev, Kamenev, and Stalin clashing as enemies. 'The explosion was absolutely unexpected by me', he said. 'During the congress I waited in uncertainty, because the whole situation changed. It appeared absolutely unclear to me.'[1]

[1] *The Case of Leon Trotsky*, pp. 322–3.

This recollection, so many years after the event, may seem quite incredible; but it is fully borne out by what its author wrote in unpublished diary notes during the congress itself.[1] To the Dewey Commission he explained that he was taken by surprise because, although he was a member of the Politbureau, the triumvirs had carefully concealed their dissensions from him and had thrashed out their differences in his absence, within the secret caucus that acted as the real Politbureau. The explanation, although true, explains little. For one thing, the crucial controversy over socialism in a single country had already been conducted in public. Trotsky could not have missed its significance if he had followed it. He evidently failed to do so. For another, Zinoviev, Kamenev, Krupskaya, and Sokolnikov had raised the demand for an open debate not within a secret caucus but at the plenary session of the Central Committee, in October. But even if they had not done so, and even if the public controversy over socialism in a single country had given no indication of the new cleavage, it would still be something of a puzzle how an observer as close, as interested, and as acute as Trotsky could have remained unaware of the trend and blind to the many omens. How could he have been deaf to the rumblings that had for months been coming from Leningrad?

His surprise, we must conclude, resulted from a failure of observation, intuition, and analysis. Moreover, it is implausible that Radek, Preobrazhensky, Smirnov, and his other friends should not have noticed what was happening and that none of them tried to bring matters to Trotsky's attention. Evidently his mind remained closed. He lived as if in another world, wrapped up in himself and his ideas. He was up to his eyes in his scientific and industrial preoccupations and literary work, which protected him to some extent from the frustration to which he was exposed. He shunned inner-party affairs. Full of the sense of his superiority and contempt for his opponents, and disgusted with the polemical methods and tricks, he was not interested in their doings. He submitted to the discipline by which they had shackled him, but he held up his head and ignored them. A few years later his biographer was told in Moscow that he used to appear dutifully at the sessions of the Central Committee, take his seat, open a book—most often a French novel—and become

[1] See the summary of these notes on pp. 255–6. The text is in *The Archives*.

so engrossed as to take no notice of the deliberations. Even if this anecdote was invented, it was well invented: it conveys something of the man's temper. He could turn his back on his adversaries, but he could not view them with detachment. He was too close to them: he saw them as the small men, the rogues, and the sharpers that they sometimes were; and he half forgot that they were also the leaders of a great state and party and that what they said and did carried immense historic weight.

Had Trotsky kept his ears open to what the Leningraders were saying, he could not have failed to realize at once that they were defending the causes he himself had defended, and attacking the attitudes he himself had attacked. As oppositionists, they started where he had left off. They argued from his premisses; they took up his arguments to carry them farther. He had criticized the Politbureau's lack of initiative, its neglect of industry, and its excessive solicitude for the private sector of the economy. So did the Leningraders. He had observed with apprehension the spirit of national narrow-mindedness which induced the party hierarchy to frame policy and think of the future in terms of self-sufficiency. Actuated by the same antagonism to 'national narrow-mindedness', Zinoviev and Kamenev were the first to come out with a critique of socialism in a single country. To Trotsky, Bukharin's and Stalin's ideas on this subject must have at first appeared as dull scholastic dogma-mongering, hardly worthy of his comment; and so he made no comment for nearly a year and a half, while socialism in a single country was becoming the new Bolshevik orthodoxy, the orthodoxy he was to fight to the end of his life. Zinoviev and Kamenev were more alive to the symptomatic meaning of the new doctrine. He could not but agree with their arguments against it for they drew them from the armoury of classical Marxist internationalism. Nor could the cry for equality that went up in Leningrad fail to strike a chord in him. Zinoviev, Kamenev, Sokolnikov, and Krupskaya had only echoed Trotsky when they protested against the stifling of party opinion. Like him, they spoke of the unholy alliance of N.E.P.-man, kulak, and bureaucrat; and like him, they called for the revival of proletarian democracy. He had warned the party against the 'degeneration' of its leadership; and now the same warning resounded even more poignantly and alarmingly in the Leningraders' outcry against the 'Thermidorian' danger. These were the ideas and the slogans that he was to take up presently and to expound in the years to come. Yet when he heard them expounded by his erstwhile adversaries, he 'waited in uncertainty' for several critical months; and his adherents waited together with him.

What contributed to his and his followers' confusion was that they had been accustomed to regard Zinoviev and Kamenev as the leaders of the party's right wing. No one had done more than Trotsky to spread this view. In *The Lessons of October* he had reminded the party of Zinoviev's and Kamenev's opposition to the October Revolution. He had argued that in 1923 Zinoviev had led the German Communists to 'capitulate' because his frame of mind had still been the same as in 1917. And when he told the party that its Old Guard might, like the hierarchy of the Second International, degenerate into a conservative, bureaucratic 'apparatus', he almost pointed an accusing finger at Zinoviev and Kamenev. No wonder that he looked at them incredulously when they appeared as the spokesmen of a new left. He suspected demagoguery. The suspicion, though not altogether groundless, made it difficult for him to grasp that the change of roles was real and that it formed part of that regroupment of men and ideas to which the extremely critical situation in the country had given rise. Zinoviev's and Kamenev's conversion was not less genuine and not less startling than that by which Bukharin, the ex-leader of the Left Communists, had become the ideologue of the new right—indeed, the two conversions supplemented each other. Official Bolshevik policy tended at present so strongly towards the right that some of those who only yesterday headed the right wing became frightened of the consequences and found themselves veering far to the left.

To be sure, personal ambitions and jealousies played their part: Zinoviev and Kamenev sought to strip Stalin of his power. But they might have had a better chance had they chosen to ride, with Bukharin, the mounting tide of isolationism and neo-Populism. Instead, they took their stand on the proletarian and internationalist traditions of Leninism which had become unpopular with the men of the party machine, on whom the outcome of the contest immediately depended. Zinoviev's and Kamenev's outlook and habits of mind as well as the moods among their followers set limits to their self-seeking. No matter how timidly or opportunistically they had behaved on important occasions, they had been Lenin's closest disciples; they were constitutionally incapable of shedding the influence that had moulded them. Others might turn their backs on the European working class and glorify, sincerely or not, the muzhik; they could not do so. Others might exalt Russia's self-sufficient socialism; to them the very thought was absurd and repugnant. The attitude to these issues, however, formed the watershed which now separated the various currents in Bolshevism.

There was yet another aspect of this change of roles. Like Trotsky and Lenin before them, Zinoviev and Kamenev grappled with the dilemma of authority and freedom, or of party discipline and proletarian democracy. They, too, felt the tension between the power and the dream of revolution. They had been the disciplinarians. Now they were tired and sick of the mechanical and rigid discipline they had enforced. Zinoviev had for years strutted the political stage, roared words of command, schemed and plotted, demoted and promoted people, built power for the revolution and for himself; he had been as if obsessed and drunken with authority. Now came the awakening, the bitter after-taste, and the yearning to find a way back to the irrecoverable clear spring of the revolution. Together with him many of the Old Guard had followed the same bents and suffered the same perplexities and disenchantments until, without knowing it, they took up attitudes indistinguishable from those of the Trotskyists whom they had just helped to defeat. Everything drove them to join hands with the men of the 1923 Opposition.

If Trotsky was to make common cause with Zinoviev and Kamenev, this was the time to do it. Till the beginning of 1926 the base from which the Leningraders operated was still intact. The administrative machinery of the city and the province was in Zinoviev's hands. He had a large body of ardent followers. He controlled influential newspapers. He possessed the material means for a long and sustained political struggle. In a word, he was still in his Northern Commune the master of a powerful fortress. He was also President of the Communist International, although Stalin was already active at its headquarters, sapping his influence. In some respects Zinoviev's position, when he came in conflict with Stalin, was much stronger than Trotsky's had ever been. Trotsky had never bothered to lay his hands on instruments of personal power; and so after his world-shaking career he began almost empty-handed the fight with the triumvirs. They found it all too easy to brand him as an alien to Bolshevism. It was far more difficult for Stalin and Bukharin to denounce Zinoviev, Kamenev, and Krupskaya as inveterate Mensheviks. The conflict was now clearly between two sections of the Bolshevik Old Guard. A coalition between Trotsky and Zinoviev, if it had come into being before Zinoviev's defeat, might have been formidable. Yet neither of them and neither of the two factions was ready. Their mutual grievances and hatreds and the memories of the knocks and insults exchanged were still too fresh to allow them to pull together.

One of the strangest moments in Trotsky's political life now followed. On 18 December the fourteenth congress, the last he was to attend, was opened. From first to last it was the scene of a political storm, the like of which the party had never witnessed in its long and stormy history. Before the eyes of the whole country the new antagonists wrestled and dealt each other mighty blows. The fate of the party and the revolution was in the balance. Nearly all the great issues which were to occupy Trotsky for the rest of his life were thrashed out. Each of the new antagonists had his eyes on Trotsky, wondering with whom he would join and waiting with bated breath for his word. Yet throughout the fortnight that the congress was in session Trotsky sat silent. He had nothing to say when to an audience convulsed with emotion Zinoviev recalled Lenin's testament and its warning against Stalin's abuse of power or when he dwelt on the danger that threatened socialism from the kulak, the N.E.P.-man, and the bureaucrat. Impassively Trotsky viewed the momentous scene when, after Kamenev had protested with great force against the establishment of autocratic rule over the party, the well-picked majority, foaming with rage and insulting the speaker, for the first time acclaimed Stalin as the Leader 'around whom the Leninist Central Committee was united'.

Nor did he rise and declare his solidarity with Krupskaya when she spoke about the stultifying effect of the Leninist cult, when she entreated the delegates to discuss the issues before them on their merits instead of swamping debates with meaningless quotations from the writings of her husband, and when, finally, she recalled by way of a warning how the campaign against Trotsky had degenerated into slander and persecution. He listened as though unconcerned to the controversy over socialism in one country, one of the greatest debates of the century. He was not provoked to make a single gesture of protest or disagreement when Bukharin built the case for socialism in one country on the party's previous rejection of Trotsky's permanent revolution, and went on to speak of building socialism at the 'snail's pace'. The triumvirs revealed the inner story of their disagreements in which Trotsky's person loomed so large: Stalin related how Zinoviev and Kamenev had asked for Trotsky's head and how he had resisted them. Zinoviev described how he and Stalin had, violating the statutes, dispersed the Central Committee of the Communist Youth after its overwhelming majority had declared for Trotsky. Speakers from all factions paid Trotsky compliments and made advances. While Krupskaya spoke, a cry: 'Lev Davidovich, you have gained new collaborators!' rose from the floor. Lashevich, hitherto one of his most embittered adversaries, acknowledged that Trotsky had not been altogether wrong in 1923. The Stalinists and Bukharinists were lavish with praise: Mikoyan held up to the new opposition the shining example of Trotsky who, when defeated, scrupulously observed party discipline. Yaroslavsky reproached the Leningraders with their rabid and still unabated anti-Trotskyism. Tomsky contrasted the 'crystal clear lucidity of Trotsky's views' and the integrity of his conduct with Zinoviev's and Kamenev's muddleheadedness and evasions. Kalinin spoke of the resentment and disgust he had always felt at their attempts to drag down Trotsky. When Zinoviev asserted his right to dissent from official policy, and complained that no opposition had ever been handled so roughly, Stalinists and Bukharinists showered on him derisive reminders of the things he had done to Trotsky. Then, winding up a great peroration, Zinoviev exhorted the congress to let bygones be bygones and to reform the party's leadership so that all sections of Bolshevik opinion should co-operate and unite. The eyes of the whole assembly were now on Trotsky: had the great and eloquent man nothing to say? His lips were sealed. He remained silent even when Andreev asked that new prerogatives be voted for the Central Committee to enable it to deal more effectively with dissenters—to enable it, that is, to break the back of the new Opposition. The latter had been heavily outvoted; but before its close the congress received with uproar and anger reports that in Leningrad turbulent demonstrations against its decisions were in progress: the Leningraders were fighting on within their fortress. And to the end not a word escaped from Trotsky's mouth.[1]

Trotsky's private papers offer us an insight into what was going on in his mind. In a note jotted down on 22 December, the fourth day of the congress, he remarked that there was 'a grain of truth'—but not more—in the view expressed by some that the Leningraders continued the work of the Trotskyist Opposition. The hue and cry, raised in 1923, about the enmity of Trotskyism towards the peasantry had paved the way for the neo-Populism which was now fashionable and against which the Leningraders reacted. It was natural that they should do so, although they had led in the drive against Trotskyism. The intense animosity of the congress towards Zinoviev's faction reflected *au fond* the hostility of the country to the town. This view, one might have thought, should have induced Trotsky to make common cause with the Leningraders at once. But the issues and the divisions did not yet seem to him as clear as they had so far appeared in his own analysis; and he entertained certain hopes which induced him to wait.

He wondered why Sokolnikov, of all people, the ultra-moderate who should have been on Bukharin's side, had joined the Leningraders. He was puzzled that the division was between Moscow and Leningrad. The artificially produced antagonism

---

[1] He made only one *Zwischenruf* in the debate. When Zinoviev explained that the year before he had asked for Trotsky's exclusion from the Politbureau because after all the accusations they had hurled at Trotsky, it was incongruous to re-elect him to the Politbureau, Trotsky interjected: 'Correct!'

Ruth Fischer, who was present in Moscow during the congress but was not admitted to it and was instead given daily reports by Bogrebinsky, Stalin's underling, 'a delegate from the G.P.U.', writes: 'Bogrebinsky was particularly interested in Trotsky. . . . Both groups feared him . . . and now both hoped to win him over; Trotsky's attitude might have been decisive among the wavering delegates from the provinces. Trotsky, Bogrebinsky noted each day, had looked well or badly; he had spoken with this person or that. "I saw Trotsky to-day in the corridors. He spoke with some of the delegates, and I could hear a little of the conversation. He said nothing on the decisive questions. He did not support the Opposition, even by hints and allusions. That is wonderful. Those dogs of Leningrad will get a thorough beating." ' R. Fischer, *Stalin and German Communism*, p. 494.

between them, he noted, veiled a deeper underlying conflict. He hoped that the organizations of the two capitals would draw together and reassert jointly the aspirations of the proletarian-socialist elements against the pro-muzhik right. He reckoned that all 'true Bolsheviks' would rise against the bureaucracy—nothing less could free the party organization of Moscow from Stalin's stranglehold. The situation was still in a state of flux. He expected something like a political landslide, of which the breach between the triumvirs was only the beginning, to shake the party and to bring about the final, far wider, and far more significant regrouping of forces. Then the lines of division would be less fortuitous and would correspond to the fundamental contradictions between town and country, worker and peasant, socialism and property. Meantime, he was not at all eager to throw in his lot with the 'vociferous, vulgar, and rightly discredited' leaders of the Leningrad Opposition. There is a whiff of *Schadenfreude* in these diary notes written as he watched Zinoviev's and Kamenev's discomfiture—as if he were saying: *Vous l'avez voulu, Vous l'avez voulu!*

Yet he could not give himself to *Schadenfreude* for any length of time; it was not in his nature to do so. Willy-nilly he had to rush to the rescue of the defeated. No sooner had the congress dispersed than the Central Committee met to consider measures for taming Leningrad. Stalin proposed to dismiss in the first instance the editorial staff of *Leningradskaya Pravda* and to turn that newspaper into the mouthpiece of official policy. Next Zinoviev was to be deposed and Kirov was to take his place at the head of the Northern Commune. The whip was to come down on the Leningraders. At this point Trotsky broke his silence—he was against the reprisals.[1] He did not contemplate an alliance with Zinoviev and Kamenev, but by trying to shield them, he at once gave offence to Stalin who had walked around him gingerly seeking to mollify him.

There was a curious scene at the session. Bukharin spoke for the course of action proposed by Stalin. Kamenev protested. It was strange, he said, that Bukharin, who had always opposed drastic reprisals against the Trotskyists, should now call for the whip. 'Ah, but he has come to relish the whip', Trotsky interjected. Bukharin, as though caught off guard, cried back: 'You

[1] N. Popov, *Outline History of the CPSU*, vol. ii, p. 255.

think I have come to relish it, but this relishing makes me shudder from head to foot.'[1] In this cry of anguish were suddenly revealed the forebodings with which Bukharin backed Stalin. From this incident dates a 'private contact', which 'after a long interval' Trotsky resumed with Bukharin—a fairly friendly, but politically fruitless and shortlived affair, the traces of which are found in their correspondence.[2] Still 'shuddering from head to foot', Bukharin did his best to persuade Trotsky not to come to Zinoviev's aid. He tried to impress on him that the party's freedom was not at stake in this case and that Zinoviev, who himself brooked no opposition, was no defender of inner-party democracy. Trotsky did not deny this, but he argued that Stalin was surely no better; and that the evil lay in a monolithic discipline and the unanimous vote which both Stalin and Zinoviev enforced—this had made it possible that on the eve of the congress the two largest organizations, those of Moscow and Leningrad, should each carry its own resolutions in 'a hundred per cent. unanimity'. He held no brief for the Leningraders; but he could not but oppose the false discipline; and he appealed to Bukharin to join him in a common effort to restore 'a healthy inner-party régime'. Bukharin, however, was afraid that by asking for more freedom they would get less; and he concluded that those who demanded inner-party democracy were in effect its worst enemies, and that the only way to save what was left of it was not to use it.

While these pathetic 'confidential' exchanges went on, Stalin lost the hope of playing Trotsky against Zinoviev and Kamenev. Earlier perhaps than Trotsky himself he realized that the two oppositions would have to join hands. He therefore gave the signal for a new drive against Trotsky. He was anxious that Trotsky should not be able to address communist gatherings in working-class districts. Uglanov, who had replaced Kamenev as leader of the Moscow organization, saw to that. Under all sorts of pretexts Trotsky was refused admission to the cells. As he was just then addressing meetings of scientists and other intellectuals, the members of the proletarian cells were told that he preferred to speak to the bourgeoisie rather than face workers. Official agitators ceased to distinguish between Trotskyists and Zinovievists, incited the rank and file against both, and hinted

[1] *The Trotsky Archives.*   [2] Ibid.

darkly that it was no matter of chance that the leaders of both were Jews—this was, they suggested, a struggle between native and genuine Russian socialism and aliens who sought to pervert it.

In another letter to Bukharin, dated 4 March, Trotsky described the vexations and the obloquy of which he had again become the object. Altogether against his inclination, he dwelt on the anti-Semitic undertones of the agitators' talk. 'I think', he wrote, attempting to arouse Bukharin, 'that what binds us, two Politbureau members, is still quite enough for us to try and check the facts calmly and conscientiously: is it true, is it possible that in *our party*, IN MOSCOW, in WORKERS' CELLS, anti-Semitic agitation should be carried on with impunity?!'[1] A fortnight later at a Politbureau meeting he asked the same astonished and indignant question. The Politbureau members shrugged, professed to know nothing, or pooh-poohed the matter. Bukharin blushed with embarrassment and shame; but he could not turn against his associates and allies. At any rate, at this stage his 'private contact' with Trotsky was coming to an end.

It was not by chance that the agitators struck the anti-Semitic note: they were briefed by Uglanov; and Uglanov took his cue from Stalin who was anything but fastidious in the choice of means. But there were means to which he could not have resorted even a year or two earlier; and playing on anti-Jewish prejudice was one of them. This had been the favourite occupation of the worst Tsarist reactionaries; and even in 1923–4 the party and its Old Guard were still too strongly imbued with internationalism to countenance such prejudice, let alone to exploit it. But the situation was changing. The new right appealed vaguely to nationalist emotions; and while these surged up, the political climate altered to such an extent that even Communists no longer frowned on anti-Semitic hints or allusions dropped in their midst. The distrust of the 'alien' was, after all, only a reflex of that Russian self-centredness, of which socialism in one country was the ideological abstract.

Jews were, in fact, conspicuous among the Opposition although they were there together with the flower of the non-

[1] *The Trotsky Archives.*

Jewish intelligentsia and workers. Trotsky, Zinoviev, Kamenev, Sokolnikov, Radek, were all Jews.[1] (There were, on the other hand, very few Jews among the Stalinists, and fewer still among the Bukharinists.) Thoroughly 'assimilated' and Russified though they were, and hostile to the Mosaic as to any other religion, and to Zionism, they were still marked by that 'Jewishness' which is the quintessence of the urban way of life in all its modernity, progressiveness, restlessness, and one-sidedness. To be sure, the allegations that they were politically hostile to the muzhik were false and, in Stalin's mouth, though perhaps not in Bukharin's, insincere. But the Bolsheviks of Jewish origin were least of all inclined to idealize rural Russia in her primitivism and barbarity and to drag along at a 'snail's pace' the native peasant cart. They were in a sense the 'rootless cosmopolitans' on whom Stalin was to turn his wrath openly in his old age. Not for them was the ideal of socialism in a single country. As a rule the progressive or revolutionary Jew, brought up on the border lines of various religions and national cultures, whether Spinoza or Marx, Heine or Freud, Rosa Luxemburg or Trotsky, was particularly apt to transcend in his mind religious and national limitations and to identify himself with a universal view of mankind. He was therefore also peculiarly vulnerable whenever either religious fanaticisms or nationalist emotions ran high. Spinoza and Marx, Heine and Freud, Rosa Luxemburg and Trotsky, all suffered excommunication, exile, and moral or physical assassination; and the writings of all were burned at the stake.

. . . .

In the first weeks of 1926 the strength of the Leningrad

[1] In 1918, while the Ukraine was under German occupation and ruled by Skoropadsky, the rabbis of Odessa pronounced anathema on Trotsky and Zinoviev. (Zinoviev, *Sochinenya*, vol. xvi, p. 224.) The White Guards, on the other hand, made much of Trotsky's Jewishness and alleged that Lenin was also a Jew. Curious echoes of this can be found in Soviet folk-lore and fiction of the early twenties. In one of Seyfulina's stories a muzhik says: 'Trotsky is one of us, a Russian and a Bolshevik. Lenin is a Jew and a Communist.' In Babel's short story 'Salt' a peasant woman says to a Red Army man: 'You don't bother your heads about Russia; you just go about saving those dirty Jews Lenin and Trotsky.' The Red Army man replies: 'We aren't talking about Jews now, you harmful citizen. The Jews haven't got nothing to do with it. By the way, I won't say nothing about Lenin, but Trotsky was the desperate son of a Governor of Tambov and went over to the working class. . . . They work like niggers, Lenin and Trotsky do, to pull us up to the path of freedom. . . .'

Opposition was broken.[1] The Leningraders could not but submit to Stalin's orders. To defy them was to challenge the authority of the Central Committee, which backed Stalin, and the legality of the congress which had elected the Committee. This Zinoviev and Kamenev, who like Trotsky still sat on that Committee, were not prepared to do. They had openly declared that Stalin had rigged the elections to the congress and that the Central Committee represented the party machine, not the party. But it was one thing to state this, and quite another to proclaim that the decisions of the congress and of the Central Committee were invalid and to refuse to submit. For Zinoviev and Kamenev in particular it would have been a dangerous undertaking to question the legitimacy of the last congress: had they not, together with Stalin, rigged the elections and packed the thirteenth congress in the same way that Stalin packed the fourteenth? By challenging the authority of the Central Committee, the Leningraders would have virtually constituted themselves into a separate party, a rival to the official All-Union Communist party. It was unthinkable that they should do so. They had all accepted the single-party system as a *sine qua non*. Nobody had shown greater zeal in asserting this principle and drawing from it the most far-reaching and absurd conclusions than Zinoviev had. Leningrad's defiance of Moscow would have amounted almost to a declaration of civil war.

And so, when Kirov appeared in Leningrad as Stalin's envoy invested with plenary power and entitled to take command of the Northern Commune, there was nothing left for Zinoviev but to yield. Almost overnight all the local branches of the party, its editorial offices, its manifold organizations, and all the resources on which the opposition had hitherto drawn, passed into the hands of Stalin's and Kirov's nominees. Two of Zinoviev's lieutenants had controlled the armed forces of Leningrad: Lashevich, as political commissar of the garrison and the mili-

[1] After the fourteenth congress the Bukharinists and the Stalinists had an increased majority in the Central Committee. The new Politbureau consisted of nine instead of seven members: Stalin, Trotsky, Zinoviev, Bukharin, Rykov, Tomsky, Kalinin, Molotov, and Voroshilov. With Kalinin and Voroshilov vacillating between the right and the centre, Stalin's faction was somewhat weaker numerically than Bukharin's. Kamenev was now only an alternate member of the Politbureau. The other alternate members were Uglanov, Rudzutak, Dzerzhinsky, and Petrovsky.

tary region, and Bakaev as head of the G.P.U. Both surrendered their offices, although Lashevich, being Vice-Commissar of Defence, remained a member of the central government. This was followed by a moral débâcle. As long as the leaders stood in the full panoply of power it seemed that they had the whole of Leningrad behind them. Now the great proletarian city appeared indifferent to their fate. The workers of Vyborg, that old rampart of Bolshevism, were the first to desert them. For years Zinoviev had bullied and browbeaten them; and so they were not moved by his latest pleas on behalf of the workers and his cry for equality, the pleas and the cry which they were to recall nostalgically a few years hence when it was too late. Humble men viewed the commotion as a brawl between bigwigs which was of no concern to them. Even those who took a less cynical view and felt with the opposition most often kept their feelings to themselves: unemployment was rampant; and the punishment for 'disloyalty' might be the loss of one's job and starvation. Thus the active following of the Leningrad opposition dwindled to a few hundred veterans of the revolution, a small and closely knit band of men, who were devoted to their ideals and leaders and who gradually found that all doors were shut upon them.

The ease and speed with which Stalin overwhelmed the Leningraders showed that the hopes to which Trotsky gave himself in the days of the fourteenth congress were unfounded. There was no sign of any further regroupment, no sign of that rallying of the communist workers against the bureaucrats he had expected. The Leningraders' struggle had caused no movement of sympathy, not even a ripple, in the cells of Moscow. The party machine worked with deadly effectiveness, breaking all resistance where it had shown itself or crushing it before it had done so. This in itself indicated the weakness of the resistance. The working class was no longer dispersed and disintegrated as it had been a few years earlier, but it lacked political consciousness, vigour, and the ability to assert itself. Yet it was on a political revival in its midst that Trotsky had reckoned when he assumed that Moscow and Leningrad would make a common stand. Zinoviev and Kamenev too had hoped for this. At the fourteenth congress they called for a return to proletarian democracy and said that the working class was no longer as splintered and demoralized as it was in the early 1920's, when the party leaders could not rely on the soundness of its political instincts and judgement. Bukharin then replied that Zinoviev and Kamenev were deluding themselves; that the working class had grown numerically by absorbing young and illiterate new-comers from the country, that consequently it was still politically immature, and that the time for a return to proletarian demo-cracy had not yet come. The void by which the Leningrad Opposition now found itself surrounded indicated that Bukharin was closer to the truth than Zinoviev and Kamenev. The work-ing class was apathetic and indifferent, although its apathy was due not merely to inherent immaturity but also to that bureau-cratic intimidation which Bukharin sought to justify. Whatever the truth of the matter, it must have become clear to Trotsky by now that he had nothing to gain by waiting. Yet, after the congress more than three months passed during which the Trotskyists and the Zinovievists did not move towards one another by one inch. Trotsky, Zinoviev, and Kamenev had not been on speaking terms since 1923; and they still said not a word to each other.

Only in April 1926 was the ice broken. At a session of the Central Committee Rykov presented a statement of economic policy. Kamenev tabled an amendment urging the Committee to take note of the ever sharper 'social differentiation of the peasantry' and to restrain the growth of capitalist farming. Trotsky tabled a separate amendment: he agreed with Kamenev's appraisal of rural conditions but added that the sluggish tempo of industrial development deprived the government of the means it needed to exercise a sufficiently strong influence on farming. In the discussion Kamenev, who as former head of the Council for Labour and Defence felt some responsibility for the in-dustrial policy which Trotsky criticized, made some barbed remarks about Trotsky. The Central Committee rejected Trot-sky's amendment. Kamenev and Zinoviev, it seems, abstained from voting. Then, when Kamenev's amendment was put to the vote, Trotsky supported it. This was the turning-point. As the session continued, they found themselves again on the same side. They unbent and moved towards each other until, at the end of the session, they acted virtually as political partners.

Only now did the three men meet in private for the first time in years. This was a strange meeting, full of heart-searchings, startling confessions, sighs of regret and of relief, forebodings, alarming warnings, and hopeful projects. Zinoviev and Kamenev were eager to make a clean breast of the past. They bemoaned the blindness which had led them to denounce Trotsky as the arch-enemy of Leninism. They admitted that they had concocted the charges against him to debar him from leadership. But had he not been mistaken also in attacking them, in reminding the party of their conflicts with Lenin in 1917, and in discrediting them rather than Stalin? They were relieved at having at last freed themselves from the net of a bizarre intrigue, the net they themselves had spun, and at having returned to serious and honest political thought and action.

As they related the various incidents of the intrigue, they made merry about Stalin, mimicking, to Trotsky's slight impatience, Stalin's behaviour and accents; but then they recalled their dealings with him with the shudder with which one remembers a nightmare. They described his slyness, perversity, and cruelty. They said that they had both written, and deposited in a safe place, letters to the effect that if they should perish suddenly and unaccountably, the world should know that this was Stalin's work; and they advised Trotsky to do the same.[1] Stalin, they maintained, had not taken Trotsky's life in 1923–4 only because he feared that some young, fervent Trotskyist might rise as an avenger. Doubtless, Zinoviev and Kamenev were eager to blacken Stalin and to advertise to Trotsky their own restraining influence on him. Trotsky himself did not take their revelations very seriously until many years later, when the Great Purges brought them back to his mind. It was indeed difficult to square what sounded like the story of a bloody court intrigue in the Kremlin of the early Tsars with the Kremlin of the Third International resounding with ideological disputes couched in Marxist terms. Had the Tsars' old fortress cast its evil spell upon Lenin's disciples? Stalin, so Zinoviev and Kamenev went on, was not interested in disputes over ideas—all he craved was power. What they failed to explain was how, if what they said was true, they could have remained in partnership with him for so long.

From these terrified and terrifying accounts and dark hints

[1] L. Trotsky, *Stalin*, p. 417.

the two men passed to plans for the future. They gave themselves to the wildest hopes. They had no doubt that all could
still be changed at a stroke. It would be enough, they said, for
the three of them to appear together in public, reconciled and
reunited, to arouse enthusiasm among Bolsheviks and bring
the party back to the right road. Rarely has the blackest gloom
yielded so easily to the most cheerful innocence.

What accounted for their optimism? It was only a few months
since they had both enjoyed the fullness of power. It was only a
few weeks since Zinoviev had lost his fief in Leningrad, and he
was still President of the Communist International. Their fall
had been so rapid and sudden that they refused to believe it was
real. They had been accustomed to see a nod from either of
them swing into motion the massive wheels of party and state.
They still had in their ears the roar of popular acclaim, a false
acclaim, which had not come from the feelings of the people
but had been artificially produced by the party machine. Suddenly a deathlike stillness surrounded them. This seemed to
them to be a delusion, a misunderstanding, or a passing incident.
What brought it about was their breach with Stalin whom they
themselves had placed, or so it seemed to them, in command of
the party. Yet, who was Stalin? A coarse, half-educated, clumsy
manipulator, a misfit, whom they had repeatedly saved from
ruin because they had found him useful in their game against
Trotsky. They had never had any doubt that as man, leader,
and Bolshevik Stalin did not reach to Trotsky's ankles. Now
that they had made common cause with Trotsky, nothing would
surely be easier than to sweep Stalin out of their way and to
bring the party back under their joint leadership.[1]

Trotsky shook his head. He did not share their optimism.
He knew better the taste of defeat. He had for years felt the full
weight of the party machine as it ran against him and drove
him into the wilderness. He had a deeper insight into the pro-

[1] Ruth Fischer describes how Zinoviev in a talk with her 'broached, almost
timidly' the subject of his alliance with Trotsky. 'This is, he said, a fight for state
power. We need Trotsky, not only because without his brilliant brain and wide
support we will not win state power, but because after we have won we need a
strong hand to guide Russia and the International back to a socialist road. Moreover, no one else can organize the army. Stalin has opposed us not with manifestoes
but with power, and he can be met only with greater power, not with manifestoes.
Lashevich is with us, and if Trotsky and we join together, we shall win.' R. Fischer,
op. cit., pp. 547–8.

cesses which had deformed the party, into that 'bureaucratic degeneration' the progress of which he had watched in impotence since 1922. And behind the party machine he perceived more clearly than they did the abysmal barbarism of old Mother Russia, not to be conjured out of existence. He was also apprehensive of the fickleness and fecklessness of his new allies. He could not forget all that had passed between him and them. Yet he was unstinting in forgiveness; and he tried to steady their nerves for a long uphill struggle.

He himself was not unhopeful. He too believed that the party would be stirred by their reconciliation. Zinoviev and Kamenev volunteered to make a public admission that Trotsky had been right all along when he warned the party against its bureaucracy. In return he was prepared to say that he had been mistaken in assailing them as the leaders of that bureaucracy when he should have concentrated his fire on Stalin. He too hoped that by joining hands the two oppositions would not merely combine their followings but multiply them. The Old Guard had, after all, looked up to Zinoviev and Kamenev. It was known that Lenin's widow was in sympathy with them. In the team which led the Leningrad opposition, although it was less outstanding than the circle around Trotsky, there were such eminent men as Lashevich, still the Deputy Commissar of Defence, Smilga, one of the most able political commissars in the civil war and a distinguished economist, Sokolnikov, Bakaev, Evdokimov, and others. With such men and with Preobrazhensky, Radek, Rakovsky, Antonov-Ovseenko, Smirnov, Muralov, Krestinsky, Serebriakov, and Yoffe, to mention only these, the Joint Opposition would command far more talent and prestige than the factions of Stalin and Bukharin had at their disposal. And, despite everything, a political revival in the working class, though delayed, would still come and would put wind into the Opposition's sails.

The partners did not have the time to make precise plans or even to define clearly the points of their agreement. A day or two after their first private meeting, Trotsky had to leave Russia for medical treatment abroad. The malignant fever from which he had suffered in the last years still persisted, rising often to over 100° F., incapacitating him during the most critical moments of the struggle and compelling him to spend many months in the Caucasus. (There he spent the winters of 1924 and 1925, and the early months of the spring.) Russian doctors were unable to diagnose and urged him to consult German specialists. The Politbureau raised no objections to his journey abroad, but insisted that he undertook it on his own responsibility. About the middle of April, accompanied by his wife and a small bodyguard, he arrived in Berlin beardless and incognito, pretending to be a Ukrainian educationist by the name of Kuzmyenko. He spent most of his time in a private clinic undergoing treatment and a minor operation; but in the intervals he moved about freely, observing the depressed Berlin of those years, so different from the Imperial capital he had known, attending a May Day parade, watching a wine festival outside the city, and so on. He was thrilled to be able, for the first time since 1917, 'to move about in a crowd without attracting anyone's attention and feeling oneself part of the nameless listening and watching mass'.[1] But somehow his incognito was discovered later and the German police warned the director of the clinic that White Russian émigrés were about to make an attempt on the patient's life. Under heavy escort Trotsky moved to the Soviet Embassy, and shortly thereafter he returned home, his temperature as high as ever. It has never been discovered whether there was any ground for the alarm about the attempt on his life.[2]

During his stay in Berlin, which lasted about six weeks, he was agitated by two political events of unequal importance. In Poland, Marshal Pilsudski, supported by the Communist party, had just carried out a *coup d'état*, which established him as dictator. In Britain the protracted strike of the coal-miners had just led to the great general strike. The absurd behaviour of the

[1] 'Only once did our companions [at the May Day parade] say to me cautiously: "There they are selling your photographs." But from these photographs no one would have recognized . . . Kuzmyenko, the official of the Ukrainian Commissariat of Education.' *Moya Zhizn*, vol. ii, p. 269.

[2] While he stayed at the Berlin Embassy, Trotsky spent many hours in discussions with Krestinsky, the Ambassador, and E. Varga, the Comintern's leading economist. The subject of his discussions with Varga was socialism in a single country. Varga admitted that as an economic theory Stalin's doctrine was worthless, that socialism in one country was moonshine, but that it was nevertheless politically useful as a slogan capable of inspiring the backward masses. Recording the discussion in his private papers, Trotsky remarked of Varga that he was 'the Polonius of the Comintern'. *The Archives*.

Polish Communists resulted in part from the tangled conditions in their country, but in part from confusion in the Comintern generated by the anti-Trotskyist campaigns: the Polish party enacted on a small scale the policy which at the same time led the Chinese Communists to back General Chiang Kai-shek and the Kuomintang. The British general strike confirmed the forecasts which Trotsky had made in *Where is Britain Going?*;[1] and at once it subjected the Comintern to new strains. The British leaders of the Anglo-Soviet Council did their best to wind up the strike before it became a revolutionary explosion; and anxious to save their own respectability, they refused to accept the aid which the Soviet trade unions offered the strikers. The Anglo-Soviet Council was thus rendered ridiculous. The British trade-union leaders still derived some advantage, however, from its existence: at the critical stage of the general strike the Communists, anxious not to embarrass the Council, were extremely reticent in criticizing their conduct. Trotsky even before he was back in Moscow assailed in *Pravda* the policy of the Anglo-Soviet Council on which Stalin and Bukharin had placed great hopes.[2]

It was only after Trotsky's return that he and the two ex-triumvirs set out in earnest to unite their factions. This was not easy. For one thing, the Trotskyist faction had been dispersed and had to be reassembled. Its strength turned out to be far less than it had been in 1923. For another, the followers of the two factions were not at all eager to unite. Their old animosities had not yet evaporated. They still distrusted each other. Among Trotsky's associates some favoured the coalition; but others, Antonov-Ovseenko and Radek, would have allied themselves with Stalin rather than with Zinoviev. Still others wished a plague on both their houses: 'Stalin will betray us', Mrachkovsky said, 'and Zinoviev will sneak away'. The rank-and-file Trotskyists in Leningrad at first refused even to disclose themselves to the Zinovievists, at whose hands they had suffered

---

[1] In his autobiography Trotsky says that the confirmation came earlier than he had expected it. *Moya Zhizn*, vol. ii, p. 272.

[2] *Pravda*, 26 May 1926. Meantime Stalin eliminated Zinoviev's followers from the Executive of the Comintern. At a session in May the Executive voted for the demotion of Fischer and Maslov, Treint, Domski, and other Zinovievist leaders of the German, French, and Polish parties.

persecution and from whom they had been accustomed to conceal their comings and goings almost as much as they had once
concealed them from the Tsarist *Okhrana*. What will happen,
they asked, if the Zinovievists change their minds and make
peace with Stalin? We shall then have delivered ourselves into
the hands of our persecutors. Trotsky had to send Preobrazhensky to Leningrad to allay these fears and persuade his re-
calcitrant followers to accept the coalition. The Zinovievists
were not less bewildered. When the news of the proposed coalition first reached Leningrad, they rushed to Moscow to remon-
strate with their leaders for their 'surrender to Trotskyism'.
Zinoviev and Lashevich had to explain that Trotskyism was a
bogy which they themselves had invented; and that they had
no use for it any longer. The admission could not but shock the
unfortunate Leningraders who had taken Zinoviev's accusations against Trotsky seriously and had repeated them after
him. But even when the mutual aversions were overcome or
subdued and the two factions began to merge, the members of
both still felt that they were entering a misalliance.[1]

Among the chiefs, too, the first elation had cooled off.
Zinoviev and Kamenev began to look over their shoulders.
They had no intention of pushing their differences with the
ruling factions to the point of an irreparable breach. They felt
uneasy at the charge that they 'surrendered to Trotskyism'.
Having admitted that they had wronged Trotsky, they still had
their record to defend; they were anxious to save for themselves
the half-spurious glory of 'pure Leninism' in which they had
walked. And so when on his return, surveying the events of the
last weeks, Trotsky began to argue that the Polish Communists
had supported Pilsudski's *coup* because the Comintern had instructed them to strive for that 'democratic dictatorship of the
workers and the peasants' which Lenin had advocated in 1905,
and not for proletarian dictatorship, Zinoviev and Kamenev
could not agree. That 'democratic dictatorship' was a taboo of
their 'old Bolshevism'; and although it was not very important
in the case of Poland,[2] it was to crop up again and again in the

---

[1] V. Serge, *Le Tournant obscur*, p. 102.

[2] Even Bukharin and Stalin disavowed the action of the Polish Communists.
See Deutscher, 'La Tragédie du Communisme Polonais' in *Les Temps Modernes*,
March 1958.

controversy over China next year. They were also taken aback by the bluntness with which Trotsky attacked the Anglo-Soviet Council, saying that it had never served any useful purpose and that it should be disbanded. Zinoviev was willing to criticize the Politbureau and the British Communists for 'hob-nobbing' with the leaders of the British trade unions; but he would not 'wreck' the Council which he had helped to sponsor. Above all, he was wary of alienating those men of the Old Guard who either backed Stalin with reservations or vacillated and urged moderation on all factions. Briefly, the two ex-triumvirs were willing to join hands with Trotsky; but they were already shrinking from an all-out attack on Stalin and Bukharin. Thus Trotsky had no sooner made the alliance with them than he had to patch up differences and make concessions. He promised Zinoviev and Kamenev to respect the taboo of the 'democratic dictatorship of workers and peasants' and to waive his demand for the disbandment of the Anglo-Soviet Council. This allowed him to establish a fairly wide measure of agreement with them on other questions.

The battle was joined, partly on Stalin's initiative, in the first days of June. Immediately after Trotsky's return, Stalin met him at the Politbureau with two fresh, incongruous yet damaging, accusations: Trotsky allegedly exhibited an impermissible 'hostility towards the British Communist Party'; and in domestic matters he gave proof of ill will and perverse defeatism when he declared that he was 'afraid of a good harvest'.[1] Trotsky refuted these charges as best he could. Then, on 6 June, he addressed a challenging letter to the Politbureau, saying that unless the party was reformed thoroughly and honestly, it would awaken one day to find itself under the undisguised rule of an autocrat.

Thus he resumed his open struggle with Stalin. He had not chosen the moment all by himself—the action and the plight of the Leningrad Opposition induced him to re-enter the fray at this time. In any case, the years of his waiting in silence or

---

[1] The first charge was based on a complaint from the British Communist party; the second on a statement in which Trotsky had said that the problem of relations between country and town would remain acute no matter whether there was a good or a bad harvest this year. If the harvest was bad, there would be a shortage of food; if it was good, the kulak would be stronger, more self-confident, and would have greater bargaining power. *The Archives.*

reticence were over. He knew that they had given him nothing: all the 'rotten compromises' with Stalin, against which Lenin had warned him, had been in vain. He was willing to compromise with Zinoviev and Kamenev in order to keep them aligned against Stalin; but he was also ready to fight it out without them. He had sized up his implacable enemy, and he knew that there was no retreat. He had lived these last years to fight another day. Now the day had come and the die was cast.

CHAPTER V

The Decisive Contest: 1926–7

THE Joint Opposition contended with the Stalinists and the Bukharinists for about eighteen months. During this time Trotsky was engaged in a political battle so intense that by comparison his earlier encounters with all the triumvirs were mere skirmishes. Tireless, unrelenting, straining every nerve, marshalling matchless powers of argument and persuasion, ranging over an exceptionally wide compass of ideas and policies, and at last supported by a large section, probably the majority, of the Old Guard which had hitherto spurned him, he made a prodigious effort to arouse the Bolshevik party and to influence the further course of the revolution. As a fighter he may appear to posterity not smaller in the years 1926–7 than he was in 1917—even greater. The strength of his mind was the same. The flame of revolutionary passion burned in him as fierce and bright as ever. And he gave proof of a force of character superior to that which he had needed and had shown in 1917. He was fighting adversaries in the camp of the revolution, not class enemies; and for such a struggle courage not merely greater but of a different kind was required. Some years later even his adversaries, when they related privately the incidents of this strife and described his mighty thrusts and his conduct under the blows, conveyed the image of a fallen Titan—rejoicing over his fall, they still recollected in awe the greatness they had struck down.[1]

Of course, the other leaders, too, brought to the contest strong passions, the resources of their uncommon intellects steeped in Marxism, tactical ingenuity, and an energy and determination which even in the weakest of them were still well above the average. The issues over which they struggled were among the greatest and the gravest over which men had ever fought: the fate of 160 million people; and the destinies of communism in Europe and Asia.

---

[1] The reference is to accounts of the struggle given to the author by many party members in Moscow in 1931.

Yet this great contest took place in a frightful void. On either side only small groups were involved. The nation was mute. Nobody knew or could know what it thought; and even to guess how its sympathies were divided was difficult. The struggle was waged over matters of its life and death; but it was waged above its head. On the face of things, nothing that the nation felt or thought could affect the outcome—the mass of the people was deprived of all means of political expression. Yet not for a moment did the antagonists take their eyes off the workers and the peasants, for inarticulate though these were, in the last instance it was their attitude that decided the issue. To win, the ruling factions needed only the passivity of the masses, while the Opposition needed for its success their political awakening and activity. Consequently, the former had the easier task: it was much simpler to confuse the masses and to breed apathy in them than to make them see the issues at stake and arouse their spirit. Furthermore, the Opposition, in its attempts to appeal to the people, was from the beginning hampered by its own inhibitions. Considering itself a section of the ruling party and continuing to acknowledge the party's unique responsibility for the revolution, the Opposition could not with a clear conscience appeal against its adversaries to the working class, the bulk of which was outside the party. Yet, as the struggle went on, mounting in bitterness, the Opposition was driven to try and seek support precisely among that mass of workers. It then came to feel the full burden of the tame and torpid popular mood. Nobody suffered from this more severely than did Trotsky: he hurled all his thunder and lightning into the void.

Nor do all the disputed questions appear in historical retrospect as real as they were to the chief actors. Some of the major issues were to lose outline and to fade soon after the disputes were over; and together with them some of the divisions which had seemed deep and unbridgeable became blurred or vanished. With cold violence Stalin denounced Trotsky as the enemy of the peasant, while Trotsky arraigned Stalin as the friend of the kulak. The sound of these recriminations still filled the air when Stalin set out to annihilate the kulak. Similarly, Stalin warned the country against the 'super-industrialization' for which Trotsky allegedly stood; but then he himself embarked with breathless precipitancy upon the course of action he had just condemned as pernicious.

As the struggle proceeds a mist envelops also most of the characters. If following this narrative we keep in mind the ultimate fate that befell Zinoviev, Kamenev, Bukharin, Rykov, Tomsky, and many others, we are struck by the inconstancy and futility of their behaviour, even though we may discern their motives. Every one of these men is completely submerged in the business of the day or of the moment, and utterly incapable of looking beyond it and forestalling next day's evil. Not only Stalin and events drive them to their doom—they drive each other; and at various times they do so with an obsessive fury which distorts their characters and contorts their minds. The imposing figures of the leaders shrink and dwindle. They become helpless victims of circumstance. The giants turn into moths rushing blindly and chasing each other madly into the flame. Only two figures seem to confront each other in irreducible reality and fixed hostility to the end—Trotsky and Stalin.

.   .   .   .

In the summer of 1926 the Joint Opposition feverishly organized its adherents. It sent out its emissaries to party branches in Moscow and Leningrad to make contact there with members who had been known to hold critical views of official policy in order to form them into Opposition groups and induce them to speak with the Opposition's voice to their party cells. Anxious to spread the network of its groups, the Opposition sent out its emissaries to many provincial towns as well, furnishing them with instructions, papers, and 'theses', dealing with its attitudes.

Soon the comings and goings of the emissaries attracted the attention of the General Secretariat, which kept track of the movements of those suspected to be in sympathy with the Opposition. Trotskyists and Zinovievists were summoned to party headquarters to explain their doings. The party committees, whenever they learned of any gatherings of oppositionists, sent their representatives on the spot to disband the meetings as illegal. When this was of no avail, they dispatched squads of zealots and ruffians to break up the meetings. The Opposition was thus driven to organize more or less clandestinely. Its supporters met stealthily in the homes of humble workers in suburban tenement blocks. When the squads of disrupters traced them there too and dispersed them, they assembled in small groups at cemeteries, in woods on the outskirts, and so on; and they posted guards and sent out patrols to protect their meetings. The long hand of the General Secretariat reached to the remoteness of these odd assembly places as well. There was no lack of grotesque incidents. One day, for instance, the sleuths of the Moscow committee discovered a clandestine meeting in a wood outside the city. The meeting was presided over by a high official of the Executive of the Comintern, one of Zinoviev's lieutenants; and it was addressed by no less a person than Lashevich, the Deputy Commissar of War. Zinoviev, as President of the Communist International, used the facilities of his office for disseminating Opposition papers and contacting groups. The headquarters of the International became the hub, as it were, of the Opposition; and this fact, too, quickly attracted Stalin's attention.

Such were the circumstances under which the Opposition managed to recruit and organize several thousand regular adherents. The estimates of its actual membership, of which approximately one half were Trotskyists and the other Zinovievists, vary from 4,000 to 8,000.[1] The remnants of the Workers' Opposition, a few hundred men at the most, also declared their accession. The Joint Opposition was anxious to rally all who were willing to join, regardless of past differences; it aspired to become the great assemblage of all Bolshevik dissenters. It may therefore be held that it suffered a decisive initial defeat when it did not succeed in recruiting a larger following. Compared with the party's total membership, which amounted to about three-quarters of a million, a few thousand oppositionists formed a tiny minority.

The strength of the factions should not, however, be seen only in the light of these figures. The great majority of the party was a jelly-like mass; it consisted of meek and obedient members, without a mind and a will of their own. It was more than four years now since Lenin had declared that the party was virtually worthless as a policy-making body, and that only the Old Guard, that 'thin stratum' which counted no more than several thousand members, was the repository of Bolshevik

[1] The lower estimate comes from Stalinist sources, the higher from Trotskyist.

traditions and principles.[1] The result of the Opposition's recruiting drive ought to be judged in the light of this statement.
The Opposition drew its support not from the inert mass but
from the thoughtful, active, and energetic elements, mostly
from the Old Guard, and partly from young Communists.
Opportunists and careerists kept aloof. The sight of broken-up
meetings and the loud threats, to which Stalinist and Bukharinist
zealots treated the adherents of the Opposition, frightened away
the timid and the cautious. The few trimmers who in 1923
had still put their stakes on the wrong horse and had described
themselves as Trotskyists now had the chance to redeem themselves by joining the ruling factions. The several thousand
Trotskyists and Zinovievists were, like the professional revolutionaries of old, men and women who felt strongly about the
great issues and braved grave personal risks. Most of them had
been prominent among the Bolshevik cadres in the most critical
times and had had many political ties with the working class.
It is doubtful whether the core of the ruling factions was
stronger even numerically. For the time being the Bukharinists
seemed to be more popular than the Stalinists; yet, two years
hence they were to be defeated far more easily than the Joint
Opposition had been, although one of their leaders presided
over the Council of the People's Commissars, another over the
trade unions, and still another over the Communist International. As to the Stalinist faction, its strength lay not in its
size, but in its leader's complete mastery of the party machine.
This allowed him to draw on all the party's resources, to rig
elections, to manufacture majorities, to veil the sectional and
personal character of his policy—in a word, to identify his own
faction with the party. At the most, only about 20,000 people
were of their own choice, directly, and actively involved in the
momentous inner-party conflict.

The Joint Opposition officially proclaimed its existence at a
session of the Central Committee in the middle of July.[2] Shortly
after the opening of the session, Trotsky read a statement of

[1] See p. 20.
[2] This was a joint session of the Central Committee and the Central Contro
Commission; it lasted from 14 to 23 July. *The Trotsky Archives, KPSS v Rezolu-
tsyakh*, vol. ii, pp. 148–69. N. Popov, *Outline History of the CPSU*, vol. ii, pp. 274 ff.
L. Trotsky, *Moya Zhizn*, vol. ii, pp. 260–75. E. Yaroslavsky, *Aus der Geschichte der
Komm. Partei d. Sowjetunion*, vol. ii, pp. 394 ff.

policy in which he, Zinoviev, and Kamenev, expressing regret over their past quarrels, declared it as their common purpose to free the party from the tyranny of its 'apparatus' and to work for the restoration of inner-party democracy, The Opposition defined its attitude as that of the Bolshevik Left, defending the interests of the working class against the wealthy peasantry, the N.E.P. bourgeoisie, and the bureaucracy. The first of its desiderata was a demand for the up-grading of industrial wages. The government had decreed a wage stop, authorizing no increase in workers' earnings unless it was justified by a rise in productivity. Against this the Opposition held that the condition of the working class was so wretched—wages were still lower than before the revolution—that in order to achieve a rise in productivity it was necessary first to improve the workers' lot. They should be free to stake out claims through the trade unions and to bargain with the industrial administration, instead of being compelled to submit to dictates and of seeing the trade unions turned into the state's obedient tools. The Opposition also demanded a reform in taxation. The government drew its revenue increasingly from indirect taxes, the brunt of which was, as always, borne by the poor. This burden, the Opposition argued, ought to be lightened, and the N.E.P. bourgeoisie ought to be made to pay higher tax rates on profits.[1]

From a parallel viewpoint the Opposition approached rural affairs. There, too, it urged a reform of taxation, claiming that the single agricultural tax, which was then in force, benefited the rich. It urged that the great mass of *bednyaks*, 30–40 per cent. of all smallholders, be exempt from taxation, and that the rest of the peasantry pay a progressive tax which would fall heaviest on

---

[1] The Opposition regarded it as scandalous that the government should obtain a high proportion of revenue from the state monopoly of vodka and acquire thereby a vested interest in the drunkenness of the masses. What the government gained as producer of vodka it lost as industrial employer through the inefficiency of drunken workers and a high rate of accidents in industry. The government excused the vodka monopoly on the ground that it combated effectively the even more disastrous mass consumption of home-brewed alcohol. This was admittedly a difficult question. The Opposition proposed that the government should tentatively, as an experiment, suspend the vodka monopoly for a year or two. The majority rejected this proposal. Within the first week of the October Revolution, we remember, the Bolsheviks had to contend with the scourge of mass drunkenness which belonged to the heritage of Mother Russia. (*The Prophet Armed*, pp. 322–4.) Ten years later the scourge was still there; it was used by the rulers as a fiscal convenience and it kept the masses politically befuddled.

the kulaks. The Opposition further pressed for the collectivization of farming. It did not advocate forced or wholesale collectivization or the 'liquidation of the kulak as a class'. It
envisaged a long-term reform to be carried out gradually, with
the peasantry's consent, and to be furthered by the government's credit policy and the use of industrial resources. None of
the Opposition's proposals went farther than the demands for a
50 per cent. rise in tax rates for kulaks and for virtually compulsory grain loans which would allow the government to step
up exports and to proceed with the import of industrial machinery.
In the face of strong official denials the Opposition maintained
that the yield of new taxation and of the grain loans should
enable the government to increase industrial investment funds,
despite rises in wages and tax reliefs for poor peasants.

The Opposition's programme culminated in the demand for
more rapid industrialization. Once again Trotsky, this time
with Zinoviev's and Kamenev's support, charged the government with inability to think ahead and plan. So timid had official policy been and so resigned to the 'snail's pace' that as a
rule industrial development ran ahead of official anticipations.
In 1925 the iron and steel industries and transport reached the
targets which the Supreme Council of the National Economy
had not expected to be attained before 1930. How much more
impetus could a far-sighted and vigorous direction impart to
the economy! The fourteenth congress had declared itself in
favour of raising the targets and accelerating the tempo. But
these resolutions had had no practical effect: they were plainly
ignored by a routine-ridden bureaucracy. To break the inertia
nothing less would do than a comprehensive and specific plan
covering five or even eight years ahead. 'Give us a real Five
Year Plan' was the Opposition's watchword.

The more firmly the Opposition pressed for the development
of the socialist sector of the economy, the more categorically did
it reject socialism in a single country. This became the central
'ideological' issue. The Opposition repudiated the idea of a
nationally self-sufficient socialism as incompatible with Leninist
tradition and Marxist principle. It held that despite all the
delays in the spread of international revolution, the party had
no reason to view the future of the U.S.S.R. in isolation and to
dismiss beforehand the prospect of revolutionary developments

abroad. The building of socialism would in any case extend over many decades, and not just a few years—why then should it be assumed that the Soviet Union would all this time stand alone as a workers' state? This was what the Stalinists and Bukharinists assumed—otherwise they would not have insisted so stubbornly that the party must accept socialism in a single country as an article of faith.

Here then was the party's entire international orientation at stake. To assume beforehand that the Soviet Union would have to build socialism alone throughout was to abandon the prospect of international revolution; and to abandon it was to refuse to work for it, even to obstruct it. The Opposition maintained that by 'eliminating' international revolution from their theoretical conception, Stalin and Bukharin tended to eliminate it also from their practical policies. Already the Comintern's strategy was strongly coloured by Bukharin's views on the 'stabilization of capitalism'; and both Stalin and Bukharin, so Trotsky and Zinoviev pointed out, were steering European communism if not towards self-liquidation then at least towards an accommodation with the parties of the Second International and the reformist trade unions. This took the form of an 'opportunist' united front, in which the Communist parties followed the Social Democratic lead and adapted themselves to the reformist attitude. Of such tactics—the very negation of directives worked out at earlier congresses of the Communist International—the Anglo-Soviet Council was the outstanding example. It arose from a pact between the leaders of the trade unions in the two countries. At no point did it or could it bring Communists in contact with the reformist masses and enable them to influence the latter. At no point, therefore, did or could the pact further the class struggle in Britain. On the contrary, the Opposition argued, by cultivating friendship with the British trade-union leaders while these curbed industrial unrest and even broke a general strike, Soviet Communists contributed to the confusion of the British workers, who could not tell friend irom foe. Trotsky and, to a lesser extent, Zinoviev and Kamenev concentrated their attack on the Anglo-Soviet Council as the epitome of that tacit abandonment of revolutionary purpose which they saw as the premiss and the corollary of socialism in a single country.

The statement which Trotsky read at the July session of the Central Committee contained little that either he or his partners had not said before. But this was the first time that they brought together the criticisms and proposals in a comprehensive declaration of policy and confronted the ruling factions with a joint challenge. The reaction was vehement. The debates were heated; and the exacerbation was heightened by a grim incident. Dzerzhinsky, highly strung and ill, delivered a long and violent speech denouncing the leaders of the Opposition, especially Kamenev. For two hours his high-pitched scream pierced the ears of the audience. Then, leaving the rostrum, he suffered a heart attack, collapsed, and died in the lobby before the eyes of the Central Committee.

Straightway the Central Committee rejected the Opposition's demand for a review of the wage scales. The leaders of the majority maintained that goods were scarce and that rises in wages, if unrelated to productivity, would cause inflation and worsen rather than improve the workers' lot. The Central Committee refused to exempt poor farmers from taxation and to impose heavier taxes on others. It resisted the demand for accelerated industrialization. It, finally, reaffirmed its support for Stalin's and Bukharin's Comintern policy and in particular for the Anglo-Soviet Council. But over all these matters the ruling factions were embarrassed and on the defensive; and it was not on grounds of policy but on those of party discipline that Stalin counter-attacked.

Stalin charged the chiefs of the Opposition with forming a regular faction within the party and thus violating the Leninist ban, now more than five years old. He aimed his blow at the weaker, the Zinovievist, section of the Opposition. He impeached Zinoviev for abusing his position as President of the Communist International and furthering the Opposition's activity from his headquarters; he arraigned Lashevich and a group of lesser oppositionists for holding the 'clandestine' meeting in the woods outside Moscow; and, finally, he brought up the case of one Ossovsky, who had expressed the view that the Opposition should constitute itself into an independent political movement and engage Stalin's and Bukharin's party in open hostility *from without* rather than act as a loyal opposition *within*. Trotsky dissociated himself and the Opposition from this view; but he pointed out that if some members came to despair of the party and saw no hope of reforming it from within, the blame lay with the leaders who had done their utmost to block every attempt at reform. The Central Committee resolved to expel Ossovsky from the party, to dismiss Lashevich from the Central Committee and the Commissariat of War, and to deprive Zinoviev of his seat in the Politbureau.[1]

Thus at this first formal encounter the Joint Opposition met with a severe reverse. The expulsion from the party of one of its adherents, even though he was a little-known 'extremist', was a menacing warning. With Lashevich's demotion the Opposition was cut off from the Commissariat of War. The worst shock was, of course, Zinoviev's dismissal from the Politbureau. As Kamenev had since the fourteenth congress been only an alternate member, both ex-triumvirs had already lost voting rights on the Politbureau; and of the chiefs of the Opposition Trotsky alone held his seat. It was because of his role in the Politbureau that Zinoviev had presided over the Communist International; it was now unthinkable that he should go on presiding. That Stalin had dared to depose the man whom only a short time ago many had considered as the senior triumvir was a sign of his extraordinary strength and self-confidence. He carried out the act with lightning dispatch and observing punctiliously all the statutory niceties. The proposal for Zinoviev's demotion had been duly tabled before the Central Committee, which alone was entitled to appoint and dismiss Politbureau members; and a massive majority voted for it.

Already at this stage there was in theory nothing to prevent Stalin from depriving Trotsky too of his Politbureau seat. He was not quite sure, however, that he would obtain the same massive majority for further reprisals; and he realized that a show of moderation could only strengthen his hand. By tackling the Opposition piecemeal he prepared party opinion all the better for the final show-down. Meantime he had little to fear from the Opposition's declarations of principle and statements of policy or from its demonstrations of protest staged at the Central Committee or the Politbureau. Little of what the

---

[1] N. Popov, *Outline History of the CPSU*, vol. ii, pp. 279–92; E. Yaroslavsky, op. cit., part ii, chapter 10; *The Trotsky Archives*; Stalin, *Sochinenya*, vol. viii, pp. 176–203; *KPSS v Rezolutsyakh*, vol. ii, pp. 160–6.

chiefs of the Opposition said there percolated to the cells down below and still less transpired in the press. As long as this was so and the ruling coalition maintained its solidarity, the verbal battles in the Politbureau and the Central Committee led the Opposition nowhere.

Precisely because of this there was nothing left for the Opposition to do but to appeal at last to the rank and file against the Politbureau and the Central Committee. In the summer of 1926 Trotsky and Zinoviev instructed their adherents to bring their common views to the notice of all party members, to disseminate policy statements, tracts, and 'theses', and to speak up in the cells. The chiefs of the Opposition themselves went into the factories and workshops to address gatherings. Trotsky made surprise appearances at large meetings held in Moscow's motor-car factory and railway workshops. But the leaders of the Opposition were no more fortunate in their efforts to shape party opinion from below than they had been in their attempts to influence policy from above. The party machine was ahead of them. Everywhere its agents, zealots and hecklers, met them with derisive booings, smothered their arguments in an infernal noise, intimidated audiences, broke up meetings, and made it physically impossible for the speakers to obtain a hearing. For the first time in nearly thirty years, for the first time since he had begun his career as revolutionary orator, Trotsky found himself facing a crowd helplessly. Against the scornful uproar with which he was met and the obsessive hissings and hootings, his most cogent arguments, his genius for persuasion, and his powerful and sonorous voice were of no avail. The insults to which other speakers were subjected were even more brutal. It was clear that the Opposition's first concerted appeal to party opinion had met with failure.

Stalin presently boasted that it was the good honest Bolshevik rank and file that had administered the Opposition the well-deserved rebuff. The Opposition replied that he had incited against it the worst elements, Lumpenproletarians and hooligans, who would not allow the decent rank and file to become acquainted with the Opposition's views. Stalin had indeed had no scruple; and the uproar with which his agents met Trotsky, Zinoviev, and their friends could hardly be mistaken for the 'voice of the people'. This, however, did not fully account for the Opposition's humiliating experience. Gangs of rowdies could disrupt the large meetings because the majority was either in sympathy or at least indifferent. An interested and self-disciplined audience usually knows how to eject or silence noisy individuals who try to prevent it from listening and collecting its thoughts. Behind the hooligans with their catcalls there were silent crowds, tame or unimpressionable enough not to think it worthwhile to exert themselves and assure order. At bottom it was the apathy of the rank and file that worsted the Opposition.

Yet the claims which the Opposition had raised on behalf of the workers, such as the demand for a rise in wages, were calculated to destroy apathy. Why then did they fail to evoke a response? On wages the ruling factions made a show of yielding. In July they had categorically refused to consider the claim, declaring that an increase in wages would greatly harm the national economy. But, in September, seeing that their adversaries were about to appeal to the rank and file, Stalin and Bukharin forestalled them and promised a rise to benefit the lowest paid and the most discontented groups of workers. The excuse for the change of policy was that the economic situation had radically improved, although no such improvement did or could occur within two months. The Opposition thus scored a partial success, but saw itself robbed of a most effective argument. Stalin further confounded it when he began to appropriate Trotsky's ideas on industrial policy. He was not by any means ready as yet for all-out industrialization; but in framing his resolutions and statements he borrowed many formulas and even entire passages from Trotsky.

The outlines of the party's rural policy were similarly blurred. Stalin insisted that the differences between the ruling factions and the Opposition were over the treatment not of the kulak, but of the middle peasant. The outcry against the kulak at the fourteenth congress had had its effect. It had aroused in the cadres a sneaking suspicion of the neo-Populist school. Bukharin could no longer afford to speak in public about the need to appease the strong farmer. The climate of Bolshevik opinion had changed: the kulak was once again recognized as *the* enemy of socialism. Although the government was still wary of antagonizing him and refused to burden him with higher taxes, it was in no mood to make new concessions either. There was now no question of any neo-N.E.P. Not that matters had improved. Caught between conflicting pressures, official policy was fixed in immobility. It had the worst of both worlds: it could count neither on the advantages which appeasement of the kulak might have yielded, nor on those which rigorous social and fiscal measures might have produced. The Opposition still had a strong case. Stalin, however, succeeded in diverting attention from it: he accused Trotsky and Zinoviev of trying to push the party to a conflict with the many millions of the middle peasants, those muzhiks *par excellence* who were no exploiters, whose attachment to private property was therefore harmless, and whose good will was essential to the alliance between proletariat and peasantry.

The Opposition had in truth no quarrel with the middle peasants.[1] It did not ask the party to turn the fiscal screws on them—and the mass of *serednyaks*, barely self-sufficient on their smallholdings, could not contribute much anyhow to the solution of the nation's food problem. However, the charge that the Opposition was out for the *serednyak*'s blood damaged its cause. Once again, as in 1923 and 1924, hosts of propagandists depicted Trotsky as the peasantry's arch-enemy; and they added that Zinoviev and Kamenev had become infected with Trotsky's hostility towards the muzhik. In the party cells people could no longer make head or tail of the charges and counter-charges. They had been apprehensive of Bukharin's pleas for the strong farmer; and now they became at least as distrustful of Trotsky's and Zinoviev's intentions. The last thing the workers, most of whom had their roots in the country, could look forward to was a conflict with the peasantry. They wished for safety first. As this was what Stalin seemed to offer them, they were wary of sticking out their necks for the Opposition.

Stalin's strength lay in the appeal he made to the popular craving for peace, safety, and stability. Trotsky once again appeared to go against that craving and to offend it. The weariness of the masses and their fear of risky experiments formed a constant background to the struggle. On that weariness and fear Stalin played even more strongly when he sought to justify his

[1] The Opposition claimed, however, that the Stalinists and the Bukharinists often played down the strength of capitalist farming by classifying the kulak as a *serednyak*.

foreign policy. Once **again he** depicted Trotsky as the Quixote of communism, who might involve the party in the most perilous ventures.

> Trotsky's policy [he said, defending the Anglo-Soviet Council] is one of spectacular gestures . . . he takes as his starting point not real men, not real and living workers . . . but some ideal and airy creatures, revolutionaries from head to foot. . . . We saw him applying this policy for the first time during the negotiations at Brest Litovsk, when he refused to sign the peace between Russia and Germany and indulged instead in a spectacular gesture, assuming that this could arouse the workers of all countries against imperialism. . . . You know well, comrades, how dearly we paid for that. Into whose hands did Trotsky's spectacular gesture play? Into the hands of . . . all those who strove to strangle the still unconsolidated Soviet Republic. . . . No, comrades, we are not going to adopt this policy of spectacular gestures, we are not going to do it today any more than we did at the time of Brest . . . we do not want our party to become a plaything in the hands of our enemies.[1]

The juxtaposition of the Brest Litovsk Peace and the Anglo-Soviet Council was altogether incongruous: even a straight breach between Soviet and British trade-union leaders—and because of Zinoviev's objections the Opposition did not press for this—could not conceivably expose the Soviet Union to dangers remotely comparable with those it had had to face during the Brest Litovsk crisis. The charge sounded even more grotesque when Bukharin made it: in 1918 he led the war party which was defeated only when Trotsky, on whose vote the issue hung, cast his vote for peace.[2] But who knew and who remembered the details of that great drama? The memory of the Bolshevik party was short; all the easier was it to arouse it to fear of Trotsky's 'heroic gestures'.

This was also the mood in which the ordinary Bolshevik listened to the debates over socialism in a single country. It was extremely difficult for him to judge the issue on its merits. The controversy, in so far as it had not bogged down in distortions and sophistries, was between two schools of economists, one conceiving the 'building of socialism' within a nationally self-contained system and the other viewing it in the context of the

---

[1] Stalin, *Sochinenya*, vol. viii, pp. 190–1.
[2] See *The Prophet Armed*, chapter xi.

broadest international division of labour. Only the most educated party members could follow the argument on this level.
The rank and file could not grasp why Zinoviev and Kamenev
insisted that Russia's internal resources, although abundant
enough to allow much progress, would not be sufficient for the
establishment of fully fledged socialism. Still less could they
absorb Trotsky's reasoning which was rooted in deeper layers of
Marxist thought. He argued that although socialist revolution
might for a time be confined to the boundaries of a single state,
socialism could not be achieved within the framework of any
nation-state, not even one as vast as the Soviet Union or as the
United States. Marxism had always envisaged socialism in
terms of an international community, because it held that historically society tended towards integration on an ever larger
scale. In the transition from the feudal to the bourgeois order
Europe had overcome its medieval particularisms. The bourgeoisie had created the national market; and on its basis the
modern nation-state had taken shape. But the productive forces
and economic energies of the advanced nations could not settle
within national boundaries; they had outgrown these even
under capitalism with its international division of labour, the
outstanding feat of progress achieved by the bourgeois West.[1]
Marx, who on this point was Smith's and Ricardo's faithful
disciple, had written in the *Communist Manifesto*:

> Modern industry has established the world market . . . [which]
> has given an immense development to commerce, navigation, and
> communication by land. . . . The need of a constantly expanding
> market for its products chases the bourgeoisie over the whole surface
> of the globe. . . . The bourgeoisie has given . . . a cosmopolitan
> character to production and consumption in every country. *To the
> great chagrin of reactionaries, the bourgeoisie has drawn from under the feet
> of industry the national ground on which it stood. . . . In place of the old local
> and national seclusion and self-sufficiency we now have the many-sided inter-
> course of nations and their universal interdependence.*[2]

How then, Trotsky asked, could one see socialism as standing
only on its national ground, in seclusion and self-sufficiency?
The high level of technology, efficiency, and abundance which

---

[1] In the 1930's Trotsky accordingly saw in the relapse of the bourgeois West into
economic nationalism (especially in the autarchy of the Third Reich) the surest
sign of its decay.   [2] My italics.

socialism presupposed, a level superior to that achieved by capitalism, could not be attained within a closed and backward economy. Socialism was even more dependent than capitalism had been on the 'many-sided intercourse of nations'. It must carry international division of labour incomparably farther than the bourgeoisie had ever dreamt of carrying it; and while the latter developed it only fitfully and undesignedly, socialism would plan it systematically and rationally. The concept of socialism in one country was therefore not merely unreal—it was reactionary as well: it ignored the logic of historic develop-ment and the structure of the modern world. Even more emphatically than before Trotsky advocated the idea of the United States of Europe as a preliminary to a socialist world community.

Whatever the merits or demerits of this reasoning, it was beyond the ken of the rank-and-file Bolshevik whose support the Opposition sought to enlist. Two years later, when he was already in exile, Radek, pondering over the reasons for the Opposition's defeat, wrote to Trotsky that they had approached their task as propagandists, dealing in great but abstract theories, not as political agitators seeking to arouse response for popular and practical ideas.[1] No doubt, Radek wrote this in a defeatist mood—he was presently to surrender to Stalin—and he did the Opposition less than justice. The practical ideas which the Opposition had advanced (its proposals about wages, taxation, industrial policy, proletarian democracy, and so forth) also failed to impress ordinary party members. There was, all the same, some truth in Radek's remark. The rank and file were weary, disillusioned, and prone to isolationism. Not for them were the sweeping historical prospects Trotsky unfolded. They craved, as Varga had put it, for a doctrine of consolation which would compensate them for the sacrifices they had made and were called upon to make. Socialism in one country was a feat of the myth-creation which was to mark the whole progress of Stalinism and which sought to conceal the gulf between Bolshevik promise and fulfilment. To Trotsky that myth-creation was a new opium for the people which the party should have refused to purvey.

[1] See Radek's memorandum 'Nado dodumat do kontsa', written in 1928 (no precise date), in *The Trotsky Archives*.

Our party [he wrote] in its heroic period looked forward unreservedly to international revolution, not to socialism in one country. Under this banner and with a programme which stated frankly
that backward Russia alone . . . could not achieve socialism, our
communist youth passed through the most strenuous years of civil
war, enduring hunger, cold, and epidemics, worked of its own
accord week-end shifts of hard labour (*subbotniki*), studied, and paid
for every step forward with numberless sacrifices. Party members
and *comsomoltsy* fought at the fronts and [on their rest days] volunteered to load logs of wood at the railway stations, not because they
hoped to build with these logs national socialism—they served the
cause of international revolution, for which it was essential that the
Soviet fortress should hold out; every log went to buttress that
fortress. . . . The times have changed . . . but the principle still
retains its full force. The worker, the poor peasant, the partisan, and
the young communist have shown by their entire conduct up to 1925
that they have no need of the new gospel. It is the official who looks
down on the masses, the petty administrator who does not wish to
be disturbed, and the hanger-on of the party machine . . . who need
it. It is they who think . . . that you cannot deal with the people
without a doctrine of consolation. . . . The worker who understands
that it is impossible to build a socialist paradise as an oasis amid the
inferno of world capitalism and who realizes that the fate of the
Soviet Republic and his own fate depend entirely on international
revolution—that worker will fulfill his duties towards the Soviet
Union much more energetically than the one who is told and believes
that we already have 'a 90 percent socialism'.[1]

Unfortunately for the Opposition and for Trotsky, the weary
and disillusioned mass, and not merely the 'petty official and
the hanger-on', responded to the doctrine of consolation more
readily than to the heroic evocation of permanent revolution.
They deluded themselves that Stalin offered them the safer, the
easier, the painless road.

Socialism in one country also stirred the people's national
pride, while Trotsky's pleas for internationalism suggested to the
simple-minded that he held that Russia could not rely on herself
and so he maintained that her salvation would ultimately have to
come from a revolutionized West. This could not but hurt the self-
confidence of a people that had achieved the greatest of revolutions—a self-confidence which, despite all the miseries of daily

---

[1] Trotsky, *The Third International After Lenin*, p. 67. The English translation has
been partly rephrased.

life, was real enough even though it was curiously blended with political apathy. Trotsky dwelt on the archaic outlook of Russia as a formidable obstacle to socialism. The Bolshevik-led masses sensed their backwardness; and the October Revolution had been their protest against it. But nations, classes, and parties, like individuals, cannot live indefinitely with an acute awareness of their own inferiority. Sooner or later they seek to suppress it. They begin to feel offended when they are reminded of it too frequently; and they are outraged when they suspect that someone is bent on reminding them of it. The apologists for socialism in one country made light of Russia's backwardness, explained it away, and even denied it.[1] They told the people that unaided they could achieve the consummation of socialism, the supreme miracle of history. It was not merely the easier and safer road that Stalin appeared to open up—it was the path of the chosen people of socialism, the path of Russia's peculiar revolutionary mission of which generations of Narodniks had dreamt. Indeed, two rival and quasi-Messianic beliefs seemed pitted against one another: Trotskyism with its faith in the revolutionary vocation of the proletariat of the West; and Stalinism with its glorification of Russia's socialist destiny. Since the impotence of Western communism had been repeatedly demonstrated, it was a foregone conclusion which of these beliefs would evoke the greater popular response.

However, for all his wishful belief in the proximity of revolution in the West, it was Trotsky rather than his adversaries who took as a rule the more sober view of current world affairs. His revolutionary idealism did not prevent him from approaching in a rigorously realistic manner specific situations either in the diplomatic field or in the communist movement. By its very nature, however, this side of his activity, his magisterial surveys and analyses of world events, could not make much impression on the rank and file, who grew or were made cynically aware of the aura of revolutionary romanticism which surrounded him.

The issues were further confused by the peculiar, scholastic style in which the controversies were conducted. For parallels

---

[1] This found a reflection even in Bolshevik history writing, especially in Pokrovsky's view of the evolution of capitalism and of the state in Russia. Pokrovsky was then the orthodox, Stalinist historian.

we should have to look to that medieval literature where theologians argued how many angels could sit on a pin-head or to the Talmudic disputes over which came first, the egg or the hen. When the ordinary Bolshevik heard Trotsky saying that the best way to advance socialism in Russia was to promote international revolution and Stalin replying that the best way to promote international revolution was to achieve socialism in Russia, the subtlety of the difference left him dizzy. Both sides argued from the canons of Leninist orthodoxy, canons which the triumvirs had first established in order to overwhelm Trotsky with them, and which they had succeeded in imposing upon him. Since then the orthodoxy had grown denser, harder, and more elaborate. Like so many orthodoxies, it served to exploit the moral authority of an inherited doctrine in the interest of the ruling group, to disguise the fact that that doctrine offered no clear answers to new problems, to reinterpret its tenets, to kill dissent or doubt, and to discipline the faithful. It was vain to search Lenin's writings for solutions to the problems of the day. A few years earlier most of the problems had not yet arisen or were only incipient; and even to the questions with which Lenin had dealt the most contradictory answers could be found for he had dealt with them in varying situations and contradictory circumstances. This did not prevent the party leaders from employing terms which with Lenin were political expressions as if they had been theological formulae. They quoted the lively epithets about his comrades which Lenin was wont to bandy about in controversy as if they had been Papal anathemas. The more independent-minded and capable of initiative any leading Bolshevik had been, the more such epithets about him could be culled from Lenin's writings or correspondence—only the trimmers and sycophants had nothing to fear from this kind of polemics. Lenin's shadow was thus conjured up to massacre his friends and disciples who now led the Opposition. The Opposition did its best to turn the shadow against the ruling factions. It alleged that it was its adversaries who were guilty of falsifying Lenin's teachings, while the Opposition strove to bring the party 'back to Leninism'.

It was true that on the central issue of the controversy—socialism in one country—the Opposition's claim to Leninist orthodoxy was extremely strong: Lenin had repeatedly spoken, as even Stalin and Bukharin had done up to 1924, about the impossibility of such a socialism.[1] If Stalin and Bukharin had been free to argue their case frankly, they might have said that in Lenin's lifetime the issue had not arisen in the form it had assumed now, that the isolation of the Russian Revolution had become much more evident since his death, that Lenin's pronouncements on this subject had therefore become irrelevant, and that they were entitled to advance their new doctrine without paying any regard to sacred texts. But Stalin and Bukharin were not free to argue thus. They, too, were swayed by the orthodoxy of their making. They could not afford to appear as the 'revisionists' of Leninism which they undoubtedly were. They had to present socialism in one country as a legitimate inference from Lenin's teachings, nay, as an idea developed by Lenin himself. Since the Leninist texts nevertheless bore strong witness in favour of the Opposition, Bukharin and Stalin had to divert the party's attention from them by turning the controversy into an endless and bizarre quibbling and hair-splitting which left the rank and file nonplussed, irritated, and finally bored to death. It is wellnigh impossible to convey in a historical narrative the obsessive repetitiveness and the unspeakable monotony of these scholastic performances. Yet the style of the controversy belongs to the very core of the events: its repetitiveness and monotony performed a definite function in the political drama. They killed in the average Bolshevik and worker every interest in the issues under debate. They gave them the feeling that those issues were of concern only to dogmatists dealing in abstruse questions, but not to ordinary people. This deprived the Opposition of its audience and enabled the ruling factions

[1] A detailed presentation and analysis of Lenin's attitude will be found in my *Life of Lenin*. Here a few brief quotations from Lenin will suffice: '. . . we put our stakes upon international revolution and were perfectly justified in doing this. . . . We have always emphasized that we look from an international viewpoint and *that in one country it is impossible to accomplish such a work as a socialist revolution.*' Lenin said this on the third anniversary of the October rising. Lenin, *Sochinenya*, vol. xxv, p. 474. (1928 edition—from later editions the italicized passage is omitted.) And again, after the final end of the civil war he declared: 'We have always and repeatedly told the workers that . . . the basic condition of our victory lies in the spread of the revolution at least to several of the more advanced countries.' At the sixth congress of the Soviets he said: 'The complete victory of the socialist revolution is unthinkable in one country, for it requires the most active co-operation of at least several advanced countries among which Russia cannot be classed. . . .' Lenin, *Sochinenya* (1950 ed.), vol. xxviii, p. 132.

to 'prove their doctrine orthodox by apostolic blows and knocks'.

The Opposition's call 'Back to Lenin!' fell similarly on deaf ears when the Opposition sought to remind the party of the freedom in which it had discussed and had managed its affairs in Lenin's time. Such reminders were double-edged, for although it was true that the Bolsheviks had enjoyed the fullest freedom of expression nearly to the end of the Lenin era, it was also true that at its end Lenin himself had severely curtailed that freedom by declaring the ban on factions and groups. Self-preservation, it might seem, should have induced the Opposition to denounce the ban as pernicious or at least obsolete and to demand its abolition. But the Opposition had by now become so entangled in the net of orthodoxy that it did not dare raise its voice against a ban which had behind it Lenin's authority. In 1924 Trotsky dissociated himself even from his friends when some of them tried to advocate freedom for inner-party groupings.[1] Two years later he still accepted the ban as valid, although he pointed out that it had been designed for a party enjoying freedom of expression and that in a muzzled party discontent and dissent tended of necessity to assume factional forms. Thus the Joint Opposition, having organized itself into a regular faction, did not have the courage to defend the act; and this half-heartedness made it doubly vulnerable. Only hypocrites, Stalin retorted, could call for a return to Lenin and flout that ban on factions and that monolithic discipline which were essential principles of Leninism. The Central Committee, he concluded, must not allow factional activity to go unpunished: there should be no room in Bolshevik ranks for those who rejected the Leninist conception of the party.

The rebuff the Opposition had received from the cells and the threat of expulsion which Stalin suspended over it caused disarray in its midst. Zinoviev and Kamenev, whose hopes had swelled with expectations of easy success, were crestfallen. Their sense of defeat was aggravated by remorse. They regretted that they had ever made the attempt to arouse the cells against the Central Committee. They were anxious to beat a retreat and to placate their adversaries. They were also uneasy over ideas that were becoming current on the ultra-radical fringe of the

[1] See p. 139.

Opposition, where many concluded that the party was completely under Stalin's and Bukharin's thumb, incapable of
absorbing any independent view, and hopelessly ossified; and
that the Opposition should learn the lesson of its defeat and at
last constitute itself as an independent party. This view, common-
ly held by those who had originally come from the Workers'
Opposition and the Decemists, began to spread also among
Trotskyists—according to Trotsky's testimony even Radek was
inclined to accept it.[1] The advocates of the 'new party' sought
to justify their attitude on broader grounds: they argued that
the old party was already in its 'post-Thermidorian' phase, that
it had 'betrayed the revolution', that it no longer spoke for the
working class, and that it had become the champion of the
bureaucracy, the kulaks, and the N.E.P. bourgeoisie. Some
held that the Soviet republic was not a workers' state any more
because its bureaucracy was a new ruling and exploiting class,
which had disinherited the toilers and appropriated the fruits
of the revolution as the French bourgeoisie had done in 1794
and after. The Opposition must therefore seek to overthrow the
bureaucracy just as Babeuf and his Conspiracy of Equals had
sought to overthrow the post-Thermidorian bourgeoisie.

Neither Zinoviev and Kamenev nor Trotsky agreed with
this. The 'Soviet Thermidor' was to them a danger to be
averted, not an accomplished fact. The revolution, they held,
had not yet come to a close. The bureaucracy was not a new
ruling or possessing class, nor an independent social force, but
merely a parasitic growth on the workers' state. Socially and
politically heterogeneous, torn between socialism and property,
the bureaucracy might eventually yield to the N.E.P. bourgeoisie
and the capitalist farmers and in alliance with these destroy
social ownership and restore capitalism. As long, however, as
this had not happened, the basic conquests of the October
Revolution were intact, the Soviet Union remained essentially
a workers' state, and the old party was still in its own way the
guardian of the revolution. Consequently, the Opposition must
not sever its links with it, but must continue to regard itself as
belonging to the party and defend with the utmost loyalty and
determination the Bolshevik monopoly of power.

From this it followed that the Opposition must not seek to

[1] Trotsky, *Écrits*, vol. i, pp. 160–3.

recruit support outside the party. Yet it was not allowed to recruit it inside either. This was an insoluble dilemma. What was immediately clear was that in order to save for itself the chance of acting further within the party, especially after Stalin had thrown out hints about expulsion, the Opposition had to yield ground. Over this Trotskyists and Zinovievists did not see eye to eye. Zinoviev and Kamenev set loyalty to the old party above all else. They wondered how they could continue the struggle while Stalin was in complete mastery of the party machine. They desired a truce. They were willing to declare that henceforth they would respect the ban on factions. They were ready to disband the organized groups they had set up, ready, that is, to demobilize the Opposition as a faction. They were anxious to separate themselves from the adherents of a 'new party'. They would have no truck with those who questioned the Bolshevik political monopoly. Indeed, they were prepared to let the main issues between themselves and Stalin and Bukharin fall into abeyance, at least for a time. Most of their supporters seemed equally anxious to beat a retreat. The Trotskyists were of a more militant spirit; and the radicals among them listened sympathetically to arguments in favour of a new party.

Amid these cross currents, Trotsky attempted to save the Opposition. To prevent Zinoviev and Kamenev from prostrating themselves before Stalin, he was prepared to go with them some way in yielding ground. They agreed that they would jointly declare their willingness to demobilize the Opposition as a faction and to dissociate themselves from the advocates of a new party; but that they would also reassert firmly the Opposition's principles and criticisms; and that they would go on opposing the ruling factions within the Central Committee and within other committees on which they sat.

On 4 October 1926 Trotsky and Zinoviev approached the Politbureau with the proposal for a truce. Stalin consented, waived the threat of expulsion, but dictated the terms. Only after much haggling did the factions agree on the statement the Opposition was to make. Without retracting any of its criticisms, indeed, after having clearly restated them, the Opposition declared that it considered the decisions of the Central Committee as binding on itself, that it ceased all factional

activity, and that it dissociated itself from Shlyapnikov and
Medvedev, the former leaders of the Workers' Opposition, and
from all those who stood for a 'new party'. On Stalin's insistence, Trotsky and Zinoviev further disavowed those foreign
groups and individuals who had declared their solidarity with
the Russian Opposition and had been expelled from their own
Communist parties.[1]

The Opposition accepted these terms with a heavy heart.
It knew that they were little short of surrender. Although it had
reaffirmed its criticisms and saved its face, the Opposition was left
without prospects and hope. Trotsky and Zinoviev had in effect
resigned their right to appeal once again to the rank and file.
They had undertaken to voice their views only within the party's
leading bodies, knowing beforehand that there they would be
regularly outvoted and that their views would have little or no
chance of reaching the rank and file. They had made the round
of a vicious circle. It was precisely because of their failure to
make any impression on the Central Committee that they had
tried to appeal to the cells; having failed to impress the cells
they were driven back into the Central Committee; and they
were trapped in it. They had weakened the Opposition by
dissociating it, for whatever reason, from Shlyapnikov's and
Medvedev's group, and by disavowing some of their own adherents abroad. By announcing the disbandment of their own
organization, they acknowledged implicitly that Stalin and
Bukharin had been justified in blaming them for forming it in
the first instance; and by declaring that they recognized the
ban on factions as valid and necessary, they blessed, as it were,
the whip with which Stalin chastised them.

Having taken on themselves all these onerous obligations and
demonstrated the Opposition's weakness, they failed to secure
the truce for which they had asked. On 16 October their statement appeared in *Pravda*. Only a week later, on 23 October, not
a trace was left of the truce. On that day the Central Committee
met to discuss an agenda for the forthcoming (fifteenth) party
conference. A more or less non-controversial agenda had already been prepared; but the Central Committee, no doubt on
Stalin's prompting, suddenly decided to add a special report

---

[1] Stalin, *Sochinenya*, vol. viii, pp. 209–13. They disavowed in particular Ruth
Fischer and Arkadi Maslov in Germany and Boris Souvarine in France.

on the Opposition, to be delivered by Stalin. This could not but reopen the wound. Trotsky protested and appealed to the majority to stand by the terms of the truce. The Central Committee nevertheless instructed Stalin to prepare his report.

Why did Stalin break the truce so soon after he had made it? He evidently wished to exploit his advantage and to rout the Opposition while it was in retreat. He was probably also provoked into fresh hostility by something that had happened two days after the truce had been announced. On 18 October the 'Trotskyist' Max Eastman published Lenin's last will in *The New York Times*—this was the first time that the full and authentic text saw the light. A year earlier he had published excerpts in *Since Lenin Died*; and Trotsky, we remember, disavowed him, and under the Politbureau's dictation denied the authenticity of the will. Stalin could not now seek to obtain another denial; but he must have suspected that Eastman had acted on Trotsky's direct or indirect inspiration. Such a suspicion was not groundless. Earlier in the year an emissary of the Opposition had indeed brought the text of Lenin's will to Paris and handed it to Souvarine who prompted Eastman to publish it. 'I think it was not only Souvarine's decision', Eastman writes, 'but the idea of the Opposition as a whole that I should be the one to publish it, one reason being that I had already got much publicity as a friend of Trotsky, another that a good many consciences in Moscow were troubled by Trotsky's disavowal of my book.'[1]

Eastman's surmise is undoubtedly correct. Among the 'troubled consciences in Moscow' none was more troubled than Trotsky's. He had denied the authenticity of the will and disavowed Eastman during that interval when neither Trotsky nor his friends wished to be drawn back into the struggle and to incur reprisals over this issue. But once he was back in the fray, after he had formed the Joint Opposition, he had every motive for trying to retrace the false step. Zinoviev and Kamenev could not but concur. It was they who had, at the fourteenth congress, raised anew the demand for the publication of the will and had repeated it at every subsequent opportunity. They, like Trotsky, would have preferred Lenin's will to be published in *Pravda*. But as this was out of the question, they could hardly have any scruple in arranging for its being broadcast by an important

[1] Quoted from Eastman's letter to the author.

bourgeois paper abroad—Lenin's will was in no sense a state
secret or an 'anti-Soviet document'. Of course, they had to act
with discretion because formally they were making themselves
guilty of a breach of discipline. The copy of the document had
been sent abroad in the heyday of the Joint Opposition, when
it was hoped that the publication would assist Oppositions in
foreign Communist parties and would have favourable repercussions in the Soviet Union as well. However, by the time the
document was published, the situation had changed: the Opposition had already suffered discomfiture, had asked for the
truce, and had dissociated itself from adherents abroad. When,
on 23 October, the Central Committee met, the newspapers all
over the world were full of the sensational disclosure; and this
doubtlessly envenomed feelings in the Central Committee. The
majority decided to disregard the truce and to give the Opposition a dressing-down.

Two days later there was a stormy scene at the Politbureau.
Stalin had just submitted his 'theses' on the Opposition which
he was to present at the fifteenth conference. He assailed the
Opposition as a 'social-democratic deviation' and demanded
that its leaders should admit the errors of their views and recant.[1] Trotsky once again protested against the breach of the
truce, spoke of Stalin's faithlessness, warned the majority that
they were embarking upon a course of action which, whether
they wished it or not, must end in wholesale ostracism. In
words charged with anger, he spoke of the fratricidal strife that
would follow, the ultimate destruction of the party, and the
mortal danger this would spell to the revolution. Then, facing
Stalin and pointing to him, he exclaimed: 'The First Secretary
poses his candidature to the post of the grave-digger of the
revolution!' Stalin turned pale, rose, first contained himself
with difficulty, and then rushed out of the hall, slamming the
door. The meeting, at which many members of the Central
Committee happened to be present, broke up in a hubbub.
Next morning the Central Committee deprived Trotsky of his
seat in the Politbureau and announced that Zinoviev would no
longer represent the Soviet Communist party on the Executive
of the Comintern, thus deposing him actually, though not

---

[1] Stalin's 'theses' appeared in *Pravda* on 22 October, the opening day of the
conference. Stalin, *Sochinenya*, vol. viii, p. 233.

nominally, from the Presidency of the International. These events overshadowed the conference which opened on the same day.

The Opposition was thrown into utter perplexity. It had yielded so much ground and gained nothing. It had renounced co-thinkers and allies, admitted itself guilty of offending against the 1921 ban, called its organizations to disband—all in order to avoid an aggravation of the struggle. What it had achieved was to get itself involved in strife more bitter than ever and, after it had tied its own hands, to bring upon itself fresh blows. The discord in its own midst grew. Zinoviev and Kamenev reproached Trotsky with having needlessly insulted Stalin and exasperated the majority just at the moment when the Opposition was seeking to soothe tempers. Even some of the Trotskyists were horrified at the vehemence with which Trotsky had assailed Stalin. Trotsky's wife describes this scene:

Muralov, Ivan Smirnov and others came to our home in the Kremlin one afternoon and waited for Lev Davidovich to come back from a Politbureau meeting. Pyatakov was the first to return. He was very pale and shaken. He poured out a glass of water, gulped it down, and said: 'You know I have smelt gunpowder, but I have never seen anything like this! This was worse than anything! And why, why did Lev Davidovich say this? Stalin will never forgive him until the third and fourth generation!' Pyatakov was so upset that he was unable to relate clearly what had happened. When Lev Davidovich at last entered the dining-room, Pyatakov rushed at him asking: 'But why, why have you said this?' With a wave of his hand Lev Davidovich brushed the question aside. He was exhausted but calm. He had shouted at Stalin: 'Grave-digger of the revolution' . . . we understood that the breach was irreparable.[1]

The scene gives a foretaste of subsequent events: a year later Pyatakov was, together with Zinoviev and Kamenev, to desert the Opposition. Even now, so Sedova affirms, he was convinced that 'a long period of reaction had opened' within Russia and without, that the working class was politically exhausted, that the party was stifled, and that the Opposition had lost. He still

[1] Quoted from Serge, *Vie et mort de Trotsky*, pp. 180–1, of which considerable fragments were written by Sedova. She describes the incident as having taken place late in 1927; but she confuses the dates. At the fifteenth conference, in October 1926, Bukharin already referred to the incident ; and he quoted Trotsky's words about the 'grave-digger of the revolution'. *15 Konferentsya VKP (b)*, p. 578.

held out against Stalin but he did so from a sense of dignity and solidarity with his comrades rather than from conviction.

With such despondency taking hold of some of them, the leaders of the Opposition decided to make another attempt to retrieve the truce: they were to refrain from attacking the ruling factions at the conference, and to speak up only in self-defence. For seven of nine days the conference lasted they did not utter a single word in reply to adversaries, who exulted throughout in their defeat, mocked at them, and tried to draw them into debate. Finally, on the seventh day, Stalin delivered a full-blast attack which lasted many hours. He gave his version of the struggle, recalling all Zinoviev had said against Trotsky as arch-enemy of Leninism and all Trotsky's strictures on Zinoviev and Kamenev, 'the strike-breakers of October', and so ridiculing the 'mutual amnesty' they had granted each other. He described with glee the Opposition's discomfiture, and said that only this had led it to sue for a truce in order to gain time and postpone its own demise. But the party must give the Opposition no respite: 'it must wage . . . a resolute struggle against the Opposition's false views . . . no matter in what "revolutionary" phraseology these may be couched', until the Opposition renounced them. He raked interminably Trotsky's life story to prove for the nth time Trotsky's inveterate antagonism to Lenin's ideas and to taunt Zinoviev and Kamenev for their 'surrender to Trotskyism'. Finally, he denounced the Opposition for inciting the party against the peasantry and urging excessive industrialization which 'would condemn millions of workers and peasants to misery', and would therefore be no better than the capitalist method of industrialization. He and his associates, so the future author of forcible industrialization and collectivization declared, favoured only such forms of economic development as would contribute immediately to the people's well-being and spare the country social convulsions; and in the name of this he called the conference to give the Opposition a 'unanimous rebuff'.[1]

When the leaders of the Opposition at last came forward, delegates noticed the very different tones in which they answered Stalin. Kamenev, who spoke first, gave a thoughtful but rather timid exposition of his views, trying in vain to blunt the edges

[1] Stalin, *Sochinenya*, vol. viii, pp. 421–63.

of the controversy. He complained about Stalin's disloyalty in launching the ferocious attack less than a fortnight after the truce. He tried to exonerate himself and Zinoviev from the charge that they had 'surrendered to Trotskyism'. They had united with Trotsky, he said, only for a definite and limited purpose as Lenin had often done. He recalled once again Lenin's testament and Lenin's fear of a split in the party; but this brought forth a howl from the floor. Then he broke into these words, part warning and part self-consolation: 'You may accuse us, comrades, of what you like, but we do not live in the middle ages! Witch trials cannot be staged now! You cannot accuse [us] . . . who call for higher taxes on the kulak and wish to help the poor peasant and together with him to build socialism—you cannot charge [us] with wishing to rob the peasantry. You cannot burn us at the stake.'[1] Exactly ten years later Kamenev was to sit in the dock at a witch trial.

Then Trotsky rose to make one of his greatest speeches, moderate in tone, yet devastating in content, masterly in logical and artistic composition, gleaming with humour—yet revealing once again the main source of his immediate weakness: his unshakeable reliance on European revolution. He spoke for the Opposition as a whole; but he also pleaded *pro domo sua*, throwing off, as if with one heave, the mountain of misrepresentation and obloquy with which he had been freshly covered at the conference. He had been accused of panic-mongering, pessimism, defeatism, and 'social-democratic deviationism'. Yet he had argued only from facts and figures; and 'arithmetic knows neither pessimism nor optimism'. To speak of the shortage of industrial goods was panic-mongering; but was there no ground for concern in the fact that in the current year industry had underproduced by 25 per cent.? Stalin had dubbed him a defeatist and made much play of his 'fear of a good harvest' because he had argued that as long as the nation suffered from a deficit of industrial goods, tension between town and country would persist, no matter whether the harvest was good or bad. Unfortunately, the last harvest was worse than they had all expected. The social differentiation of the peasantry was growing apace. None of these difficulties were as yet disastrous; but the omens had to be noticed in time. The Opposition had asked

[1] *15 Konferentsya VKP (b)*, p. 486.

that the well-to-do pay higher taxes and that the poor be granted reliefs. This demand may have been justified or not; but 'what is there in it that is social-democratic?' The Opposition was against a credit policy which favoured the kulak—was this social-democratic? It favoured a modest rise in wages—was this social-democratic? It did not share Bukharin's view that capitalism had regained stability—was that social-democratic? Was the Opposition's criticism of the Anglo-Soviet Council perhaps 'social-democratic?'

He recalled his service in the Comintern, his intimate co-operation with Lenin, and especially the support he had given Lenin in the transition to N.E.P., the N.E.P. he allegedly wished to disrupt. He was charged with 'disbelief' in the building of socialism. Yet had he not written that 'the sum total of the advantages *vis-à-vis* capitalism which we possess gives us, if we use the advantages properly, the chance to raise the coefficient of industrial expansion in the next few years not only to twice but even three times the pre-war 6 per cent. per year and perhaps even higher'.[1] It was true that he did not believe in socialism in one country and that he had been the author of the theory of permanent revolution. However, permanent revolution had been dragged in artificially: he alone, not the Opposition, was responsible for that theory. As a sop to Zinoviev and Kamenev he added: 'and I myself consider this issue to have been deposited in the archives long, long ago.' But what had his critics to say? They held it against him that in 1906 he had forecast that after the revolution urban collectivism would inevitably clash with peasant individualism. Had they not lived to see that prediction come true? Had they not proclaimed N.E.P. precisely because of such a clash? Had not 'the middle peasants talked with the Soviet Government through naval guns' at Kronstadt and elsewhere in 1921? The critics held it against him that he had predicted a collision between revolutionary Russia and conservative Europe. Had they overslept the years of intervention? 'If we, comrades, are alive, this is because, after all, Europe has not remained what it was.'

---

[1] This was indeed the rate at which Soviet industry expanded later under the Five Year Plans. (Trotsky quoted here a passage from his booklet *Towards Socialism or Capitalism?* published in 1925.) In 1930 Stalin was to ask for an annual increase of 50 per cent.! See my *Stalin*, p. 321.

However, the fact that the revolution had survived did not guarantee it against a repetition of conflicts with the peasantry and the capitalist West; nor did it argue in favour of socialism in one country. Indeed, they would have to face new conflicts and face them in worse conditions if they were to advance only at a 'snail's pace' and turn their backs on international revolution. Bukharin had written that 'the controversy is over this: can we build socialism and complete the building if we leave aside international affairs . . ?'. 'If we leave aside international affairs', Trotsky retorted, 'we can; but the whole point is that we cannot leave them aside (laughter). You can go for a walk naked in the streets of Moscow in the month of January, if you leave aside the weather and the militia (laughter). But I am afraid, neither the weather nor the militia will leave you aside. . . . Since when has our revolution acquired this . . . self-sufficiency?'

Here Trotsky came to the 'core of the problem': What would happen in Europe while Russia would be building socialism? So far they had all agreed with Lenin in assuming that Russia would need 'a minimum of thirty to fifty years' to achieve socialism.[1] What would the world look like in the course of these years? If within that time revolution were to win in the West, the question over which they argued would lapse. The adherents of socialism in one country evidently presupposed that this would not happen. They must then be starting from one of the following three possible assumptions: Europe—this might be the first assumption—would economically and socially stagnate, its bourgeoisie and proletariat keeping each other in a precarious balance. But such a situation could hardly last forty or even twenty years. The next assumption might be that European capitalism was capable of a new ascendancy. In that case, 'if capitalism were to flourish and if its economy and culture were to be on the ascendant that would mean that we had come too early', that is that the Russian Revolution was doomed. '. . . an advancing capitalism will . . . have the appropriate military, technical, and other means to stifle and crush us. This dark prospect is, in my view, ruled out by the entire condition of the world economy.' In any case, one could not base on such an assumption the vista of socialism in Russia.

---

[1] Stalin denied that this had been Lenin's view (*Sochinenya*, vol. ix, p. 39), but he had little ground for the denial.

Finally, one might assume that in the course of thirty to fifty years European capitalism would decline but that the working class would prove incapable of overthrowing it. 'Can you imagine this?' Trotsky asked.

I am asking you why I should accept this assumption which is nothing but black and groundless pessimism about the European proletariat; and why should we at the same time cultivate an uncritical optimism about the building of socialism by the isolated forces of our country? In what sense is it my ... duty as a communist to assume that the European working class will not be able to take power in the course of forty or fifty years. ... I see no theoretical or political reason for thinking that we with our peasantry will find it easier to achieve socialism than the European proletariat will find it to seize power. ... Even to-day I believe that the victory of socialism in our country can be safeguarded only together with a victorious revolution of the European proletariat. This is not to say that what we are building is not socialism, or that we cannot or should not go ahead with it full steam. ... If we did not think that ours is a workers' state, even though it is bureaucratically deformed ...; if we did not think that we are building socialism; if we did not think that we have enough resources in our country to promote the socialist economy; if we were not convinced of our full and final victory, then, of course, there should be no place for us in the ranks of the Communist Party. ...

Then also the Opposition would have to build another party and seek to arouse the working class against the existing state. This, however, was not its purpose. But let them beware: Stalin's disloyal and unscrupulous methods, freshly exemplified by the manner in which he had turned the truce into a scrap of paper, might produce a real split in the party and lead to a struggle between two parties.[1]

The assembly listened to Trotsky in breathless suspense and respectful hostility, even though repeatedly he had to interrupt his speech at the most dramatic moments and beg to be allowed to go on; again and again the conference prolonged his speaking time. Restrained and persuasive, he showed no sign of vacillation or weakness. Larin, who took the platform immediately after Trotsky, thus expressed the mood of the majority:

[1] *15 Konferentsya VKP (b)*, pp. 505–35.

'This was one of the dramatic episodes of our revolution . . . the revolution is outgrowing some of its leaders.'[1]

It was in a very different mood that the delegates listened to Zinoviev as he made a plaintive apology and tried to ingratiate himself with them. They treated him with rough contempt and hatred, drove him from the platform, and did not allow him to speak even on the affairs of the Comintern, for which he had been responsible; and this despite the fact that they were about to vote on his 'withdrawal' from the Comintern's Executive.[2]

As one looks back upon these congresses and conferences and compares the tenor of their debates, one is struck by the venom and the violence with which the ruling factions treated the Opposition; and one senses almost palpably how, from assembly to assembly, the coarse brutality rises to an ever higher pitch and turns into a fury. An utterly grotesque effect is created by the fact that some of the most churlish and vindictive assaults on the Opposition and some of the most fulsome tributes to Stalin came from people who only a few years hence were to become disgusted with him, turn into his belated critics, and perish as his helpless victims. Among those who at this conference distinguished themselves by their zealotry were Gamarnik, the future chief Political Commissar of the Red Army, who was to be denounced as traitor and was to commit suicide on the eve of Tukhachevsky's trial; Syrtsov, Chubar, Uglanov, who

---

[1] Ibid., p. 535. Larin had stood on the extreme right wing of the Mensheviks up to 1914, joined the Bolsheviks in the summer of 1917, and was then in friendly relations with Trotsky. His attitude towards the 1923 Opposition was ambiguous; later he joined the Stalinists.

[2] This, according to the verbatim report, is the conclusion of Zinoviev's speech: 'Comrades, I would like to say a few words about the bloc [i.e. the Joint Opposition]. I would like to say (*interruptions: You have talked enough. . . . Enough! Noise.*) I would like to say a few words about the bloc and the Comintern . . . (*voices: enough, enough! You should have spoken about this earlier and not about other things!*) Now, this is not right. Would you say that the problem of socialism in one country [about which Zinoviev had spoken] is not an important one? Why then did Stalin speak about it for three hours . . .? (*Noise, protests.*) I am asking for ten to fifteen minutes, so that I may say something about the bloc and the problems of the Comintern. (*Noise, Voices: enough!*) You know, comrades, that the party is now deciding that I should cease to work in the Comintern. (*Exclamation from the floor: This has already been decided!*) Such a decision is absolutely inevitable in the present circumstances, but will it be fair on your part not to grant me five minutes to enable me to speak on Comintern problems? (*Noise. Shouts: Enough! The chairman rings the bell.*) I beg you, comrades, leave me another ten-fifteen minutes to cover these two points.' (*The chairman orders a vote; and an overwhelming majority is against prolonging Zinoviev's time by ten minutes.*) Ibid., p. 577.

were all to die as 'saboteurs and plotters'; and even Ossinsky, the former Decemist, who now professed his faith in socialism in one country, but was also to end as 'wrecker and enemy of the people'. None, however, excelled Bukharin. Only a few months earlier he still appeared to be in amicable intercourse with Trotsky. Now he stood by Stalin's side, as Zinoviev had stood there two years earlier, and assailed the Opposition with reckless virulence, exulting in its plight, bragging, threatening, inciting, sneering, and playing up to the worst elements in the party. The kindly scholar was as if transfigured suddenly, the thinker turned into a hooligan and the philosopher into a thug destitute of all scruple and foresight. He praised Stalin as the true friend of the peasant smallholder and the guardian of Leninism; and he challenged Trotsky to repeat before the conference what he had said at the Politbureau about Stalin 'the grave-digger of the revolution'.[1] He jeered at the restraint with which Trotsky had addressed the conference, a restraint due only to the fact that the party had 'seized the Opposition by the throat'. The Opposition, he said, appealed to them to avert the 'tragedy' that would result from a split. He, Bukharin, was only amused by the warning: 'Not more than three men will leave the party—this will be the whole split!', he exclaimed amid great laughter. 'This will be a farce not a tragedy.' He thus scoffed at Kamenev's apology:

When Kamenev comes here and . . . says: 'I, Kamenev, have joined hands with Trotsky as Lenin used to join hands with him and lean on him', one can only reply with homeric laughter: what sort of a Lenin have they discovered! We see very well that Kamenev and Zinoviev are leaning on Trotsky in a very odd manner. (*Prolonged laughter and applause.*) They 'lean' on him in such a way that he has saddled them completely (*giggling and applause*), and then Kamenev . . . squeals: 'I am leaning on Trotsky'. (*Mirth*) Yes, altogether like Lenin! (*Laughter*).

Within barely two years Bukharin would try to 'lean' on a broken and prostrate Kamenev and whisper in terror into his ears that Stalin was the new Genghiz Khan.[2] But now, self-assured and complacent, juggling and jingling with quotations from Lenin, he returned to the attack on permanent revolution, on Trotsky's 'heroic postures', hostility towards the muzhik, and

---

[1] *15 Konferentsya VKP* (*b*), pp. 578–601.   [2] See pp. 440–3.

'fiscal theory of building socialism'; and again and again he extolled the steadfastness, the reliability, and the caution of his own and of Stalin's policies which secured the alliance with the peasantry. When the Opposition 'screamed' about the strength of the kulak and the danger of peasant strikes and of famine in the towns, it was trying to frighten the people with bogies. The party should not forgive them this and the 'chatter about the Soviet Thermidor', unless they came with their heads bowed, to repent, confess, and beg: 'Forgive us our sins against the spirit and the letter and the very essence of Leninism!' Amid frantic applause he went on:

Say it, and say it honestly: Trotsky was wrong when he declared that ours was not a *fully* proletarian state! Why don't you have the plain courage to come out and say so? . . . Zinoviev has told us here how well Lenin treated oppositions. Lenin did not expel any opposition even when he was left with only two votes for himself in the Central Committee. . . . Yes, Lenin knew his job. Who would try and expel an opposition when he could muster two votes only? (*Laughter.*) But when you get all votes and you have only two against you and the two shriek about Thermidor, then you may well think about expulsion.

The conference was delighted with this display of cynicism and shook with merriment. From the floor Stalin shouted: 'Well done, Bukharin. Well done, well done. He does not argue with them, he slaughters them!'[1]

What accounted for Bukharin's strange, almost macabre performance? No doubt he was genuinely frightened of the policies advocated by the Opposition. He dreaded the collision with the peasantry which they might provoke; and he did not see that it was his and Stalin's policy that led to this. The Opposition, although far too weak to replace the ruling group, was strong enough to compel Stalin's faction to shift its ground. True, at this conference it looked as if the Bukharinists had gained the upper hand within the ruling coalition: Bukharin, Rykov, and Tomsky presented the three main reports on behalf of the Central Committee. However, even they had to reckon with the Opposition. Bukharin himself had now to tread cautiously in matters of rural policy—he could no longer appeal frankly to the strong farmer. He saw Stalin's faction increasingly

[1] *15 Konferentsya VKP (b),* p. 601.

sensitive to Trotsky's and Zinoviev's criticisms and inclined to
steal page after page from their book. Stalin was already yielding to the demands for more rapid industrialization; this showed
itself even in the resolutions voted by this conference. Bukharin
would have preferred the ruling coalition to stand fast and to
defeat its adversaries without having to borrow their ideas and
confuse the issues. He wondered how far the Opposition's pressure might not push the party. He 'shuddered from head to
foot' at the thought that it might drive it to a bloody conflict
with the peasantry. And so, he was, at this moment, even more
anxious than Stalin to free official policy from the Opposition's
indirect influence. He desperately clung to Stalin so as to keep
him from yielding further ground; and he countenanced and
instigated Stalin's violence and trickery hoping that the defeat
of the Opposition would assure peace in the country. No
sacrifice of tact, taste, and decency was for him too high a price
to pay for this.

The ferocity of the attacks sprang also from embarrassment
and perplexity. Stalin's faction shrank from the enormity of the
step it was to take two years hence. Its speakers, too, imputed
to Trotsky and Zinoviev that they were prompting the party
to embark upon the forcible collectivization of the peasantry.
Kaganovich, for instance, who was to play a very prominent
part in the destruction of private farming, exclaimed: 'Theirs is
the road of plundering the peasantry, a pernicious road, no
matter how much Trotsky and Zinoviev may protest against
this—such indeed are their slogans.'[1] Once more also the Opposition had run up against the wall of the single-party system.
When it asked for freedom within that system, it was charged
with jeopardizing the system itself: Bukharin and Stalin claimed
that it tended to constitute itself into another party. Molotov,
in his inarticulate manner, hit the mark: Opposition speakers,
protesting against suppression, had recalled that even during
the Brest Litovsk crisis Lenin had allowed the Left Communists
to publish their own paper attacking him without fear or
favour: and to this Molotov replied: 'But in 1918 . . . the Mensheviks and the Social Revolutionaries too had their newspapers.
Even the Cadets had theirs. Somehow the present position is
not at all like that.'[2] Once again: the Bolsheviks could not enjoy

---

[1] *15 Konferentsya VKP (b)*, p. 637.   [2] Ibid., p. 671.

the freedom they had denied to others. Kaganovich recalled the words Trotsky had spoken at the eleventh congress when he presented the case against the Workers' Opposition. It was inadmissible, Trotsky then said, for party members to speak about their comrades and leaders in terms of 'We' and 'They', for if they did, they would, whatever their intentions, oppose themselves to the party, seek to exploit its difficulties, and assist those who had raised the banner of Kronstadt. 'Why then', Kaganovich asked, 'did you, comrade Trotsky, have the right to say this to Medvedev and Shlyapnikov when they made a mistake (and these comrades had been old Bolsheviks), and why can we not tell you that you are going on the Kronstadt road? . . .'[1]

It was not only the ghosts of Kronstadt and of the Workers' Opposition that were pressed into the onslaught on Trotsky. Shlyapnikov and Medvedev joined it in their own persons. After the Opposition had, on Stalin's insistence, declared that it would have no truck with them, Stalin managed by threat and cajolery to persuade Shlyapnikov and Medvedev to admit the error of their ways, to repent, and to denounce the Opposition. With jubilation the Central Committee then broadcast their recantation and announced that it granted them a pardon. The two men had urged the Joint Opposition to give up allegiance to the single-party system, to form itself from a faction within the old party into a new party. But confronted with the threat of their own expulsion from the old party and incensed by the fact that the Joint Opposition had disavowed them, they surrendered to Stalin. Theirs was the first recantation Stalin succeeded in extorting—a precedent and an example for many others. Before the end of the conference, Stalin surprised the Opposition by yet another stroke: he announced that Krupskaya had severed her connexion with Trotsky and Zinoviev.[2] It was whispered in Moscow that Stalin had blackmailed her with hints of indiscretions about Lenin's private life—'I shall appoint', he allegedly said, 'someone else to be Lenin's widow'. It is more plausible that Krupskaya withdrew from the Opposition because she was horrified to see the party founded by her husband split and torn asunder. As she had been among Stalin's and Bukharin's most outspoken critics, her defection did the Opposition much harm.   ·

[1] Ibid., p. 638.   [2] Ibid., pp. 754-5.

Finally, Stalin played against Trotsky and Zinoviev the leaders of foreign Communist parties. On their behalf Klara Zetkin, the veteran German Communist, who had at the Fourth Comintern Congress, when Lenin was already ill, paid in the name of the whole International a great and solemn tribute to Trotsky, now dissociated herself from him and Zinoviev, charging them with provoking a crisis in the International and supplying grist to the mills of all enemies of communism. '. . . even the lustre which attaches to the names of the leaders of the Opposition', she declared with a show of dignity, 'is not enough to redeem them. . . . The merits of these comrades . . . are imperishable. They will not be forgotten. Their deeds have entered into the history of the revolution. I am not forgetting them. However, . . . there exists something greater than deeds and merits of individuals.'[1]

The Opposition was routed; and the conference sanctioned the expulsion of the three chiefs of the Opposition from the Politbureau, threatening them with further reprisals if they dared to reopen the controversy.

Thus the Joint Opposition reached a point similar to that at which the 1923 Opposition had arrived after its defeat. The formal verdicts having gone against it, it had to decide what to do next: whether to go on with the struggle and risk wholesale and final expulsion, or accept defeat, at least temporarily? Each of the two sets of the Opposition reacted in a different manner. The Zinovievists were inclined to lie low. This was no easy matter, because the official attacks on them went on unabated, despite the formal closing of the controversy. The newspapers, purporting merely to comment on the resolutions of the conference, filled pages with the most virulent polemics, and gave the attacked no chance to answer. Rank-and-file Oppositionists paid for the courage of their convictions: they lost jobs, were subjected to ostracism, and were treated as little better than outcasts. Zinoviev and Kamenev resigned themselves to the mildest forms of passive resistance. Anxious to protect their adherents, they advised them to keep their views to themselves and, if need be, even to deny their association with the Opposition. Such advice could not but discredit the Opposition and demoralize those to whom it was given; they began to desert and recant.

[1] *15 Konferentsya VKP* (b), pp. 698–707.

The Trotskyists, on the other hand, who had already gone through a similar trial, knew that they had nothing to gain from inaction and nothing to hope for from half measures. Trotsky himself reviewed the recent experience in diary notes written towards the end of November.[1] For himself he defined the Opposition's predicament with greater candour than he could afford to use in public or in the Central Committee. He acknowledged defeat. He attributed it not merely to Stalin's disloyalty and to bureaucratic intimidation, but to the lassitude and disillusionment of the masses, who had expected too much from the revolution, had found their hopes cruelly deceived, and reacted against the spirit and the idea of early Bolshevism. The young, finding themselves under a tutelage from the moment they entered politics, could not develop any critical faculties and political judgement. The ruling factions played on popular weariness and craving for security and frightened people with the bogy of permanent revolution. Speaking for the record, Trotsky usually dwelt on the antagonism between the ruling group and the rank and file. Off the record he admitted that the ideas and slogans of the ruling group met an emotional need in the rank and file, that this overlaid their antagonism, and that the Opposition was at variance with the popular temper.

What then was to be done? It is not the business of the Marxist revolutionary, Trotsky reflected, to bow to the reactionary mood of the masses. At times when their class consciousness is dimmed, he must be prepared to become isolated from them. The isolation need not last long, for the time was one of transition and crisis; and, within the Soviet Union and without, the forces of revolution might yet surge up. This was, in any case, no time for the Opposition to flag or waver, even if the odds were against it. The revolutionary has to fight no matter whether he is destined to end as Lenin did—to live and see his cause triumph—or to suffer Liebknecht's fate who served his cause through martyrdom. In his private notes and in talks with friends, Trotsky hinted at this alternative more than once; and although he did not give up hope that he might 'end like Lenin', he seemed already more and more resigned inwardly to 'Liebknecht's fate'.

[1] See his notes of 26 November 1926 in *The Archives*.

I did not believe in our victory [Victor Serge recollects] and at heart I was even sure that we would be defeated. When I was sent to Moscow with our group's messages for Lev Davidovich, I told him so. We talked in the spacious office of the Concessions Committee . . . he was suffering from a fit of malaria; his skin was yellow, his lips were almost livid. I told him that we were extremely weak, that we, in Leningrad, had not rallied more than a few hundred members, that our debates left the mass of workers cold. I felt that he knew all this better than I did. But he, as a leader, had to do his duty and we, as revolutionaries, had to do ours. If defeat was inevitable, what else could one do but meet it with courage . . .?[1]

.   .   .   .

The winter of 1926–7 passed in relative calm. The Opposition was debilitated by internal discord. Trotsky did his best to prevent his partnership with Zinoviev from dissolving; and as Zinoviev was in near panic, the Joint Opposition paid for its unity with irresolution. In December its leaders even protested to Stalin against attempts made in the party cells of Moscow to draw them into fresh debates.[2] In the same month the Executive of the Comintern discussed the situation in the Russian party; and willy-nilly the Opposition had to restate its attitude. Once again Trotsky had to defend his own record and, protesting against the 'biographical method' used in inner party controversy, he went through the story of his relations with Lenin in order to demonstrate to an audience whose mind was closed that the 'irreconcilable antagonism between Trotskyism and Leninism was but a myth'.[3] The Executive confirmed the expulsion from foreign Communist parties of Trotskyists and Zinovievists on the ground that they denied the proletarian character of the Soviet state. Trotsky declared that the Opposition would combat any of its supposed foreign supporters who

---

[1] V. Serge, *Le Tournant obscur*, p. 116.

[2] Trotsky's and Zinoviev's 'Letter' to Stalin and the Politbureau of 13 December 1926. *The Archives.*

[3] On this occasion Trotsky gave an illuminating account of his attitude towards Lenin up to 1917. He spoke of the 'inner resistance' with which he drew nearer and nearer to Lenin. All the more was his eventual acceptance of Leninism wholehearted and complete. He compared his case with that of Franz Mehring who had embraced Marxism only after he had struggled against it as a leading Liberal. In spite or rather because of this Mehring's conviction was unshakeable and in his old age he paid for it with his freedom and life, whereas Kautsky and Bernstein, and the other men of the 'old guard' of Marxism, deserted the banner. See *The Trotsky Archives* statement of 9 December. See also *The Stalin School of Falsification*, p. 85.

took such a view. Half resigned to Souvarine's expulsion, he stood up for Rosmer and Monatte, who had been his political friends since the First World War and had founded and led the French Communist party, from which they were now banished.[1] But apart from such minor political interventions, he spent the winter in reserve, editing volumes of his *Works* and 'carrying out a more thorough theoretical examination of many questions'.

The 'theoretical question' which, apart from the economic argument against socialism in one country, preoccupied him most strongly was the 'Soviet Thermidor'. In the ranks of the Opposition and among its sympathizers abroad there was a great deal of confusion over this. Some argued that the Russian Revolution had already passed into the Thermidorian phase. Those who held this view also spoke of the bureaucracy as of the new class which had destroyed the proletarian dictatorship and exploited and dominated the working class. Others, and Trotsky most of all, hotly contested this opinion. As often happens when an historical analogy becomes a political shibboleth, neither of the disputants had a clear view of the precedent to which they referred; and Trotsky was to revise repeatedly his own interpretation of it. At this stage he defined the 'Soviet Thermidor' as a decisive 'shift to the right' which might occur within the Bolshevik party against the background of general apathy and disillusionment with the revolution, and result in the destruction of Bolshevism and the restoration of capitalism. From this definition Trotsky concluded that it was at least premature to speak of a Soviet Thermidor, but that the Opposition was justified in raising an alarm. One element of a 'Thermidorian situation' had been in evidence all too strongly: the masses were weary and disillusioned. But the decisive 'shift to the right' leading to restoration had not occurred, although the 'Thermidorian forces' working towards it had gathered strength and momentum.

---

[1] *Inter alia* Trotsky intervened with the Politbureau when the latter planned to send Pyatakov on a trade mission to Canada. He pointed out that because of the presence of many Ukrainian émigrés in Canada such a mission might be dangerous for Pyatakov who had led the Bolsheviks in the Ukraine during the civil war. Pyatakov had just been refused admission to the U.S.A. as one 'who sentenced to death worthy citizens of Russia'. Trotsky's letter to Ordjonikidze, 21 February 1927. See *The Archives*.

There would be no compelling need to go here into this rather abstruse argument if it were not that the view which Trotsky now formulated determined in part his own behaviour and the Opposition's fate in subsequent years and that the controversy over it generated indescribable heat and passion in all factions. This was indeed one of the seemingly most irrational phenomena in the struggle. It was enough for an Oppositionist to utter the word 'Thermidor' at any party meeting, and at once tempers flared up and the audience fumed and raged, although many had only the faintest idea what it was about. It was enough that they knew that the Thermidorians had been the 'grave-diggers' of Jacobinism and that the Opposition charged the ruling group with being engaged in some deep plot against the revolution. This curious historical slogan enraged even educated Bukharinists and Stalinists, who knew that its meaning was far less simple. The Opposition argued that the men of the Thermidor had not been out to destroy Jacobinism and to put an end to the First Republic—they had done this unwittingly from weariness and confusion. In the same manner the Soviet Thermidorians, not knowing what they were doing, might do the same. The analogy preyed on the thoughts of many a Stalinist and Bukharinist and sapped their confidence. It brought to their minds the uncontrollable element in revolution, of which they were increasingly if dimly aware; it made them feel that they were or might become playthings in the hands of vast, hostile, and unmanageable social forces.

Uneasily many a Bolshevik felt that this might be true. To whatever faction he belonged, he was terrified by the ghosts which the Opposition had conjured up. This was a case of *le mort saisit le vif*. When the Bukharinist or the Stalinist disclaimed any affinity with the Thermidorian, he did so not with calm self-assurance, but with that resentment, born of inner uncertainty, with which Bukharin spoke at the fifteenth conference of the Opposition's 'unforgivable chatter about Thermidor'.[1] His fury against the Opposition helped him to smother his own fears. The Oppositionist saw the ghost stalking the streets of Moscow, hovering over the Kremlin, or standing among the Politbureau members at the top of the Lenin Mausoleum on days of national

[1] See p. 305.

celebrations and parades. The uncannily violent passions which the bookish historical reminiscence aroused sprang from the irrationality of the political climate in which the single-party system had grown up and developed. The Bolshevik felt alienated from his own work—the revolution. His own state and his own party towered high above him. They appeared to have a mind and a will of their own which bore little relation to his mind and his will and to which he had to bow. State and party appeared to him as blind forces, convulsive and unpredictable. When the Bolsheviks made of the Soviets 'organs of power' they were convinced, with Trotsky, that they had established 'the most lucid and transparent political system' the world had ever seen, a system under which rulers and ruled would be closer to one another than ever before and under which the mass of the people would be able to express and enforce its will as directly as never before. Yet nothing was less 'transparent' than the single-party system after a few years. Society as a whole had lost all transparency. No social class was free to express its will. The will of any class was therefore unknown. The rulers and the political theorists had to guess it, only to be more and more often taught by events that they had guessed wrongly. The social classes therefore appeared to act, and up to a point did indeed act, as elemental forces, unpredictably pressing on the party from all sides. Groups and individuals within the party seemed to be unknowingly pushed in the most unexpected directions. On all sides cleavages arose or reappeared between what men thought (of themselves and others), what they willed, and what they did—cleavages between the 'objective' and the 'subjective' aspects of political activity. Nothing was now more difficult to define than who was the foe and who was the friend of the revolution. Both the ruling group and the Opposition moved in the dark, fighting against real dangers and against apparitions, and chasing one another and one another's shadows. They ceased to see one another as they were and saw each other as mysterious social entities with hidden and sinister potentialities which had to be deciphered and rendered harmless. It was this alienation from society and from one another that prompted the ruling factions to declare that the Opposition worked as an agency of alien social elements and the Opposition to claim that behind the ruling men stood Thermidorian forces.

Which then were those forces? The wealthy peasants, the N.E.P. bourgeoisie, and sections of the bureaucracy, Trotsky replied—briefly, all those classes and groups which were interested in a bourgeois restoration. The working class remained attached to the 'conquests of October' and was implicitly hostile to the Thermidorians. As for the bureaucracy, Trotsky expected that in a critical situation it would split: one section would back the counter-revolution; another would defend the revolution. He saw the divisions within the party as an indirect reflection of that cleavage. The right wing stood closest to the Thermidorians; but it was not necessarily identical with them. Bukharin's defence of the men of property savoured of a Thermidorian aspiration; but it was not clear whether the Bukharinists were the actual Thermidorians or merely their unwitting auxiliaries, who would in danger rally to the revolution. The left, i.e. the Joint Opposition, alone, according to this view, represented within the party the proletarian class interest and the undiluted programme of socialism; it acted as the vanguard of the anti-Thermidorians. The centre, the Stalinist faction, had no programme; and although it controlled the party machine, it had no broad social backing. It balanced between right and left and spawned on the programmes of both. As long as the centre was in coalition with the right, it helped to pave the way for the Thermidorians. But it had nothing to gain from a Thermidor which would be its own undoing; and so, when faced with the menace of counter-revolution, the centre or a large section of it would rally to the left in order to oppose, under the left's leadership, the Soviet Thermidor.

There is no need to run ahead of our story and to point out to what extent events confirmed or falsified this view.[1] Here it will be enough to indicate one important practical conclusion which Trotsky drew from it. This was, briefly, that under no conditions must he and his associates enter into an alliance with Bukharin's faction against Stalin. In certain circumstances and under certain conditions, Trotsky urged, the Opposition must even be prepared to form a united front with Stalin against Bukharin. The conditions were those that applied in any united front: the Opposition must not give up its independence, its

---

[1] A further analysis of the problem is found in chapter VI and in *The Prophet Outcast*.

right of criticism, and its insistence on inner-party freedom. According to a well-known tactical formula, left and centre should march separately and strike jointly. True, for the time being the Opposition had no opportunity to apply this rule: the Stalinists and the Bukharinists shared power and maintained unity. But Trotsky had no doubt that they would fall out presently. His tactical rule was designed to drive a wedge between them and assist in bringing about a realignment which would allow the Opposition to take command of all 'anti-Thermidorians', including the Stalinists. In the next few years the whole conduct of the Opposition was to be governed by this principle: 'With Stalin against Bukharin?—Yes. With Bukharin against Stalin?—Never!'

When this tactical decision, for which Trotsky was mainly responsible, is viewed in the grim light of the end which befell all anti-Stalinist factions and groups, it cannot but appear as an act of suicidal folly. The Thermidorian spirit which Trotsky saw as if embodied in the ineffectual Bukharin appears to have been the figment of an imagination overfed with history. And as one ponders, with the full knowledge of the after-events, Trotsky's many anxious alarms about the 'danger from the right', i.e. from Bukharin's faction, and his evident underestimation of Stalin's power, one may marvel at the short-sightedness or blindness which in this instance characterized the man so often distinguished by prophetic foresight. However, a view taken only from the angle of the *dénouement* would be one-sided. Trotsky's decision has to be seen also against the background of the circumstances in which he took it. N.E.P. was at its height, the forces interested in a bourgeois restoration were still alive and active, and nobody dreamt as yet of the forcible suppression of N.E.P. capitalism and of the 'liquidation of the kulaks as a class'. Trotsky could not take for granted the outcome of the contest between the antagonistic forces of Soviet society. The phantom of Thermidor, as he saw it, was still half real. Eight or even ten years after 1917 the possibility of a restoration could not be ruled out. As Marxist and Bolshevik, he naturally felt it to be his prime duty to strain all strength and to mobilize all energies against it. This determined his inner-party tactics. If anything at all could still pave the way for restoration, then it was Bukharin's rather than Stalin's policy. Within this context Trotsky could not but conclude that the Opposition must lend conditional support to the latter against the former. Such a conclusion was in line with Marxist tradition which approved alliances between left and centre against the right, but considered any combination of left and right directed against the centre as unprincipled and inadmissible. Thus, seen in its contemporary setting and judged in Marxist terms, Trotsky's attitude had its logic. It was his misfortune that subsequent events were to transcend that logic and to show it up as the logic of the Opposition's self-effacement. It was Trotsky's tragedy indeed that in the very process by which he defended the revolution he also committed political suicide.

. . . .

In the spring of 1927 the inner-party struggle flared up again in connexion with an issue which had hitherto played almost no part in it, but which was to remain at its centre to the end, until the final expulsion and the dissolution of the Joint Opposition.

That issue was the Chinese Revolution.

It was about this time that the Chinese Revolution entered upon a grave crisis which had been prepared by developments dating back to the close of the Lenin era. The Bolsheviks had very early set their eyes on the anti-imperialist movements among the colonial and semi-colonial nations, believing that these movements constituted a major 'strategic reserve' for proletarian revolution in Europe. Both Lenin and Trotsky were convinced that Western capitalism would be decisively weakened if it were cut off from the colonial hinterland which supplied it with cheap labour, raw materials, and opportunities for exceptionally profitable investment. In 1920 the Comintern proclaimed the alliance of Western communism and the emancipatory movements of the East. But it did not go beyond the enunciation of the principle. It left open the forms of the alliance and the methods by which it was to be promoted. It acknowledged the struggles of the nations of Asia for independence as the historic equivalent of bourgeois revolutions in Europe; and it recognized the peasantry and, up to a point, even the bourgeoisie of those nations as allies of the working class. But the Leninist Comintern did not yet attempt to define clearly the relationship between the anti-imperialist movements and the struggle for socialism in Asia itself, or the attitude of the Chinese

THE DECISIVE CONTEST: 1926–7

and Indian Communist parties towards their own 'anti-imperialist' bourgeoisie.

It was too early to resolve these questions. The impact of the
October Revolution on the East was still too fresh. Its strength
and depth could not yet be gauged. In the most important
countries of Asia the Communist parties were only beginning
to constitute themselves; the working classes were numerically
weak and lacked political tradition; even bourgeois anti-im-
perialism was still in a formative period. Only in 1921 did the
Chinese Communist party, based on small propagandist circles,
hold its first congress. But no sooner had it done so and set out to
formulate its programme and shape its organization than Moscow
began to urge it to seek a *rapprochement* with the Kuomintang.
The Kuomintang basked in the moral authority of Sun Yat-sen
which was then at its height. Sun Yat-sen himself was eager for
an agreement with Russia which would strengthen his hands
against Western imperialism; and in his vague, 'classless', Populist socialism, he was prepared to co-operate with the Chinese
Communists as well, but only if they accepted his leadership
unreservedly and supported the Kuomintang. He signed a pact
of friendship with Lenin's government, but found it more
difficult to get the Chinese Communists to co-operate with him
on his terms.[1]

The Communists were led by Chen Tu-hsiu, one of the
intellectual pioneers of Marxism in Asia, its first great propagandist in China, and the most outstanding figure of the Chinese
Revolution up to the advent of Mao Tse-tung, to whom he was
inferior as tactician, practical leader, and organizer, but superior,
it seems, as thinker and theorist. Chen Tu-hsiu had been the
initiator of the great campaign against the privileges the Western
powers enjoyed in China: the campaign starting from Pekin
University, of which Chen Tu-hsiu was a professor, assumed
such power that under its pressure the Chinese government
refused to sign the Versailles Treaty which sanctioned the
privileges. It was largely under Chen Tu-hsiu's influence that

[1] The account given in these pages is based *inter alia* on Brandt, Schwartz,
Fairbank, *A Documentary History of Chinese Communism*; Mao Tse-tung, *Selected
Works*; M. N. Roy, *Revolution und Konterrevolution in China*; Chen Tu-hsiu, 'An Open
Letter to the Party' (*Militant*, 1929); Stalin, *Works*; Trotsky, *Problems of the Chinese
Revolution*; Isaacs, *The Tragedy of the Chinese Revolution*; Tang Leang-Li, *The Inner
History of the Chinese Revolution*; files of *Bolshevik*, *Inprekor*, and *Revolutsionnyi Vostok*.

the Marxist propagandist circles had developed which formed the Communist party. He remained the party's undisputed leader from the moment of its foundation till late in 1927, throughout the crucial phases of the revolution. From the beginning he viewed with apprehension the political advice his party received from Moscow. He admitted the need for Communists to co-operate with the Kuomintang, but was afraid of too close an alliance which would prevent communism from establishing its own identity; he preferred his party to stand on its own feet before it marched with the Kuomintang. Moscow, however, insistently urged him to drop his scruples; and he possessed none of the strength of character and none of the slyness of Mao Tse-tung, who in similar situations never raised objections to Moscow's advice, always pretended to accept it, and then ignored it and acted according to his own lights, without ever provoking a genuine breach with Moscow. Chen Tu-hsiu was straightforward, soft, and lacked self-confidence; and these qualities made him a tragic figure. At every stage he frankly stated his objections to Moscow's policy; but he did not stick to them. When overruled, he submitted to the Comintern's authority, and against his better knowledge carried out Moscow's policy.

As early as in 1922–3 two men who were later prominent in the Trotskyist Opposition, Yoffe and Maring-Sneevliet,[1] played a crucial part in associating the young Chinese Communist party with the Kuomintang and in preparing the ground for the policy which Stalin and Bukharin were to pursue. Yoffe, as Ambassador of Lenin's government, negotiated the pact of friendship with Sun Yat-sen. Eager to facilitate his task and, no doubt, going beyond his terms of reference, he assured Sun Yat-sen that the Bolsheviks were not interested in promoting Chinese communism and that they would use their influence to ensure that the Chinese Communists co-operated with the Kuomintang on Sun Yat-sen's terms. Maring attended, as delegate of the Communist International, the second congress of the Chinese Communist party in 1922. It was on his initiative that the party

[1] Maring-Sneevliet, a Dutch Marxist, had been closely associated with the beginnings of communism in Indonesia, and represented the Dutch party in Moscow. In later years, especially throughout the 1930's, he was Trotsky's ardent follower. During the Second World War he led a resistance group in occupied Holland and was executed by the Nazis.

established contact with the Kuomintang and began to discuss the conditions of adherence to it. But Sun Yat-sen's terms were stiff; and the negotiations broke down.

Later in the year Maring returned to China and told Chen Tu-hsiu and his comrades that the Communist International firmly instructed them to join the Kuomintang, regardless of terms. Chen Tu-hsiu was reluctant to act on this instruction, but when Maring invoked the principle of international communist discipline, he and his comrades submitted. Sun Yat-sen insisted, like Chiang Kai-shek later, that the Communist party must refrain from criticizing openly the Kuomintang's policy and must observe its discipline—otherwise he would expel the Communists from the Kuomintang and consider his alliance with Russia null and void. By the beginning of 1924 the Communist party had joined the Kuomintang. It did not at first take Sun Yat-sen's terms to heart: it maintained its independence; and it pursued distinctly communist policies, incurring the Kuomintang's displeasure.

Communist influence grew rapidly. When in 1925 the great 'movement of 30 May' spread over southern China, the Communists were in its vanguard, inspiring the boycott of Western concessions and concerns and leading the general strike of Canton, the greatest so far in China's history. As the momentum of the movement increased, the Kuomintang leaders became frightened, tried to curb it, and clashed with the Communists. The latter sensed the approach of civil war, were anxious to untie their hands in time, and made representations to Moscow. In October 1925 Chen Tu-hsiu proposed to prepare his party's exodus from the Kuomintang. The Executive of the Communist International, however, vetoed the plan and admonished the Chinese party to do its utmost to avoid civil war. Soviet military and diplomatic advisers, Borodin, Blucher, and others, worked at Chiang Kai-shek's headquarters, arming and training his troops. Neither Bukharin nor Stalin, who by now effectively directed Soviet policy, believed that Chinese communism had any chance of seizing power in the near future; and both were anxious to maintain the Soviet alliance with the Kuomintang. The growth of communist influence threatened to disrupt that alliance and so they were determined to keep the Chinese party in its place.

Moscow thus urged Chen Tu-hsiu and his Central Committee to refrain from class struggle against the 'patriotic' bourgeoisie, from revolutionary-agrarian movements, and from criticism of Sun Yat-senism, which had since Sun Yat-sen's death become canonized as the ideology of the Kuomintang. To justify their attitude in Marxist terms, Bukharin and Stalin evolved the theory that the revolution which had begun in China, being bourgeois in character, could not set itself socialist objectives; that the anti-imperialist bourgeoisie behind the Kuomintang was playing a revolutionary role; and that it was consequently the duty of the Communist party to maintain unity with it, and do nothing that could antagonize it. Seeking further to substantiate their policy on doctrinal grounds, they invoked the view Lenin had expounded in 1905 that in the 'bourgeois' Russian Revolution, directed against Tsardom, socialists must aim at a 'democratic dictatorship of workers and peasants', not at a proletarian dictatorship. This precedent had little or no relevance to the situation in China: in 1905 Lenin and his party did not seek an alliance with the Liberal bourgeoisie against Tsardom—on the contrary, Lenin preached untiringly that the bourgeois revolution could conquer in Russia only under the leadership of the working class, in irreconcilable hostility to the Liberal bourgeoisie; and even the Mensheviks, who did seek an alliance with the bourgeoisie, did not dream of accepting the leadership and discipline of an organization dominated by it. Bukharin's and Stalin's policy was, as Trotsky later pointed out, a parody not merely of the Bolshevik but even of the Menshevik attitude in 1905.

However, these doctrinal sophistries served a purpose: they embellished Moscow's policy ideologically and soothed the conscience of Communists who felt uneasy over it. The opportunism of that policy showed itself startlingly when, early in 1926, the Kuomintang was admitted to the Communist International as an associate party and the Executive of the International elected with a flourish General Chiang Kai-shek as honorary member. With this gesture Stalin and Bukharin demonstrated their 'good will' to the Kuomintang and browbeat the Chinese Communists. On 20 March, only a few weeks after the 'General Staff of World Revolution' had elected him its honorary member, Chiang Kai-shek carried out his first anti-communist coup. He barred communists from all posts at the headquarters of the Kuomintang, banned their criticisms of Sun Yat-sen's political philosophy, and demanded from their Central Committee that it should submit a list of all party members who had joined the Kuomintang. Pressed by Soviet advisers, Chen Tu-hsiu and his comrades agreed. But, convinced that Chiang Kai-shek was preparing civil war against them, they were anxious to organize communist-led armed forces to match, if need be, his military strength; and they asked for Soviet assistance. The Soviet representatives at Canton categorically vetoed this plan and refused all assistance. Once again Chen Tu-hsiu bowed to the Comintern's authority.[1] The newspapers of Moscow made no comment on Chiang's coup—they did not even report the event. The Politbureau, fearing complications, sent Bubnov, the ex-Decemist, to China to enforce its policy and persuade the Chinese communists that it was their revolutionary duty to 'do coolie service' to the Kuomintang.[2]

During all these events the Chinese issue remained as if outside the Russian inner party controversy. The fact deserves to be underlined: it disposes of one of the legends of vulgar Trotskyism which maintains that the Opposition had from the beginning unremittingly resisted Stalin's and Bukharin's 'betrayal of the Chinese Revolution'. No doubt, Trotsky himself had had his misgivings as early as at the beginning of 1924. He had then expressed at the Politbureau a critical view of the adherence of the Chinese Communists to the Kuomintang; and in the following two years he restated his view on a few occasions. But he did it almost casually. He did not dwell on the matter and did not go to its heart. When he found that at the Politbureau he stood alone—all other members backed the Chinese policy—he did not try to repeat his objections before the wider forum of the Central Committee. Not once, it seems, in these years, 1924-6, did he speak about China in the Executive or the commissions of the Comintern. Not once, at any rate, did he allude in public to any difference of opinion in this matter. He appears to have given to it far less attention and far less weight than he

[1] Chen Tu-hsiu relates that the Chinese Central Committee requested the Soviet military advisers in Canton to supply from the munitions which had arrived for Chiang Kai-shek at least 5,000 rifles to the communists to enable them to arm insurgent Kwantung peasants. The request was refused.
[2] Quoted from Chen Tu-hsiu's 'Open Letter'.

gave British or even Polish communist policies. He was evidently not clearly aware of the force of the tempest breaking over China and of the magnitude and gravity of the approaching crisis in communist policy.

Early in 1926 he was still concerned more closely with the conduct of Soviet diplomacy towards China than with the direction of communist affairs there. He presided over a special commission—Chicherin, Dzerzhinsky, and Voroshilov were its members—which was to prepare recommendations for the Politbureau on the line that Soviet diplomacy ought to pursue in China. Of the commission's work little is known, apart from its report which Trotsky submitted to the Politbureau on 25 March 1926.[1] As he did not dissociate himself from the report, it must be assumed that he was in basic agreement with it. The commission made its recommendations in strictly diplomatic terms, without reference to the objectives of the Chinese Communist party. While that party strove, in co-operation with the Kuomintang, to abolish the *status quo* in China, the commission offered instructions for the Soviet diplomatic services on the attitudes they should adopt within the *status quo*. Both the Communist party and the Kuomintang called for the political unification of the country, that is for the overthrow of Chang Tso-lin's government, whose writ ran in the north, and for the spread of revolution from south to north. Trotsky's commission reckoned with China's continued division; and its recommendations were as if calculated to prolong it. At this time Chiang Kai-shek was already preparing his great military expedition against the north. Amid the confusion which reigned across the Soviet Far Eastern frontier, Trotsky's commission sought not to promote revolution but to secure every possible advantage for the Soviet government. Thus the commission suggested that Soviet diplomatic agencies should seek a *modus vivendi* and a division of spheres between Chiang Kai-shek's government in the south and Chang Tso-lin's in the north.

Trotsky later maintained that at the Politbureau, during the discussion on the report, Stalin tabled an amendment that Soviet military advisers should dissuade Chiang Kai-shek from undertaking his expedition. The commission rejected the amendment, but in more general terms it advised Soviet agencies in China to

[1] *The Archives.*

'urge moderation' on Chiang Kai-shek. The Politbureau's main concern was with safeguarding Russia's position in Manchuria against Japanese encroachments. The commission therefore advised that Russian envoys in northern China should encourage Chang Tso-lin to pursue a policy of balancing between Russia and Japan. Moscow, too weak to eliminate Japanese influence from Manchuria and not believing in the Kuomintang's ability to do so, was ready to reconcile itself to Japan's predominance in southern Manchuria, provided that Russia, remaining in possession of the North Eastern Chinese Railway, maintained her hold on the northern part of the province. The commission urged Soviet envoys to prepare public opinion 'carefully and tactfully' for this arrangement, which was likely to hurt patriotic feelings in China. The Politbureau's motives were mixed and tangled. It was concerned over Manchuria. But it also feared that Chiang Kai-shek's expedition against the north might provoke the Western powers to intervene in China more energetically than hitherto. And it also suspected that Chiang was planning the expedition as a diversion from revolution, a means to absorb and disperse the revolutionary energies of the south.

In April the Politbureau accepted the report of Trotsky's commission. At this point, however, Trotsky raised the problem of the strictly communist policy in China. This, he held, should remain independent of Soviet diplomatic considerations: it was the diplomats' business to make deals with existing bourgeois governments—even with old-time war-lords; but it was the revolutionaries' job to overthrow them. He protested against the admission of the Kuomintang to the Comintern. Sun Yatsenism, he said, extolled the harmony of all classes; and so it was incompatible with Marxism committed to class struggle. In electing Chiang Kai-shek an honorary member, the Executive of the Comintern had played a bad joke. Finally, he repeated his old objections to the adherence of the Chinese Communists to the Kuomintang.[1] Once more, all members of the Politbureau, including Zinoviev and Kamenev, who were now on the point of forming the Joint Opposition, defended the official conduct of Chinese communist affairs. This exchange, too, was incidental. It occurred within the closed doors of the Politbureau; and it had no consequences.

[1] Stalin, *Sochinenya*, vol. x, pp. 154-5.

Then, for a whole year, from April 1926 till the end of March
1927, neither Trotsky nor the other leaders of the Opposition
took up the issue. (Only Radek, who since May 1925 had
headed the Sun Yat-sen University in Moscow and had to
expound party policy to perplexed Chinese students, 'pestered'
the Politbureau for guidance. This he failed to obtain and he
expressed mild misgivings.) Yet this was the most crucial and
critical year in the history of the Chinese Revolution. On 26
July, four months after the Politbureau had discussed the report
of Trotsky's commission, Chiang Kai-shek, ignoring Soviet
'counsels of moderation', issued his marching orders for the
northern expedition. His troops advanced rapidly. Against Moscow's expectation, their appearance in central China acted as a
tremendous stimulus to a nation-wide revolutionary movement.
The northern and central provinces were astir with risings
against Chang Tso-lin's administration and the corrupt war-
lords who supported it. The urban workers were the most active
element in the political movement. The Communist party was
in the ascendant. It led and inspired the risings. Its members
stood at the head of the trade unions, which had sprung into
being overnight and found enthusiastic mass support in liberated
cities and towns. All along the route of Chiang Kai-shek's advance the peasantry welcomed his troops and, counting on their
support, rose against war-lords, landlords, and usurers, ready to
dispossess them.

Chiang Kai-shek was frightened by the tide of revolution and
sought to contain it. He forbade strikes and demonstrations,
suppressed trade unions, and sent out punitive expeditions to
subdue the peasants and to requisition food. Intense hostility
developed between his headquarters and the Communist party.
Chen Tu-hsiu, reporting these events to Moscow, demanded
that his party should at last be authorized to make its exodus
from the Kuomintang. He was still for a united front of Communists and Kuomintang against the northern war-lords and
the agencies of the Western powers; but he held that it was
imperative for his party to shake off the Kuomintang's discipline, regain freedom of manœuvre, encourage the proletarian
movement in the towns, back the peasantry's struggle for land,
and get ready for open conflict with Chiang Kai-shek. A rebuff
was once again the answer Chen Tu-hsiu received from the

Executive of the International. Bukharin rejected his demand as dangerous 'ultra left' heresy. As the Central Committee's *rapporteur* at the party conference in October, Bukharin reasserted the need 'to maintain a single national revolutionary front' in China where 'the commercial industrial bourgeoisie was at present playing an objectively revolutionary role . . .'.[1] It might be difficult, he went on, for Communists to satisfy in these circumstances the peasantry's clamour for land. The Chinese party had to keep a balance between the interests of the peasantry and those of the anti-imperialist bourgeoisie which was opposed to agrarian upheaval. The Communists' overriding duty was to safeguard the unity of all anti-imperialist forces; and they must repudiate all attempts at disrupting the Kuomintang.[2] Patience and circumspection were the watchwords— all the more so as the revolutionary atmosphere was affecting the Kuomintang too, bringing about its 'radicalization', and 'reducing its right wing to impotence'.

Somewhat later Stalin also, speaking at the Comintern's Chinese commission, extolled Chiang Kai-shek's 'revolutionary armies', demanded from the Communists complete submission to the Kuomintang, and warned them against any attempts at setting up Soviets at the height of a 'bourgeois revolution'.[3]

On the face of it, Stalin's and Bukharin's predictions about a 'leftward shift in the Kuomintang' presently came true. In November the Kuomintang government was reconstructed into a broad coalition, in which leftish groupings led by Wang Ching-wei, Chiang's rival, came to the fore, and which included two Communist Ministers in charge of agriculture and labour. The new government moved from Canton to Wuhan. The Kuomintang right, however, was far from being 'reduced to impotence'. Chiang Kai-shek remained in supreme command of the armed forces and was busy setting the stage for his dictatorship. It was rather the Communists within the government who were reduced to impotence. The Minister of Agriculture exerted himself to stem the tide of agrarian revolt; and the Minister of Labour had to swallow Chiang Kai-shek's anti-labour

---

[1] *15 Konferentsya VKP (b)*, p. 27.
[2] Ibid., pp. 28-29.
[3] Stalin, *Sochinenya*, vol. viii, pp. 357-74.

decrees.[1] From Moscow ever new envoys arrived to calm the Communists: after Bubnov's departure, the eminent Indian communist leader M. N. Roy appeared with this mission at Wuhan towards the end of the year 1926.

The Politbureau was still preaching unity with the Kuomintang when in the spring of 1927 Chiang Kai-shek, still honorary member of the Executive of the Comintern, carried out another coup by which he initiated open counter-revolution. The scene was Shanghai, China's largest city and commercial centre, dominated by the extra-territorial enclaves of the Western powers and their warships anchored in the harbour. Shortly before Chiang Kai-shek's troops had entered, the workers of Shanghai rose, overthrew the old administration, and took control of the city. Once again the hapless Chen Tu-hsiu appealed to Comintern headquarters seeking to impress them with the significance of the event—the greatest proletarian rising Asia had seen—and to disentangle his party from its commitments to the Kuomintang. Once again he and his comrades were pressed to reaffirm their allegiance to the Kuomintang and also to yield control over Shanghai to Chiang Kai-shek. Bewildered but disciplined, rejecting assistance offered them by Chiang's own detachments, the Communists on the spot accepted these instructions, laid down arms, and surrendered. Then, on 12 April, only three weeks after their victorious rising, Chiang Kai-shek ordered a massacre in which tens of thousands of Communists and of workers who had followed them were slaughtered.

Thus the Chinese Communists were made to pay their tribute to the sacred egoism of the first workers' state, the egoism that the doctrine of socialism in one country had elevated to a principle. The hidden implications of the doctrine were brought out and written in blood on the pavements of Shanghai. Stalin and Bukharin considered themselves entitled to sacrifice the Chinese Revolution in what they believed to be the interest of the consolidation of the Soviet Union. They sought desperately to avoid any course of action which might turn the capitalist powers against the Soviet Union and disturb its hard-won and precarious peace and equilibrium. They conceived their Chinese

[1] M. N. Roy, *Revolution und Konterrevolution in China*, pp. 413 ff. Harold Isaacs, *The Tragedy of the Chinese Revolution*, chapters 14 and 15.

policy in the same mood in which they shaped their present domestic policies, believing that it was wisdom's first commandment to keep to the safe side and to proceed cautiously, step by step, in the conduct of all affairs of state. The same logic which had induced them to placate the 'strong farmer' at home led them to woo so excessively the Kuomintang. They had indeed expected the Chinese Revolution to develop at the snail's pace at which Bukharin thought that socialism could progress in Russia.

As so often in history, this kind of weary and seemingly practical realism was but a pipe-dream. It was impossible to ride the dragons of revolution and counter-revolution at the snail's pace. But the Bolsheviks had for years exerted themselves to gain a breathing space for the Soviet Union. Having gained it, they sought to draw it out indefinitely; and they reacted with sore resentment against anything which might conceivably interrupt it or shorten it. At home a policy risking conflict with the peasantry might interrupt it. Abroad a forward communist policy might interrupt it. The ruling factions were determined that this should not happen; and so, almost without turning a hair, they made the Chinese Revolution prolong with its dying breath the breathing space for the workers' first state.[1]

It was only on 31 March 1927, after a year's silence and barely a fortnight before the Shanghai massacre, that Trotsky attacked the Politbureau's Chinese policy.[2] That he had been implicitly opposed to that policy and its premisses cannot be doubted. His earlier protests against the entry of the Chinese party into the Kuomintang and against the honour the Comintern had bestowed on Chiang Kai-shek had shown it. His own conceptions, developed consistently over more than twenty years, made it impossible for him to accept even for a moment the ideological arguments with which Stalin and Bukharin endeavoured to justify their political strategy. Nothing was farther from the exponent of permanent revolution than their view that because the upheaval in China was bourgeois in character, the Communists

---

[1] Stalin attempted to treat the next Chinese Revolution (1947-9) in the same way, but the momentum of that revolution was too great for that; and Mao Tse-tung had learned his lesson from the experience of Chen Tu-hsiu.

[2] See his letter to the Politbureau and the Central Committee in *The Archives.*

there must forgo their socialist aspirations for the sake of an alliance with the Kuomintang bourgeoisie. It was inherent in Trotsky's whole way of thinking that he should take the view that the bourgeois and the socialist phases of the revolution would merge, as they had merged in Russia; that the working class would be the chief driving force throughout; and that the revolution would either win as a proletarian movement ushering in a proletarian dictatorship or would not win at all.

Why then did he keep silent during the decisive year? He was, of course, ill much of the time; he was up to his neck in domestic issues and the affairs of European communism; he was engaged in an unequal struggle; and he had to reckon with the Opposition's delicate tactical situation. His attention—so his private papers suggest—did not become focused on the Chinese problem before the early months of 1927. He had not been aware how far the Politbureau's opportunism and cynicism had gone. He had no knowledge of the reluctance with which the Chinese Communists had acted on its instructions. He had no inkling of Chen Tu-hsiu's many appeals and protests—Stalin and Bukharin had locked them up in secret files; nor was he acquainted with other confidential communications that had passed between Moscow and Canton or Wuhan. When at last, having little more to go by than news generally accessible, he became alarmed and raised the issue within the Opposition's leading circle, he found that even there he was almost isolated.

Up to the end of 1926 Zinoviev and Kamenev had had little with which to reproach official policy. Sticking to the 'old Bolshevik' ideas of 1905, they too held that the Chinese Revolution must of necessity limit itself to its bourgeois and anti-imperialist objectives. They approved the party's entry into the Kuomintang. In his heyday at the Comintern Zinoviev himself must have played his part in implementing this policy and in overruling Chen Tu-hsiu's objections. But even the most important Trotskyists, Preobrazhensky, Radek, and also, it seems, Pyatakov and Rakovsky, were taken aback when Trotsky applied the scheme of permanent revolution to China.[1] They did not think that proletarian dictatorship could be established and that the Communist party could seize power in a country even more

---

[1] See Trotsky's 1928 correspondence with Radek and Preobrazhensky in *The Archives*.

retarded socially than Russia had been. Only when Trotsky threatened to raise the issue on his own responsibility and virtually to split the Opposition over it, and only after it had become abundantly clear that the workers were in fact the 'chief driving force' of the Chinese Revolution and that in obstructing it Stalin and Bukharin had long since gone beyond the point at which 'old Bolshevik' theory and dogma had any meaning, did the leaders of the Opposition consent to open a controversy over China in the Central Committee. Even then they were prepared to turn against the official policy but not against its premisses. They were willing to attack the excessive zeal with which Stalin and Bukharin had made of the Chinese party Chiang Kai-shek's accomplice in subduing strikes, demonstrations, and peasant risings; but they still held that the Communists should remain within the Kuomintang, and that this 'bourgeois' revolution could not usher in a proletarian dictatorship for China. This was a self-contradictory and self-defeating attitude, for once it had been conceded that the Communists must stay within the Kuomintang, it was inconsistent to expect them not to pay for it.

Trotsky contented himself with opening the new controversy inside the limits within which Zinoviev, Kamenev, Radek, Preobrazhensky, and Pyatakov were prepared to conduct it. In the early months of the year the chiefs of the Opposition were still seeking to adjust their differences; only towards the end of March did they define the common ground from which they would start the attack. They now embarked upon a new and dangerous venture. Trotsky was conscious of its bleak prospects. On 22 March, the very day when the workers of Shanghai were up in arms and Chiang Kai-shek's troops were entering the city, he remarked in his private papers that there was 'the danger that at the Central Committee they would turn the matter into a factional squabble instead of discussing it seriously'. Regardless of this, the issue had to be posed, for 'how can one keep silent when nothing less than the head of the Chinese proletariat is at stake?'[1]

The fact that the Opposition applied itself to China so late and with so many mental reservations weakened its stand from the beginning. The policy which was in the next few weeks to

[1] *The Trotsky Archives.*

produce the débâcle had been pursued for at least three long
years. It could hardly have been reversed within two or three
weeks. Even as Trotsky was resolving that he could not keep
silent when 'the head of the Chinese proletariat was at stake'
that head was already under Chiang Kai-shek's hammer-blow.
When the Opposition then denounced Stalin and Bukharin as
those responsible, they retorted by asking where the Opposition
had been and why it had kept silent during three long years.[1]
They plausibly suggested that the critics' indignation was
spurious, that the Opposition had been on the look-out for a
debating point, and that it grasped at the Chinese issue 'as a
drowning man grasps at a straw'. The rejoinders were not
wholly undeserved. Stalin further brought to light the inconsistencies in the Opposition's attitude and exploited to the utmost
the differences between Trotsky and his colleagues. This does not
alter the fact that the Opposition's criticisms, even if belated and
half-hearted, were justified. As to Trotsky—throughout these
fateful weeks, day after day he struggled with all his courage and
energy for a last-minute revision of policy. His analyses of the
situation were of crystalline clarity; his prognostications were
faultless; and his warnings were like mighty alarm bells.

Posterity can only marvel at the malignant complacency and
wilfulness with which the ruling factions shut their ears during
these weeks, and throughout the rest of the year, when, amid
many rapid shifts in China, Trotsky ceaselessly tried to induce
them to salvage at least the wreckage of Chinese communism.
At every stage they spurned his promptings, partly from political calculation and partly because they were bent on proving
him wrong. When events proved him right and brought fresh
disasters, they steered frantically and yet half-heartedly in the
direction which he had favoured but which it was already too
late to take; and invariably they sought to justify themselves
by heaping denunciation and abuse upon Trotskyism.

It will not be out of place to survey here at least some of
Trotsky's interventions. In his letter to the Politbureau, of 31
March, complaining that he had no access to reports from
Soviet advisers and Comintern envoys, he pointed to the up-
surge in China of the workers' movement and of communism as
the dominant feature of this phase of the revolution. Why, he

---

[1] Stalin, *Sochinenya*, vol. x, pp. 17, 21, 25 and *passim*.

asked, did the party not call upon the workers to elect Soviets, at least in the main industrial centres such as Shanghai and Hankow? Why did it not encourage agrarian revolution? Why did it not seek to establish the closest co-operation between insurgent workers and peasants? This alone could save the revolution which, he insisted, was already confronted with the danger of a counter-revolutionary military coup.

Three days later, on 3 April, he came out against an editorial statement in the *Communist International* to the effect that the crucial issue in China was 'the further development of the Kuomintang'.[1] This was exactly what was not the crucial issue, he replied. The Kuomintang could not lead the revolution to victory. Workers and peasants must be urgently organized in Councils. Day in, day out, he protested against speeches by Kalinin, Rudzutak, and others, who asserted that all classes of Chinese society 'look upon the Kuomintang as upon *their* party and should give the Kuomintang government their wholehearted support'. On 5 April, a week before the Shanghai crisis, he wrote emphatically that Chiang Kai-shek was preparing a quasi-Bonapartist or fascist coup and that only Workers' Councils could frustrate him. Such councils, Soviets, should first act as a counterbalance to the Kuomintang administration, and then, after a period of 'dual power', become the organs of insurrection and revolutionary government. On 12 April, the day of the Shanghai massacre, he wrote a scathing refutation of a eulogy of the Kuomintang which had appeared in *Pravda*—its author, Martynov, had for twenty years been the most rightwing Menshevik, had joined the Communist party only some years after the civil war, and was now a leading light in the Comintern. In the next few days Trotsky wrote to Stalin, once more asking in vain to be shown the confidential reports from China. Grotesquely, on 18 April, a week after the Shanghai massacre, the eastern secretariat of the Comintern invited him to autograph, as other Soviet leaders did, a picture for Chiang Kai-shek as a token of friendship. He refused and rebuked with angry contempt the Comintern officials and their inspirers.[2]

By this time reports about the Shanghai butchery had reached Moscow. Stalin's and Bukharin's pleadings were still fresh in

---

[1] *Kommunisticheskii Internatsional*, 18 March 1927.
[2] All this correspondence is quoted from *The Archives*.

everyone's memory. Fortunately for them the criticisms of the Opposition had not become public knowledge—only some party cadres, Comintern officials, and Chinese students in Moscow were aware of the controversy. Stalin and Bukharin did their best to belittle the events and presented them as an episodic set-back to the Chinese Revolution.[1] They were compelled, how-ever, to modify their policy. The 'alliance' with Chiang Kai-shek having lapsed, they instructed the Chinese Communists to attach themselves all the closer to the 'left Kuomintang', i.e. the Wuhan government, headed by Wang Ching-wei. The left Kuomintang was temporarily in conflict with Chiang Kai-shek and anxious to benefit from communist support. Moscow readily granted it and pledged that Chen Tu-hsiu and his comrades would refrain as before from 'provocative' revolutionary action and submit to Wang Ching-wei's discipline.[2]

Trotsky asserted that the new policy merely reproduced the old mistakes on a smaller scale. The Communists should be encouraged to adopt at last a forward policy, to form Workers' and Peasants' Councils, and to support with all their strength the rebellious peasantry in the south of China, where Chiang Kai-shek's writ did not run and they could still act. True, he saw the possibilities of revolutionary action as greatly reduced: Chiang Kai-shek's coup was, despite official attempts to belittle it, a 'basic shift' from revolution to counter-revolution and a 'crushing blow' to the urban revolutionary forces. But he as-sumed that Chiang Kai-shek had not succeeded in overwhelm-ing the scattered and elusive agrarian movements; that the peasants' struggle for the land would go on; and that it might in time provide the stimulus for a revival of revolution in the towns.[3] Communists should throw their full weight behind the agrarian movements; but to be able to do this they must at last break with the Kuomintang, 'left' as well as right, and pursue their own aims. On this point again the Zinovievists disagreed. They still preferred the Chinese party to remain within the left Kuomintang; but they wished it to pursue there an independent

---

[1] Stalin, *Sochinenya*, vol. ix, pp. 259–60 and *passim*.

[2] See Stalin's 'theses' in *Sochinenya*, vol. ix, p. 221. This attitude was reluctantly adopted by the Chinese Communist party during its congress at the end of April. See Chen Tu-hsiu's 'Open Letter'.

[3] See 'The Situation in China After Chiang's Coup and the Prospects' (written on 19 April 1927) in *The Archives*.

policy, in opposition to Wang Ching-wei. Along these lines the Opposition argued its case in many statements, none of which saw the light.

The Opposition's return to the attack over China threw the ruling factions into a fever. Their predicament was grave, for never before had the futility of their policy been so glaringly revealed and never before had their leaders so outrageously and ridiculously disgraced themselves. About the same time another setback, minor by comparison, added to their embarrassment. The Anglo-Soviet Council broke up: the leaders of the British trade unions contracted out. In the diplomatic field, there was high tension between Britain and the Soviet Union. Yet another of the great hopes of official policy had vanished into thin air. The ruling factions made the most of this circumstance, however, precisely to divert attention from China and to block all discussion. They raised an outcry about the danger of war and intervention and created a state of public nervousness and national alarm, in which it was all too easy to damn the Opposition as unpatriotic. Stalin cracked the whip, threw out fresh threats of expulsion, and used every means of moral pressure to silence his critics. At his prompting Krupskaya begged Zinoviev and Kamenev not to make a 'row over China' and to remember that they might find themselves 'criticizing the party from the outside'. The Opposition wished to avoid the 'row'. Trotsky and Zinoviev proposed that the Central Committee should meet and thrash out the differences in private so that the discussion should not be reported even in the confidential bulletin which the Central Committee issued for the 'activists'. Stalin, however, would have no debate even off the record, and the Politbureau refused to call the meeting.[1]

Then, in the last week of May, Trotsky forced a debate at a session of the Executive of the Comintern. He appealed from the

---

[1] On 7 May Trotsky wrote a letter to Krupskaya. Hurt by her talk about the 'row over China' he asked her not to evade a great issue. 'Who is right, we or Stalin?' He recounted all that the Opposition had done to secure an off-the-record discussion, and he reminded Krupskaya that until recently she had been with the Opposition against Stalin's 'brutality and disloyalty'. Had Stalin's régime become any better since? He wrote to Lenin's widow in grief and disappointment, mingled with a warm sentiment—this was in a way his farewell to her—and he hesitated over the conclusion of the letter: 'From all my heart I wish you good health and good . . . confidence in the integrity of that line which . . . .' He deleted, rewrote, and deleted again the last two lines. The draft of the letter is in *The Archives*.

Russian party to the International. In doing so he acted within his rights. The Executive of the International was nominally the court of appeal before which any communist was entitled to lodge a complaint against his own party. *Pravda*, however, denounced the appeal beforehand as an act of disloyalty and a breach of discipline. The Opposition, nevertheless, used the opportunity to subject to criticism the whole of the official policy at home as well as abroad, in Asia as well as in Europe. To strengthen its hands and protect itself against reprisals, or, as Trotsky put it, 'to spread the expected blow over many shoulders', the Opposition staged a political demonstration similar to that the Forty Six had made in 1923: on the eve of the session a group of Eighty Four prominent party men declared their solidarity with Trotsky's and Zinoviev's views.[1] Stalin could not indeed apply disciplinary measures immediately against Trotsky and Zinoviev without applying them also to the Eighty Four and then to the Three Hundred, who signed the statement of solidarity. But their joint *démarche* enabled Stalin to claim that the Opposition had broken its pledge and reconstituted itself as a faction.[2]

On 24 May Trotsky addressed the Comintern Executive. Ironically, he had to begin with a protest against the Executive's treatment of Zinoviev, its erstwhile President who not so long ago had indicted him before this very Executive—Zinoviev was now not even admitted to the session. Trotsky spoke of the 'intellectual weakness and uncertainty' which led Stalin and Bukharin to conceal from the International the truth about China and to denounce the Opposition's appeal as a crime. The Executive should publish its proceedings—'the problems of the Chinese revolution could not be stuck in a bottle and sealed

[1] The document is sometimes referred to as the Statement of the Eighty Three and sometimes of the Eighty Four. It was presented to the Central Committee between 23 and 26 May. Later the number of signatories rose to 300.

[2] See Trotsky's letter of 12 July 1927 addressed to one of the leaders of the Opposition who held ambassadorial posts abroad (either to Krestinsky or to Antonov-Ovseenko). His correspondent thought that the *démarche* of the Eighty Four needlessly aggravated the struggle. Trotsky admitted that such doubts had been entertained by Oppositionists in Moscow as well, but said that they decided in favour of the *démarche* as a measure of self-protection. He did not believe that matters were aggravated because the Opposition spoke out. He thought that his correspondent had through a long absence become cut off from Russia and invited him to make a trip to Moscow to get the feeling of the atmosphere there. *The Archives.*

up'. It should beware of the grave dangers lurking in the 'régime' of the International modelled on that of the Russian party. Some foreign communist leaders were impatient with the Opposition and imagined that the Russian party and the International would resume normal life once Trotsky and Zinoviev had been disposed of. They were deluding themselves. 'The contrary will happen. . . . On this road there will be only further difficulties and further convulsions.' No one in the International trusted himself to speak up for fear that criticism might harm the Soviet Union. But nothing was as harmful as lack of criticism. The Chinese débâcle had proved it. Stalin and Bukharin were concerned mainly with self-justification and with covering up their disastrous mistakes. They claimed that they had foreseen it all and provided for all. Yet only a week before the crisis in Shanghai Stalin had boasted at a party meeting that 'we would use the Chinese bourgeoisie and then throw it away like a squeezed lemon'. 'This speech was never made public because a few days later the "squeezed lemon" seized power.' Soviet advisers and Comintern envoys, especially Borodin, behaved 'as if they represented some sort of a *Kuom*intern':

they hindered the independent policy of the proletariat, its independent organization and especially the arming of the workers. . . . Heaven forbid that arms in hand the workers should frighten away that great chimera of a national revolution which should embrace all classes of Chinese society. . . . The Communist Party of China is a shackled party. . . . Why has it not had and why does it not have to this day its own daily newspaper? Because the Kuomintang does not want it. . . . But in this way the working class has been kept politically disarmed.[1]

While the Executive was in session, the tension between Britain and the Soviet Union reached a critical point: the British police had raided the offices of the Soviet trade mission in London and the British government broke off relations with Russia. Stalin exploited this circumstance. 'I must state, comrades', he told the Executive, in conclusion of his speech, 'that Trotsky has chosen for his attacks . . . an all too inopportune moment. I have just received the news that the English Conservative government has resolved to break off relations with the U.S.S.R. There is no need to prove that what is intended is

[1] Trotsky, *Problems of the Chinese Revolution*, pp. 91–92.

a wholesale crusade against communists. The crusade has already started. Some threaten the party with war and intervention; others with a split. There comes into being something like a united front from Chamberlain to Trotsky. . . . You need not doubt that we shall be able to break up this new front.'[1] He staked everything on the left Kuomintang as confidently as he had previously staked it on the right Kuomintang: 'Only blind people can deny the left Kuomintang the role of the organ of revolutionary struggle, the role of the organ of insurrection against feudal survivals and imperialism in China.'[2] He demanded, in effect, that the Opposition should keep silent on pain of being charged with bringing aid and comfort to the enemy.

This was not the first time that Stalin had made hints about 'a united front from Chamberlain to Trotsky'. A few months earlier *Pravda* had done it anonymously.[3] But for the first time now vague and anonymous insinuation was replaced by direct accusation. Here is Trotsky's rejoinder:

It would be manifestly absurd to believe that the Opposition will renounce its views. . . . Stalin has said that the Opposition stands in one front with Chamberlain and Mussolini. . . . To this I reply: Nothing has facilitated Chamberlain's work as much as Stalin's false policy, especially in China. . . . Not a single honest worker will believe the insane infamy about the united front of Chamberlain and Trotsky.

In reply to Stalin's appeal in favour of the left Kuomintang Trotsky said:

Stalin assumes and wants the International to assume responsibility for the policy of the Kuomintang and of the Wuhan government as he repeatedly assumed responsibility for the policy of . . . Chiang Kai-shek. We have nothing in common with this. We do not want to assume even a shadow of responsibility for the conduct of the Wuhan government and the leadership of the Kuomintang; and we urgently advise the Comintern to reject this responsibility. We say directly to the Chinese peasants: the leaders of the left

[1] Stalin, *Sochinenya*, vol. ix, pp. 311–12.   [2] Ibid., p. 302.
[3] *The Archives* contain the draft of a sharp protest against this, written on 6 January 1927, addressed to the Politbureau. Zinoviev objected to its sharpness and produced another draft begging the Politbureau to protect the Opposition from slander.

Kuomintang . . . will inevitably betray you if you follow them . . . instead of forming your own independent Soviets. . . . [They] will unite ten times with Chiang Kai-shek against the workers and peasants.[1]

These exchanges were still going on in the Kremlin when in the remote south of China Trotsky's prediction was already coming true. In May there occurred the so-called Chan-Sha coup. The Wuhan government in its turn set out to suppress trade unions, sent out troops to quell peasant risings, and struck at the Communists. For almost a month the Soviet press kept silent about these events.[2] The resolutions of the Executive, dictated by Stalin and Bukharin, were grotesquely out of date even before they had come off the printing presses; and Stalin hastened to frame new instructions for the Chinese party. He still ordered it to remain within the left Kuomintang and go on supporting the Wuhan government; but he instructed it to protest against the employment of troops against the peasants and to advise the Wuhan government to seek the assistance of Peasant Councils in restraining the agrarian movement instead of resorting to arms. By now, however, the left Kuomintang was expelling Communists from its ranks. Throughout June and July the breach between them deepened; and the stage was set for a reconciliation between the left Kuomintang and Chiang Kai-shek.

At once the repercussions were felt in Moscow. Almost daily Trotsky protested against the suppression of information. Zinoviev asked that a party court should try Bukharin, who as Editor of *Pravda* was responsible for the suppression. At last Zinoviev and Radek agreed to demand with Trotsky that the Communists should contract out of the left Kuomintang. This was now pointless, for since the left Kuomintang had broken with the Communists, there was nothing that even Stalin could do but to advise them to . . . break with it.

Stalin was in fact already preparing to carry out one of his major turns of policy and switch over to that 'ultra left' course which, towards the end of the year, was to lead the Chinese Communists to stage, at the ebb of the revolution, the futile and

[1] *The Archives; Problems of the Chinese Revolution*, pp. 102-11.

[2] The chiefs of the Opposition learned about them from a confidential bulletin of the Soviet News Agency.

bloody rising of Canton. In July he withdrew Borodin and Roy from China and sent out Lominadze, a secretary of the Soviet Comsomol, and Heinz Neumann, a German communist, both without an inkling of Chinese affairs and both with a bent for 'putschism', to carry out a coup in the Chinese party. They branded Chen Tu-hsiu, the reluctant but loyal executor of Stalin's and Bukharin's orders, as the 'opportunist' villain of the piece and made him the scapegoat for all failures.

At home Stalin went on playing up the danger of war and of an anti-communist crusade; and he intensified the drive against the Opposition. He sent many chiefs of the Opposition abroad on the pretext that they were needed for various diplomatic missions. Pyatakov, Preobrazhensky, and Vladimir Kossior had joined Rakovsky at the Embassy in Paris. Kamenev was appointed Ambassador to Mussolini—there could be no more frustrating and humiliating assignment for the former Chairman of the Politbureau. Antonov-Ovseenko was in Prague; Safarov, the Zinovievist leader of the Comsomol, was posted to Constantinople; others were detailed to Austria, Germany, Persia, and Latin America. Thus the leading group of the Opposition was largely dispersed. One after another the Eighty Four were demoted, penalized, and, on the pretext of administrative appointments, shifted to remote provinces. Repression was all the less disguised and blunter the lower down it reached: men of the rank and file were sacked and dispatched into the wilderness without any pretext.

The Opposition was exasperated and tried to defend itself, protesting against the veiled forms of deportation and exile. This was of no avail. The ruling factions saw in every one of the Opposition's attempts at self-defence a new offence justifying fresh reprisals. Every complaint was received as another sign of malignant insubordination; and every cry or even whimper of protest as a call to revolt. So persistent were the Stalinists and Bukharinists in twisting the Opposition's intentions and in making even its most timorous gestures appear as acts of unheard of defiance that in the end every such gesture did become an act of defiance, that the Oppositionist had to be full of stubborn insubordination to lodge any complaint whatsoever, and that even a whimper of protest resounded like the clarion call of rebellion. Any incident, however trivial in itself, was now liable to arouse furious passions in the factions, to make their blood boil, and to shake party and government.

One such incident was 'the meeting at the Yaroslavl station'. About the middle of June Smilga was ordered to leave Moscow and to take up a post at Khabarovsk, on the Manchurian frontier. The leader of the Baltic Fleet in the October Revolution, a distinguished political commissar in the civil war, and an economist, Smilga was one of the most respected and popular chiefs of the Zinovievist faction. On the day of his departure from Moscow several thousand Oppositionists and their friends gathered at the Yaroslavl railway station to see him off and to demonstrate against the surreptitious victimization. The crowd was angry. The demonstration was unprecedented. It was staged in a public place, amid all the traffic normal at a great railway junction. Travellers and passers-by, non-party men, mingled with the demonstrators, overheard their unflattering remarks about the party leaders, and picked up their agitated exclamations. They also listened to Trotsky and Zinoviev who made speeches. Because of these circumstances, the farewell to Smilga became the Opposition's first public, though only half premeditated, demonstration against the ruling group. Trotsky, aware of the delicacy of the situation, addressed the crowd in a restrained manner. He made no reference to the inner-party conflict. He did not, it seems, even allude to the cause of the manifestation. Instead, he spoke gravely about the international tension and the threat of war and about the allegiance which all good Bolsheviks and citizens owed the party.

The ruling group nevertheless charged Trotsky and Zinoviev with making an attempt to carry the inner-party controversy beyond the party. Humble Oppositionists who were found to have been present at the Yaroslavl station were expelled from the cells without ado. The excitement over the incident lasted throughout the summer—against the background of a continuing war scare which led to a run on food shops.

'This is the worst crisis since the revolution', Trotsky declared in a letter to the Central Committee on 27 June.[1] He referred to the war scare and its adverse effects; and he pointed out that if the Central Committee believed the danger to be as imminent as its agitators made it out, then this was one reason more why

[1] *The Archives.*

the Committee should review its policy and restore normal relations, 'the Leninist régime', within the party. The opportunity for this, he pointed out, was close at hand: the Central Committee was preparing a new congress of the party—let it then open a free pre-congress debate and bring back all the virtually·banished adherents of the Opposition and allow them to take part. Even before his appeal had reached its destination, the press spoke again about the Opposition's collusion with foreign imperialists. On the next day Trotsky again addressed himself to the Central Committee, saying *inter alia* that Stalin evidently aimed at the physical annihilation of the Opposition: 'The further road of the Stalinist group is mechanically predetermined. To-day they falsify our words, to-morrow they will falsify our deeds.' 'The Stalinist group will be compelled, and it will be compelled very soon, to use against the Opposition all those means that the class enemy used against the Bolsheviks in July 1917', during the 'month of the great slander', when Lenin had to flee from Petrograd—they would speak about 'sealed cars', 'foreign gold', conspiracies, &c. 'This is whither Stalin's course is leading—to this and all the consequences. Only the blind do not see it; only the Pharisee will not admit it.'[1]

Stalin indignantly denied that he was out to annihilate his critics. Shortly thereafter, however, he made up his mind to have the leaders of the Opposition arraigned before the Central Committee and the Central Control Commission—these two bodies acted jointly as the party's supreme tribunal. A demand for the exclusion of Zinoviev and Trotsky from the Central Committee was placed before them—this was to be the penultimate disciplinary measure before their expulsion from the party. In principle only a congress which elected the members of the Central Committee could deprive them of their seats; but the 1921 ban on factions gave that power also to the party's supreme tribunal, enabling it, in the intervals between congresses, to depose members who had infringed the ban. About the end of June the indictment of the two Opposition chiefs, by Yaroslavsky and Shkiryatov, had been lodged. It contained only two counts: Trotsky's and Zinoviev's appeal

---

[1] *The Archives.* The Zinovievists were so horrified at the idea of the guillotine descending upon them or so incredulous that they implored Trotsky to tone down his warning.

from the Russian party to the Executive of the International; and the demonstration at the Yaroslavl station. Both charges were so flimsy that in four months the tribunal, fervent Stalinists and Bukharinists to a man, could find no sufficient ground for a verdict.

As the proceedings dragged on, Stalin's impatience grew. He was bent on extracting a verdict of expulsion before he convened the fifteenth congress. As long as the chiefs of the Opposition sat on the Central Committee they were *ex officio* entitled to present to the congress full criticisms of official policy and even to offer formal counter reports, as Zinoviev and Kamenev had done at the last congress. They could therefore reveal the whole truth about China and place it in the centre of an open debate conducted in the hearing of the nation and the world. Stalin could not afford to take this risk. For this and for other reasons—events compelled him again to shift his ground in domestic policy as well and thus to admit implicitly his failures —Stalin had to do his utmost to deny Trotsky and Zinoviev the platform of the congress. For this he had first to expel them from the Central Committee. Once he had done so he could be sure that the excited attention of the congress would be absorbed by the inner party cabal rather than by the Chinese débâcle and other issues of policy, and that the chiefs of the Opposition would appear at the congress, if at all, only as defendants appealing against a degrading verdict. The congress was convened for November. He had to make the most of his time.

On 24 July Trotsky appeared for the first time before the Presidium of the Central Control Commission to answer the charges. It was five years since he himself had indicted the Workers' Opposition before the same body. The man who had then been in the chair—Solz, an old and respected Bolshevik, whom, in Lenin's days, some described as the 'party's conscience'—now sat as a Stalinist among Trotsky's judges. Presiding over the proceedings was the hot-tempered, yet in his way honest and even generous, Ordjonikidze, Stalin's countryman and friend, to whose expulsion from the party Trotsky had objected when Lenin insisted on it because of Ordjonikidze's behaviour in Georgia in 1922.[1] Yaroslavsky and Shkiryatov, Trotsky's accusers, were also among the members of the

[1] See p. 91.

Presidium. Another judge was one Yanson, whom the Control Commission had in the past censured for an excess of anti-Trotskyist zeal. The rest were likewise stalwarts of the ruling factions. Trotsky could not expect them to consider his case fairly. Indeed, he began his plea by denouncing their partiality and demanding that at least Yanson be disqualified. Yet even these men came to their job dispirited, their hearts in their mouths. They as well as the defendant went back in their thoughts to the French Revolution and were haunted by memories of Jacobin purges. Across 130 years the sepulchral cry of the condemned Danton: 'After me it will be your turn, Robespierre!' rang in their ears.

Shortly before the opening of the proceedings Solz, conversing with one of Trotsky's associates and trying to show him how pernicious was the Opposition's role, said: 'What does this lead to? You know the history of the French Revolution—and to what this led: to arrests and to the guillotine.' 'Is it your intention then to guillotine us?' the Oppositionist asked, to which Solz replied: 'Don't you think that Robespierre was sorry for Danton when he sent him to the guillotine? And then Robespierre had to go himself. . . . Do you think he was not sorry? Indeed he was, yet he had to do it. . . .'[1] Judges and defendants alike saw the giant and bloody blade above their heads; but as if gripped by fatality they were unable to avert what was coming; and each, hesitantly or even tremblingly, went on doing what he had to do to hasten its descent.

Trotsky replied briefly to the two formal charges laid against him. He denied the court the right to sit in judgement over him for a speech he had delivered before the Executive of the International. He would similarly deny any 'district commission' the right to try him for anything he had said at the Central Committee—his judges, the leading bodies of the party, acknowledged themselves to be subject to the International's authority. As to the second charge, the farewell demonstration for Smilga, the ruling group denied that it had intended to penalize Smilga. But 'if Smilga's assignment to Khabarovsk was a matter of administrative routine then how dare you say that our collective farewell to him was a collective demonstration against the Central Committee?' If, however, the assign-

[1] *The Archives*; Trotsky, *The Stalin School of Falsification*, pp. 126–48.

ment was a veiled form of banishment, then 'you are guilty of duplicity'. These trifling accusations were mere pretexts. The ruling group was determined 'to hound the Opposition and prepare its physical annihilation'. Hence the war scare produced in order to intimidate and silence critics. 'We declare that we shall continue to criticize the Stalinist régime as long as you have not physically sealed our lips.' That régime threatened to 'undermine all the conquests of the October Revolution'. The Oppositionists had nothing in common with those old-time 'patriots' for whom Tsar and Fatherland were one. Already they had been accused of giving aid and comfort to the British Tories. Yet they had every right to turn the accusation against the accusers. Stalin and Bukharin, by backing the Anglo-Soviet Council, had indeed indirectly helped Chamberlain; and their 'allies', the leaders of the British trade unions, had in all essentials backed Chamberlain's foreign policy, including the rupture of relations with the U.S.S.R. At the party cells official agitators asked suggestive questions, 'worthy of the Black Hundreds', about the sources from which the Opposition obtained means for carrying on with its activity. 'If you were really a Central Control Commission you would feel bound in duty to put an end to this dirty, abominable, contemptible, and characteristically Stalinist campaign. . . .' If the ruling group were genuinely concerned with the nation's security, they would not have dismissed the best military workers, Smilga, Mrachkovsky, Lashevich, Bakaev, and Muralov, only because they were adherents of the Opposition. This was a time to assuage the conflicts within the party, not to aggravate them. The drive against the Opposition had its origin in a rising tide of reaction.

Having surveyed the major questions at issue, Trotsky wound up with a forceful evocation of the French Revolution. He referred to the conversation, quoted above, between Solz and an Oppositionist. He said that he agreed with Solz that they all ought to consult anew the annals of the French Revolution; but it was necessary to use the historical analogy correctly:

During the great French Revolution many were guillotined. We, too, brought many people before the firing squad. But there were two great chapters in the French Revolution: one went like this (*the speaker points upwards*); the other like that (*he points downwards*). . . . In the first chapter, when the revolution moved upwards, the

Jacobins, the Bolsheviks of that time, guillotined the Royalists and Girondists. We, too, have gone through a similar great chapter when we, the Oppositionists, together with you shot the White Guards and exiled our Girondists. But then another chapter opened in France when . . . the Thermidorians and the Bonapartists, who had emerged from the right wing of the Jacobin party, began to exile and shoot the left Jacobins. . . . I would like Comrade Solz to think out his analogy to the end and to answer for himself first of all this question: which chapter is it in which Solz is preparing to have us shot? (*Commotion in the hall.*) This is no laughing matter; revolution is a serious business. None of us is scared of firing squads. We are all old revolutionaries. But we must know who it is that is to be shot and what chapter it is that we are in. When we did the shooting, we knew firmly what chapter we were in. But do you, Comrade Solz, see clearly in which chapter you are preparing to shoot us? I fear . . . that you are about to do so in . . . the Thermidorian chapter.

He went on to explain that his adversaries were mistaken in imagining that he called them names. The Thermidorians were not deliberate counter-revolutionaries—they were Jacobins, but Jacobins who had 'moved to the right'.

Do you think that on the very next day after 9 Thermidor they said to themselves: we have now transferred power into the hands of the bourgeoisie? Nothing of the kind. Look up the newspapers of that time. They said: We have destroyed a handful of people who disturbed the peace in the party, and now after their destruction the revolution will triumph completely. If comrade Solz has any doubts about it . . .

Solz: You are practically repeating my own words.

Trotsky: . . . I shall read to you what was said by Brival, a right Jacobin and Thermidorian, when he reported on that session of the Convention which had resolved to hand over Robespierre and his associates to the revolutionary tribunal: 'Intriguers and counter-revolutionaries draping themselves with the togas of patriotism, they had sought the destruction of liberty; and the Convention decreed to place them under arrest. They were: Robespierre, Couthon, St. Just, Lebas, and Robespierre the Younger. The Chairman asked what my opinion was. I replied: Those who had always voted in accordance with the principles of the Mountain . . . voted for imprisonment. I did more . . . I am one of those who proposed this measure. Moreover, as secretary, I hasten to sign and to transmit to you this decree of the Convention.' That is how the report was made by a Solz . . . of that time. Robespierre and his associates—these were the counter-revolutionaries. 'Those who had always voted in accordance with the principles of the Mountain' meant in the language of that time 'those who had always been Bolsheviks'. Brival considered himself an old Bolshevik. 'As secretary I hasten to sign and to transmit to you this decree of the Convention.' To-day, too, there are secretaries who hasten to 'sign and transmit'. To-day, too, there are such secretaries. . . .[1]

The Thermidorians too, Trotsky went on, had struck at the left Jacobins amid cries of *La Patrie en danger*! Convinced that Robespierre and his friends were only 'isolated individuals', they did not understand that they struck against 'the deepest revolutionary forces of their age', forces opposed to the Jacobin 'neo-N.E.P.' and Bonapartism. They branded Robespierre and his friends as aristocrats—'and did we not hear to-day this same cry "aristocrat" from the lips of Yanson addressed to me?' They stigmatized the left Jacobins as Pitt's agents just as the Stalinists denounced the Opposition as the agents of Chamberlain, 'that modern pocket edition of Pitt'.

The odour of the 'second chapter' now assails one's nostrils . . . the party régime stifles everyone who struggles against Thermidor. The worker, the man of the mass, has been stifled in the party. The rank and file is silent. [Such had also been the condition of the Jacobin Clubs in their decay.] An anonymous reign of terror was instituted there; silence was compulsory; the 100 per cent. vote and abstention from all criticism were demanded; it was obligatory to think in accordance with orders received from above; men were compelled to stop thinking that the party was a living and independent organism, not a self-sufficient machine of power. . . . The Jacobin Clubs, the crucibles of revolution, became the nurseries of Napoleon's future bureaucracy. We should learn from the French Revolution. But is it really necessary to repeat it? (*Shouts.*)

Not all was lost yet, however. Despite grave differences, a split could still be avoided. There was still 'a gigantic revolutionary potential in our party', the stock of ideas and traditions inherited from Lenin. 'You have squandered a great deal of this capital, you have replaced much of it with cheap substitutes . . . but a good deal of pure gold still remains.' This was an age of stupendous shifts, of sharp and rapid turns, and the scene might

---

[1] Loc. cit.

yet suddenly become transformed. 'But you dare not hide the facts, for sooner or later they will become known anyway. You cannot hide the victories and the defeats of the working class.' If only the party were allowed to ponder the facts and to form its opinion freely, the present crisis could be overcome. Let the ruling group therefore take no rash and irreparable decision. 'Beware, lest you should find yourself saying later: we parted company with those whom we should have preserved and we preserved those from whom we should have parted.'

It is impossible to read these words without recalling the 'cold shiver running down one's spine' of which the young Trotsky spoke in 1904 when, at the threshold of his career, he thought of the future of Lenin's party and compared it with the fate of the Jacobins. The same cold shiver ran down his spine twenty-three years later. In 1904 he had written that 'a Jacobin tribunal would have tried under the charge of *modérantisme* the whole international labour movement, and Marx's lion head would have been the first to roll under the guillotine'. Now he himself fought with a lion's courage for his own head before the Bolshevik tribunal. In 1904 he had been disgusted with Lenin's 'malicious and morally repulsive suspiciousness—a flat caricature of the tragic Jacobin intolerance'. Now he invoked Lenin's ideas against the intolerance and the 'malicious and morally repulsive suspiciousness' of Lenin's successors. But his view of Jacobinism was almost diametrically opposed to the one he had voiced in his youth. Then he saw Jacobinism as incompatible with Marxist socialism—these were 'two opposed worlds, doctrines, tactics, mentalities . . .', for Jacobinism meant an 'absolute faith in a metaphysical idea and absolute distrust of living people', while Marxism appealed in the first instance to the class consciousness of the working masses. And so in 1904 he demanded a clear choice between the two, because the Jacobin method, if revived, would consist in 'placing above the proletariat a few well-picked people . . . or one person invested with the power to liquidate and degrade'. Now he confronted these few well-picked people and the one person acquiring the power to liquidate and degrade. But his main charge against them was not that they acted in the Jacobin spirit, but, on the contrary, that they worked to destroy it. Now he dwelt on the affinity of Marxism and Jacobinism; and he identified himself and his adherents with Robespierre's group; and it was he who turned the charge of *modérantisme* against Stalin and Bukharin.

Thus the 'conflict of the two souls in Bolshevism, the Marxist and the Jacobin', a conflict which we first noticed in 1904,[1] and which underlay all Bolshevik affairs in recent years, now caused Trotsky to view Jacobinism from an angle altogether opposed to that from which he had first approached it. This conflict was in varying degrees characteristic for all Bolshevik factions. Curiously, they all appeared to identify themselves with the same aspect of Jacobinism. While Trotsky compared his own attitude with Robespierre's and saw his adversaries as 'moderantists', Solz and others with him saw Stalin as the new Robespierre and Trotsky as the new Danton. In truth, as events were to show, the alignments and divisions were far more complex and confused. What Jacobinism and Bolshevism had in common was—substitutism. Each of the two parties had placed itself at the head of society but could not rely for the realization of its programme on the willing support of society. Like the Jacobins, the Bolsheviks 'could not trust that their *Vérité* would win the hearts and the minds of the people'. They too looked around with morbid suspicion and 'saw enemies creeping from every crevice'. They too had to draw a sharp dividing line between themselves and the rest of the world because 'every attempt to blur the line threatened to release inner centrifugal forces'; and they too were drawing it 'with the edge of the guillotine', and having destroyed their enemies outside their own ranks began to see enemies in their own midst. Yet, as a Marxist Trotsky reiterated now what he had first said in 1904: 'The party must see the guarantee of its stability in its own base, in an active and self-reliant proletariat, and not in its top caucus, which the revolution . . . may suddenly sweep away with its wings. . . .' He cried out again that 'any serious group . . . when it is confronted by the dilemma whether it should, from a sense of discipline, silently efface itself, or, regardless of discipline, struggle for survival—will undoubtedly choose the latter course . . . and say: perish that "discipline" which suppresses the vital interests of the movement'.

·   ·   ·   ·   ·   ·   ·   ·   ·   ·

[1] See *The Prophet Armed*, pp. 91–96.

Before the end of July the party tribunal dispersed without passing a verdict on Trotsky and Zinoviev. The majority of the judges still appeared to be as sorry for them as 'Robespierre had been sorry for Danton'. Stalin, however, pressed for a decision. With every day the consequences of his 'colossal blunders' were becoming clearer. The final collapse of the Chinese Revolution threatened to discredit him. The Anglo-Soviet Council had finally ceased to exist: its British members had not uttered even a word of protest against the rupture of relations between Britain and Russia. At home the war scare and the run on shops had led to another goods famine. The peasantry was restless. There was reason to fear that it would not deliver enough food to the towns in the autumn. So far Stalin had been able to conceal his responsibility: he had managed to suppress all the warnings and predictions made by his opponents. Almost every one of Trotsky's recent speeches might have exploded his laboriously acquired and still precarious authority; but he had not allowed Trotsky's voice to penetrate through the thick walls of the Kremlin and gain resonance outside. However, the date of the fifteenth congress was approaching; and with it Trotsky's and Zinoviev's opportunity to state their case. The whole country would listen. It would be impossible to suppress speeches made at a congress in the way criticisms uttered at the Central Committee were concealed. At any cost Stalin had to deprive them of this opportunity.

He had yet another reason for haste. He had to reckon with strains within the ruling coalition. The rightist policy of recent years was nearing exhaustion. It was increasingly difficult to maintain it abroad, in the Comintern. At home also everything pointed to the need for a change of policy; and although its possible scope was far from evident, it was clear that the change would require of the party a more radical attitude towards the peasantry and a bolder course in industry. On all these issues Stalinists and Bukharinists had hitherto patched up their differences so as to be able to present a common front to the Opposition. But the moment was drawing close when it might become difficult to patch them up any longer and a breach might follow. Yet Stalin could not turn against Bukharin, Rykov, and Tomsky as long as he had not concluded his struggle against Trotsky and Zinoviev. He could not confront two oppositions simultaneously, especially as a change of policy would appear to many a vindication of Trotsky's and Zinoviev's views. He had to crush the Joint Opposition and free his hands as soon as possible.

He lashed out with redoubled vehemence after Trotsky had made his so-called Clemenceau statements, first on 11 July, in a letter to Ordjonikidze, and then again, before the end of the month, in an article which he submitted to *Pravda*. Referring to the war scare, Trotsky had repeatedly declared that if war came the leaders of the ruling factions would prove themselves incompetent and unequal to their task and that the Opposition would in the interest of defence continue to oppose them and seek to take over the direction of the war. These declarations brought on Trotsky charges of disloyalty and defeatism. In refutation he explained that the Opposition stood for the 'unconditional defence' of the U.S.S.R. and that in war it would seek to replace the ruling factions precisely in order to pursue hostilities with the utmost vigour and clear-headedness which could not be expected from those who now led the party. Only 'ignoramuses and scoundrels' could from 'their rubbish heaps' blame this attitude as defeatist. This was, on the contrary, an attitude dictated by genuine concern over defence—'victory is not obtained from the rubbish heap'. And here came the much disputed 'Clemenceau statement':

One may find examples, and quite instructive ones, in the history of other social classes [Trotsky wrote to Ordjonikidze]. We shall cite only one: at the beginning of the imperialist war [i.e. of the First World War] the French bourgeoisie had at its head a blundering government, a government without rudder and sail. Clemenceau and his group were in opposition to it. Regardless of war and military censorship, regardless even of the fact that the Germans stood 80 kilometres from Paris (Clemenceau said: 'Precisely because of this'), he waged a furious struggle against the government's petty bourgeois irresolution and flabbiness—for the prosecution of the war with truly imperialist ferocity and ruthlessness. Clemenceau did not betray his class, the bourgeoisie; on the contrary, he served it more faithfully, firmly, resolutely, and wisely than did Viviani, Painlevé and company. The further course of events proved this. Clemenceau's group took office and by means of a more consistent policy—his was a policy of imperialist robbery—secured victory. ... Did any French journalistic scribes label Clemenceau's group defeatist? Of course, they did: fools and slanderers trail along in the camp of every social class. Not always, however, do they have the same opportunity to play important roles.[1]

This then was the example which Trotsky declared he would follow—an example, it may be added, which early in the Second World War Churchill followed in his opposition to Chamberlain. The retort came at once. The Stalinists and Bukharinists raised an outcry that Trotsky threatened to stage a *coup d'état* in the middle of war, while the enemy might be standing less than eighty kilometres from the Kremlin—what other proof of his disloyalty was needed? About the same time a group of army leaders addressed to the Politbureau a secret statement expressing solidarity with the Opposition and criticizing Voroshilov, the Commissar of War, for military incompetence. Among the signatories were, apart from Muralov, until recently chief Army Inspector, Putna, Yakir, and other generals, who were to perish in the Tukhachevsky purge ten years later.[2] The ruling factions took the *démarche* of the military as an earnest of the Opposition's intentions.

The hue and cry about the Clemenceau statement lasted till the end of the year, till Trotsky's banishment; and it was to echo for many years afterwards: it was cited whenever Trotsky's treachery was to be shown up. Very few were those party men who knew what the Clemenceau statement was about; most understood it in fact as Trotsky's threat to turn the next war into civil war, if not as an actual prelude to a *coup*. That he had meant to utter no such threat and that the precedent he invoked had not implied one did not matter. Few, very few, Bolsheviks had any idea what it was that the French 'Tiger' had done and by what means he had seized power. To Trotsky the reference to Clemenceau occurred naturally—it was in Paris he himself had watched Clemenceau's struggle ten years earlier. But the precedent was remote, obscure, and therefore sinister to the public, to the great majority of the Central Committee, and even to the members of the new Politbureau (among whom hardly any one, apart from Bukharin, had any knowledge of French affairs). This is how Trotsky himself describes satirically the

---

[1] Quoted *in extenso* by Stalin in *Sochinenya*, vol. x, p. 52.
[2] Tukhachevsky himself did not sign the statement, and he was at no time involved with the Joint Opposition.

dazed nescience with which the Central Committee received his analogy:

From my article . . . Molotov had first learned many things which he then reported to the Central Committee as terrible *prima facie* evidence of these insurrectionist designs. And so Molotov learned that during the war there was in France a politician called Clemenceau, that that politician waged a struggle against the French government of the time in order to force a more determined and ruthless imperialist policy. . . . Stalin then explained to Molotov and Molotov expounded to us the real sense of that precedent: on the example set by Clemenceau's group the Opposition intends to fight for another policy of socialist defence—and that means an insurrectionist policy like that which the Left Social Revolutionaries adopted [in 1918].[1]

It was all too easy to frighten with the mysterious conundrum the cells, first in Moscow and then in the provinces, whence the clamour rose that it was time to render the Opposition harmless.

On 1 August the Central Control Commission and the Central Committee again considered the motion calling for Trotsky's expulsion. Once again Stalin, Bukharin, and others railed in good set terms and read out interminable indictments in which they brought up every detail of Trotsky's political past, beginning with 1903, and showed it in the darkest colours. Even the long-forgotten charges, made once, in 1919, by the Military Opposition, namely, that during the civil war Trotsky was the enemy of the Communists in the army and ordered brave and innocent commissars to be shot, were brought up afresh.[2] This time, however, the Clemenceau statement provided the gravamen of the indictment which was that the Opposition could not be depended upon to behave loyally in war and to contribute to the defence of the Soviet Union.

In reply Trotsky recalled the paramount responsibility he had

[1] See Trotsky's note 'Clemenceau' dated 2 August 1927, in *The Archives*.
[2] This particular charge was made by Yaroslavsky, but it shocked even Stalinists, and Ordjonikidze expressly dissociated himself from it. *The Archives*. Yaroslavsky belonged to the Military Opposition in 1919. The charges against Trotsky were then brought before the Politbureau by Smilga and Lashevich and the commissars whom Trotsky allegedly persecuted were Zalutsky and Bakaev—all four now prominent in the Opposition. For the account of the incident see *The Prophet Armed*, pp. 420-1, 425-6, 432.

borne over many years for the party's defence policy and for
formulating the Communist International's views on war and
peace. He attacked Stalin's and Bukharin's reliance for defence
on broken reeds, or, as he put it, on 'rotten ropes' and 'rotten
props'. Had they not hailed the Anglo-Soviet Council as a bul-
wark against intervention and war; and had this not turned out
a rotten prop? Was their alliance with the Kuomintang not a
rotten rope? Had they not weakened the Soviet Union by
sabotaging the Chinese Revolution? Voroshilov had stated that
'the peasant revolution [in China] might have interfered with
the northern expedition of the generals'. But this was precisely
how Chiang Kai-shek viewed the matter. 'For the sake of a
military expedition you have put a brake on the revolution . . .
as if the revolution were not . . . itself an expedition of the oppressed against the oppressors.' 'You came out against the
building of Soviets in the "army's rear"—as if the revolution
were the rear of an army!—and you did it in order not to dis-
organize the hinterland of the very same generals who two days
later crushed the workers and the peasants in their rear.' Such
a speech by Voroshilov, the Commissar of Defence and a member of the Politbureau, was in itself 'a catastrophe—equivalent
to a lost battle'. In case of war 'the rotten ropes will fall to
pieces in your hands'—and this was why the Opposition could
not forbear criticizing the Stalinist leadership.

But would criticism not weaken the moral standing of the
U.S.S.R.? To pose the question thus was 'worthy of the Papal
Church or of feudal generals. The Catholic Church demands
unquestioning recognition of its authority by the faithful. The
revolutionary gives his support while criticizing; and the more
undeniable his right to criticize the greater in time of struggle is
his devotion to the creative development and strengthening of
that in which he is a direct participant.' 'What we need is not a
hypocritical *union sacrée* but honest revolutionary unity.' Nor was
victory in war primarily a matter of arms. Men had to wield the
arms and men were inspired by ideas. What then was the idea
underlying Bolshevik defence policy? It might be possible to
secure victory in one of two ways: either by waging war in a
spirit of revolutionary internationalism, as the Opposition proposed to do, or by waging it in the Thermidorian style—but this
meant victory for the kulak, further suppression of the worker,

and 'capitalism on the instalment plan'. Stalin's policy was neither the one nor the other; he vacillated between the alternatives. But war would brook no irresolution. It would force the Stalinist group to make a choice. In any case, the Stalinist group, itself not knowing whither it was going, could not secure victory.

At this point of Trotsky's speech the record notes that an exclamation of eager assent came from Zinoviev, but that Trotsky stopped to correct himself: instead of saying that 'Stalin's leadership was *incapable* of assuring victory' he stated that it 'would make victory *more difficult*'. 'But where would the party be?' Molotov interjected. 'You have strangled the party'— Trotsky thundered back; and he repeated again with deliberation that under Stalin victory would prove 'more difficult'. The Opposition, therefore, could not identify the defence of the U.S.S.R. with the defence of Stalinism. 'Not a single Oppositionist will renounce his right and duty to fight for the correction of the party's course on the eve of war or during war . . . therein lies the most important prerequisite of victory. To sum up: for the socialist fatherland? Yes! For the Stalinist course? No!'[1]

After the Second World War these prophecies were as if lost in the blaze of Stalin's triumphs. Stalin did, after all, secure Russia's victory; and the sequel showed no similarity to 'capitalism on the instalment plan'. However, Trotsky spoke at the height of N.E.P. when Russia was still one of the industrially most backward nations; when private farming predominated in the country; when the kulak was growing in strength; and when the party was still a whirlpool of conflicting trends; and he spoke conditionally about a danger of war which the ruling factions claimed to be imminent. One can only speculate about the course which a war fought against this background might have taken and how Stalin might have fared in it. At any rate, against such a background Trotsky's assessment of the prospects was far more plausible than it appears when it is related to the Soviet Union of the years 1941–5. Yet, even after the Second World War, Stalinism endeavoured to overcome the tensions inside the Soviet Union by a forcible expansion of its rule into eastern and central Europe. It might be argued that the alterna-

[1] *The Archives*; *The Stalin School of Falsification*, pp. 161–77.

tive to expansion was precisely that 'capitalism on the instalment plan' inside the Soviet Union of which Trotsky had spoken. And even in the light of victory, Trotsky's strictures on Stalin's and Voroshilov's incompetence do not appear altogether groundless. In 1941, in the very first months of Russo-German hostilities, Voroshilov fumbled and bungled so badly that as a general he could never raise his head again. As to Stalin, the General Secretary of 1927 had as yet little of that empirical military knowledge and experience that the dictator of the later period had gathered in long years of absolute rule. And, although Stalin's role in the Second World War is, and will for a long time to come remain, the subject of historical controversy, it seems nevertheless established that victory was indeed '*more difficult under Stalin*' than it need perhaps have been, that under a leadership more far-sighted than his the Soviet Union might not have suffered as severe initial defeats as those of 1941–2, and that it might not have needed to pay for its final triumph the prodigious price in human life and wealth which it did pay.[1]

The weakness of Trotsky's attitude lay not in what he said against his adversaries. It lay elsewhere—in the manner in which he envisaged the Opposition's action in war. That there was not a shred of defeatism in it is obvious. But how did he imagine himself acting the Soviet Clemenceau? He returned to this question on 6 August, when the Central Committee and the Central Control Commission continued to debate the motion for his expulsion. It was preposterous, he said, to charge him with incitement to insurrection: Clemenceau had not staged any insurrection or coup, or acted in any unconstitutional manner; he overthrew the government he opposed and himself assumed office in a most lawful way, using for this purpose the machinery of parliament. But the Soviet Union, it might be said, had no such parliamentary machinery? 'Yes,' Trotsky rejoined, '*fortunately*, we do not have it.' How then could any opposition overthrow any government constitutionally? 'But we do have', Trotsky went on, 'we do have the machinery of our party.' The Opposition, in other words, would act within the party statutes and seek to overthrow Stalin by a vote at the Central Committee or perhaps at a congress. But had not

[1] See the appraisal of Stalin's role in the war in my *Stalin, a Political Biography*, pp. 456–60; and chapters xii–xiv, *passim*.

Trotsky himself repeatedly said and demonstrated that the party's nominal constitution was a sham and that its real constitution was Stalin's bureaucratic absolutism? Did not events daily prove this to be true? That, Trotsky replied, was why the Opposition strove to reform the inner-party régime: '... in case of war, too, the party ought to preserve or rather to restore a more supple, a sounder, and a healthier inner régime, which would make it possible to criticize in time, to warn in time, to change policy in time.' The ruling factions, however, made no bones about it: they would allow no such reform and permit no change in the leadership by any constitutional method. With this in mind they viewed Trotsky's declaration; and they concluded that as he would not be able to overthrow Stalin by means of any parliamentary procedure or vote, he would have to resort to a *coup d'état*. From their viewpoint they were in a sense consistent in regarding his Clemenceau statement as a proclamation of the Opposition's right to insurrection. Even though he had not in fact proclaimed that right—he was to do it in exile eight or nine years hence; and the ruling factions realized that it was inherent in the situation, which they had created, that he should proclaim it.

With even greater logic Trotsky charged that it was they who threatened to perpetuate their hold on the party and maintain themselves in power by measures of civil war; and that they were preparing to use such measures against the Opposition. And indeed, in raising the clamour against the Clemenceau statement, Stalin was out to establish indirectly the principle, which Bolshevik tradition did not allow him to proclaim openly, that his rule was inviolable and inalienable, and that any attempt to replace it was tantamount to counter-revolution. This was *the* issue which underlay the affair. The storm over the Clemenceau statement revealed the width, the depth, and the unbridgeability of the gulf between the ruling group and the Opposition: by force of circumstances the language in which they addressed each other was already the language of civil war.

Yet even now the party tribunal, deliberating for the second month over Trotsky's expulsion, still demurred at passing the verdict. For once Stalin had run ahead of his followers and associates. They were not yet quite ready to do his bidding.

Still entangled in shreds of old loyalties, still thinking of their adversaries as comrades, still worried about niceties of party statutes, and anxious to preserve appearances of Bolshevik decorum, they sought once again to come to terms with the Opposition. The latter were only too glad to meet them half-way; and so Trotsky and Zinoviev tried to calm the emotions aroused by the Clemenceau statement with a declaration of the Opposition's loyalty to party and state and of its commitment to the unconditional defence of the Soviet Union in any emergency. A new 'truce' was arranged; and on 8 August the Central Committee and the Central Control Commission concluded their deliberations, ignoring the motion for expulsion and content to pass no more than a vote of censure on the chiefs of the Opposition.

For the moment it looked as if the Opposition would be able to participate in the fifteenth congress and there make another appeal to the party. The leaders prepared a full and systematic statement of policy, a *Platform*, such as they had never before been able to present. The *Platform* was thrashed out in Opposition circles, carefully amended, and supplemented.[1] However, matters had long since gone beyond the point at which 'normalization' might have been possible. This was the last 'truce'; and it was even more shortlived than the previous one. The ruling factions had hesitantly agreed to it on the tacit understanding that the leaders of the Opposition, having narrowly escaped punishment, would hold their fire. This was not how the latter understood their obligation. They felt entitled to go on with what to them was normal expression of opinion and criticism, especially in the months preceding a congress, the season for an all-party debate. Stalin and his closest henchmen did what they could to nullify the truce. He exacerbated the Opposition by continuing, with and without pretext, to penalize and banish its adherents. He laid the blame at the Opposition's door, saying that it had broken the truce by preparing its *Platform*, by refusing to join in a condemnation of its sympathizers in Germany, &c. Seeing that he was behind schedule in his drive, he delayed the fifteenth congress by a month.

On 6 September Trotsky and his friends approached the

---

[1] The *Platform* is known under the title *The Real Situation in Russia* under which Trotsky published it later in exile.

Politbureau and the Central Committee and pointed out that the General Secretariat was pursuing its own policy, one which did not even accord with that of the Stalinist-Bukharinist majority; and they submitted a detailed report on the new persecution and a protest against the postponement of the congress. Trotsky asked once again for a loyal pre-congress debate with the participation of the banished Oppositionists. He also demanded that the Central Committee should, in accordance with a custom honoured in the past, publish the Opposition's *Platform* and circulate it, together with all the official documentation, among the party electorate. After fierce and relentless interventions by Stalin, the Central Committee rejected the Opposition's complaints and refused to publish the *Platform* as part of the papers for discussion. Moreover, it forbade the Opposition to circulate the document by its own means.

This was, of course, a new bone of contention. For the Opposition to observe the fresh ban was to surrender ignominiously, perhaps for ever. But to snap their fingers at it was also risky, for the *Platform* would then have to be produced and circulated clandestinely or semi-clandestinely. The Opposition resolved to take the risk. To protect itself against reprisals—to 'spread the blow' once again—and also to impress the congress, Trotsky and Zinoviev called on their adherents to sign the *Platform en masse*. The collection of signatures was to reveal the size of the Opposition's following; and so the campaign was from the outset a trial of strength in a form the Opposition had not hitherto dared to undertake.

Stalin could not allow this to go on undisturbed. On the night of 12–13 September the G.P.U. raided the Opposition's 'printing shop', arrested several men engaged in producing the *Platform*, and announced with a flourish that it had discovered a conspiracy. The G.P.U. maintained that they had caught the Oppositionists red-handed, working hand in glove with notorious counter-revolutionaries; and that a former officer of Wrangel's White Guards had set up the Opposition's printing shop. On the day of the raid Trotsky had left for the Caucasus; but several leaders of the Opposition, Preobrazhensky, Mrachkovsky, and Serebriakov, attempted to come forward with a refutation and declared that they assumed full responsibility for the 'printing shop' and the publication of the *Platform*. All three were immediately expelled from the party and one of them, Mrachkovsky, was imprisoned. This was the first time that such punishment was inflicted on prominent men of the Opposition.

The incident foreshadowed the 'amalgams' on which the great purges of the next decade were to be based. The G.P.U. revelations were calculated to impress all those who had listened incredulously to Stalin's asseverations about the 'united front from Chamberlain to Trotsky'. If the consciences of such people were uneasy and if they wondered whether that 'united front' was not a figment of Stalin's imagination, the story of the un- covered plot was there to reassure them. The native figure of the 'Wrangel officer' appeared as a link between the Opposition and the dark forces of world imperialism. The doubting and the confused received a clear warning. They were shown the net in which they might get entangled once they undertook or merely condoned any form of activity directed against the official leaders, no matter how innocent that activity appeared at first sight.

The blow was well aimed. By the time the Opposition managed to show up the G.P.U. revelations as a fake, the harm had been done. Zinoviev, Kamenev, and Trotsky—he had interrupted his stay in the Caucasus and returned to Moscow—intervened with Menzhinsky, the head of the G.P.U. since Dzerzhinsky's death, and cleared up the farcical circumstances of the plot. The G.P.U. had caught members of the Opposition duplicating typewritten copies of the *Platform*. The Opposition, it transpired, did not even possess a clandestine printing shop of the kind that every underground group had operated in Tsarist days. A few young men had volunteered to do the typing and duplicating. True, some of them were not party members; but this was their only fault—Stalin was later unable to find for them a label more damaging than 'bourgeois intellectuals'. A former Wrangel officer had indeed helped in the work and promised to assist in circulating the *Platform*; but Menzhinsky admitted, first to Trotsky and Kamenev and then to the Central Committee, that the G.P.U. had employed the officer as an *agent provocateur*; and that his particular job had been to spy on the Opposition. Stalin himself confirmed the disclosure and said: 'But what is wrong with this former Wrangel officer helping the Soviet government to unearth counter-revolutionary plots? Who can deny Soviet authorities the right to win over ex-officers and use them to uncover counter-revolutionary organizations?'[1] Thus Stalin first pointed to the Wrangel officer as proof positive of the counter-revolutionary character of the Opposition's activity, and then went on to say that he saw no reason why he should not have used the officer to provide the proof. The Opposition cried out: 'Our enemies, persecutors, and slanderers!' But it did not recover from the effects of the slander.

Trotsky had returned in haste to Moscow not only because of this affair. While he was in the Caucasus the Presidium of the Comintern unexpectedly announced that it was due to meet before the end of September and that it had placed on its agenda Trotsky's expulsion from the Executive of the International. He appeared before the Executive on 27 September to speak for the last time—with scorn and passion—to the envoys of all Communist parties. These were grotesque assizes. The foreign Communists who sat in judgement over one of the founding fathers of their International and denied him all the merits of a revolutionary, were almost to a man pathetic failures as revolutionaries: instigators of abortive risings, almost professional losers of revolution, or heads of insignificant sects all basking in the glory of that October in which the accused man had played so outstanding a part. Among them were Marcel Cachin, who had during the First World War, while Trotsky was being expelled from France as the author of the Zimmerwald Manifesto, gone as agent of the French government to Italy to back Mussolini's pro-war campaign; Doriot, the future fascist and Hitler's puppet;[2] Thälmann, who was to lead German communism in its capitulation to Hitler in 1933 and then perish in Hitler's concentration camp; and Roy, who had just returned from China where he had done his best to induce the Chinese party to lick the dust before Chiang Kai-shek. J. T. Murphy, an insignificant envoy of one of the most insignificant foreign Communist parties, the British, was chosen to table the motion of expulsion. The disdain which Trotsky hurled at this conventicle was proportionate to the insult they inflicted on him.

'You are accusing me', he told the Executive, 'of a breach of

---

[1] Stalin, *Sochinenya*, vol. x, p. 187.

[2] Doriot, it seems, was not present at the session; but he was an alternate member of the Executive and one of Trotsky's most vehement accusers.

discipline. I have no doubt that even your verdict is ready.'[1]
None of the members of the Executive dared to make up his
mind for himself—they were all only carrying out orders. Such
was their servility that the General Secretary of the Russian
party had the insolence to assign an envoy of a foreign Communist party to the job of a minor official in a remote Russian
province—the reference was to Vuyovich, the Yugoslav representative with the Comintern, a Zinovievist, who was now
also to be expelled. He, Trotsky, had been called to account for
having appealed from the Russian party to the International—
'just as in Tsarist days so now the *pristav* (the bailiff) still beats
up any one who dares to complain against him higher up'. The
supposed leaders of international communism did not even have
the dignity to try and save appearances: in their sycophancy
they had forgotten to expel Chiang Kai-shek and Wang Ching-
wei from their Executive, and the Kuomintang was still af-
filiated to the International; but they sat in judgement over
those who were of the flesh and the blood of the Russian
Revolution.[2]

In the course of four fateful years, Trotsky went on, they had
convened no congress of the International; in Lenin's days a
congress was held every year, even during civil war and blockade.
There had been no discussion of any of the grave issues that had
emerged, for all these issues were taboo—in all of them Stalin's
policy had suffered shipwreck. 'Why does the press of the
Communist parties keep silent? Why does the press of the International keep silent?' The Executive trampled almost daily
on the statutes of their organization; and then they charged the
Russian Opposition with breaches of discipline. 'The Opposition's . . . only guilt', he confessed, 'is that it has been too
amenable to the schemes of the Stalinist Secretariat which have
been calamitous to the revolution.' 'The manner in which the
congress of the Russian party is being prepared is a mockery . . .
Stalin's favourite weapon is slander.' 'Whoever knows history
knows that every step on the usurper's road is always marked
by such false accusations.' The Opposition could not give up
the right to speak out against a régime which was the deadliest

[1] *The Archives.*
[2] *L'Humanité* had hailed Chiang Kai-shek as the 'hero of the Shanghai commune', Trotsky said.

danger to the revolution: 'When a soldier's hands are tied, the chief danger is not the enemy but the rope which ties the soldier's hands.'

'He launched the attack', so Murphy, the author of the motion for expulsion, recollects, 'with all the vigour and power of which he was capable. He challenged us on every aspect of the problems that had been under discussion during the last three years . . . a forensic effort such as only he could make'; and, turning his back on the Executive of the organization on which he had once placed his highest hopes, he 'marched out with head erect'.[1] The Executive was not troubled even by such scruples as still beset the Russian Central Committee—its verdict had indeed been ready.

At this point the struggle in Moscow led to a diplomatic incident which aroused a flurry of international excitement. Since the breach between Britain and Russia Soviet relations with France had also deteriorated. The French government and press raised again the old clamour about the unpaid loans, the clamour which had been first heard after Lenin's government had repudiated all Tsarist debts to foreign creditors. The Politbureau and the Central Committee discussed the question on and off. In 1926 Trotsky was in favour of conciliating the French. Britain was then in the throes of industrial unrest; the Chinese Revolution was on the ascendant; France was labouring under the effects of inflation; and the Soviet Union was in a position of strength in which he held it to be advisable to make a concession to the French and remove a grievance of the small *rentiers*. At that time, however, Trotsky relates, Stalin was in an over-confident mood and would not hear of any settlement. Then, in the autumn of 1927, when the issue again became topical, Stalin was anxious to go some way to meet the French demands. Now, however, Trotsky and his friends were opposed to this. He argued that after the defeat of the Chinese Revolution, the breakdown of the Anglo-Soviet Council, and the

[1] J. T. Murphy, *New Horizon*, pp. 274-7. Murphy relates that before the sitting he met Trotsky in the corridor. 'Everybody had their heavy overcoats and fur hats, and the hat and coat rack in the hall was full. Trotsky was looking around, when . . . [Murphy's secretary] asked: "Can I help you, Comrade Trotsky?" Quick as thought he answered smartly: "I am afraid, not. I am looking for two things—a good communist and somewhere to hang my coat. They are not to be found here." ' The meeting lasted from 9.30 p.m. to 5 a.m.

rupture with Britain, the Soviet government was too weak to yield; and that any concession on its part would be taken as a further sign of weakness.

For the Opposition the situation was complicated by the fact that Rakovsky, as Ambassador, conducted the negotiations in Paris and became the butt of French attacks. Already in August the French envoy in Moscow had expressed his government's displeasure because of Rakovsky's connexion with the Trotskyist Opposition.[1] At the Central Committee, on the other hand, Stalin then made an attempt to play off Rakovsky against Trotsky: he asserted that it was Rakovsky, a 'loyal Oppositionist', who was urging Moscow to yield to the French. Trotsky wrote to Rakovsky and asked him to keep in mind that his role in Paris had become an issue in the inner-party struggle.[2] Rakovsky's devotion to the Opposition and to Trotsky personally was such that the reminder could not but impress him. But even before he had received it, he took a step which caused one of the great diplomatic scandals of the time. He put his signature to a Manifesto, calling workers and soldiers in capitalist countries to defend the Soviet Union in case of war. In these years of 'stabilization' and 'normalcy' in diplomatic relations with bourgeois governments, it was not customary for Soviet Ambassadors to make such revolutionary appeals. The French press fulminated. The French government declared Rakovsky *persona non grata*. Aristide Briand, the Foreign Minister, declared that the Soviet government should be all the more willing to recall its unruly Ambassador because it was anyhow improper that an adherent of the Opposition should represent it in Paris.

Moscow's reply was ambiguous. Chicherin, as Commissar of Foreign Affairs, defended his Ambassador, but the French ministry had reason to think that its attacks on Rakovsky were not altogether unwelcome to Chicherin's superiors. Trotsky maintained that Stalin played a disloyal game over Rakovsky's recall, and that the Soviet Foreign Office should have told Briand bluntly not to meddle in the inner affairs of the Bolshevik party. However, since the French government had declared Rakovsky *persona non grata*, Moscow had no choice but to

---

[1] See Degras (ed.), *Soviet Documents on Foreign Policy*, vol. ii, pp. 247–55.

[2] Trotsky's letter to Rakovsky of 30 September 1927 is in *The Archives*.

recall him. Rakovsky, distinguished though he was as diplomat, had chafed at his foreign assignments and was eager to plunge back, after an interval of four years, into the struggle at home. Trotsky, too, was glad to have his old friend back at his side. The Opposition derived some credit from the circumstances of Rakovsky's recall: the fact that one of its chiefs had drawn on himself the enmity of a bourgeois government, because he had appealed to foreign workers and soldiers to defend the Soviet Union, strikingly refuted the charges about the Opposition's defeatism and the 'united front from Chamberlain to Trotsky'.

Stalin, realizing that it was not enough to heap accusations on his adversaries, now endeavoured to enhance his popularity in a more positive way. The Opposition had, in its *Platform*, renewed the demands which it had made the year before and which the ruling factions then pretended to fulfil. It asked that wage increases be granted to poorly paid workers, that the eight-hour day be strictly observed, that tax reliefs be accorded to *bednyaks*, and so on. The *Platform* asserted that the ruling factions had honoured none of their promises; and that the conditions of the proletarian and semi-proletarian masses had gone from bad to worse. In reply to this Stalin made a startling move: he announced that the government would shortly introduce a seven-hour working day and a five-day week, and that the workers would receive the same wages as before. The occasion for the promulgation of the reform was to be the approaching tenth anniversary of the October Revolution, on which the Polit-bureau was to address the nation in a solemn Manifesto hailing the seven-hour day as the greatest achievement of socialism so far—the consummation of the first decade of revolution.

This was sheer duplicity. The Soviet Union was too poor to afford such a reform—even thirty years later, after it had become the world's second industrial power, its workers still worked eight hours a day and six days a week.[1] However, Stalin was not concerned with the economic realities of the case.

---

[1] The seven-hour day and five-day week were nominally in force for about thirteen years, but were not honoured in practice. At the beginning of the Second World War the return was decreed to the normal week and the eight-hour day and these remained obligatory for nearly two decades. Only in 1958 was a *gradual* return to the seven-hour day (but not yet to the five-day week) initiated.

He produced his sensational piece of legislation without discussing it beforehand with the trade unions, the Gosplan, and
even the Central Committee. The Bukharinists had misgivings.
Tomsky, who led the trade unions, did not hide his dislike of
the stunt. Stalin, nevertheless, forced it through; and for the
middle of October a special session of the Central Executive
Committee of the Soviets was convened to Leningrad to give its
formal and solemn sanction.

At the session, on 15 October, after Kirov had presented an
official report, Trotsky demonstrated the spuriousness of the
scheme. He recalled that when the Opposition had urged a
modest increase in wages, the claim had been indignantly rejected as threatening to strain the nation's economic resources.
How then could the economy now sustain a seven-hour day?
The Opposition maintained that even the eight-hour day was
not properly observed in state-owned industry—why then did
Stalin suddenly pull this grand reform out of his hat? Would it
not have been more honest to offer the workers some more
modest but real gains? It was a shame to commemorate the
revolution with such conjuring tricks. Trotsky pointed out that
none of the blueprints for the first Five Year Plan, which had
just been completed after years of preparation, contained as
much as a hint of the shorter working day. How then could they
genuinely shorten it when they had planned for years ahead on
the assumption of a longer working day? The whole reform,
he concluded, had one purpose only—to assist the ruling group
in its show-down with the Opposition.

In this dispute reason, truth, and honesty were all on Trotsky's side; and not for the first or the last time they led him
immediately into a trap. Nothing suited Stalin better than
Trotsky's protests. The Stalinists flocked to the factories to tell
the workers of Trotsky's latest indignity. He wanted, they said,
to rob the workers of the boon the party bestowed on them; he
was obstructing the epoch-making reform in which all could
see the dawn of socialism. What was the good of all his professions of Bolshevik loyalty and of all his posturing as a champion of the working class? To the men in the factories Trotsky's
arguments were unknown. Old and cool-headed workers may
have guessed them and have had their own thoughts about
Stalin's dubious gift. But the great credulous mass hailed it and

was impatient with the critics. The Opposition had mostly argued about issues which were high above the workers' head: Kuomintang, the Anglo-Soviet Council, permanent revolution, Thermidor, Clemenceau, &c. The one point at which the Opposition's language had not been abstruse was its demand for an improvement in the workers' lot. The demand had gained it wide, though passive, sympathy. Much of that sympathy was now dissipated. The wall of indifference and hostility closed around the Opposition.

Yet—so strong is sometimes in men 'the wish for that for which they only faintly hope'—precisely at this moment a freakish event brought the leaders of the Opposition solace and encouragement. During the session at which the seven-hour day was debated official demonstrations were staged in Leningrad to honour the event. There was the usual pomp and circumstance. Party leaders reviewed a parade and a march past of vast multitudes. Trotsky and Zinoviev were not to be seen among the leaders. By chance or choice, as if to demonstrate their divorce from officialdom, they stood on a lorry some distance away from the official stands, in a place where the demonstrators were due to pass on the way from the parade. Behind Trotsky's back was the Tauride Palace where ten years ago he had thundered against Kerensky and stirred the workers of the capital to enthusiasm, action, revolt. The columns of demonstrators, having marched past the official stands, approached. They recognized the two chiefs of the Opposition, halted, moved on, and halted again, stared mutely, raised hands, gesticulated, waved caps and handkerchiefs, moved on, and stopped again. The throngs around the lorry grew and traffic was blocked, while the space around the official stands became empty. It was as if an echo of the enthusiastic acclaim and clamour of the crowds of 1917 had come back. In truth, the crowd in front of Trotsky and Zinoviev, although visibly agitated, was subdued and timid. Its behaviour was ambiguous. If it intended to demonstrate sympathy with the Opposition, the demonstration was little more than a dumb show. It expressed the crowd's respect or pity for the defeated, not any readiness to fight on their side.

But the chiefs of the Opposition mistook the temper of the demonstrators. 'This was a silent, vanquished, stirring acclamation', so an eye-witness describes the scene. But 'Zinoviev and Trotsky received it with definite joy, as a manifestation of strength. "The masses are with us!" they said the same evening.'[1] The incident had a sequel quite out of proportion to its importance. It was largely under its impression, hoping that the masses were indeed with them at last, that the leaders of the Opposition decided to make a direct 'appeal to the masses' on the anniversary of the revolution, three weeks later. The ruling factions, on the other hand, saw in the ambiguous behaviour of the crowd a warning: they realized that they must take no chances with the temper of the people.

Stalin presently returned to the attack. On 23 October he once again asked for Trotsky's and Zinoviev's expulsion from the Central Committee. At length after four months he had worn down hesitation and resistance in the men who constituted the party's supreme tribunal. At last they were ready to do his bidding. But their fears and misgivings were still with them and showed in an extraordinary nervousness and the violence of the proceedings. There was a morbid tenseness in the air, such as might be felt at an execution where hangman and accomplices view their victim with deep hatred but also with deep awe and with gnawing uncertainty about the justice of the deed and the consequences. Whatever the victim says or does stirs in them these contradictory emotions which rise to a pitch of fury. All are convinced that the victim must die if they are to live; and all shudder at the thought of the horrors which may follow. They try to rout their own qualms by urging on the hangman, demanding haste, and hurling shameless abuse and heavy stones at the condemned. Such was the behaviour of the Stalinists and the Bukharinists at this session. They constantly interrupted Trotsky's last pleas with bursts of hatred and vulgar vituperation. They shut their ears to his arguments; and they urged the chairman to shut his mouth. From the chairman's table inkpots, heavy volumes, and a glass were flung at Trotsky's head while he spoke. Yaroslavsky, Shvernik, Petrovsky, President of the Ukraine, and others loudly incited Stalin and prompted him to get on with the job. There was no end to the

---

[1] Victor Serge, *Mémoires d'un Révolutionnaire*, p. 239. Trotsky's own description of the same scene in *Moya Zhizn*, vol. ii, p. 278, seems to reflect something of the wishfulness with which he first viewed the demonstration.

threats, gibes, and curses, which made this assembly look like a meeting of damned souls.[1]

Of the ruling group Stalin alone spoke with self-possession, with coarse and cold hatred, and without a trace of any qualm. He went over the familiar list of charges; and his speech—it was in it that he justified the employment of *agents provocateurs* (the 'Wrangel officer') against party members—was a feat of cynicism even for him.[2] Only Trotsky spoke with equal self-possession. His voice rose above the delirium for a last challenge before his departure. He warned the factions that Stalin's aim was nothing less than the extermination of all Opposition; and, amid cries of mockery, he forecast the long series of bloody purges in which not only his adherents but many Bukharinists and even Stalinists would be engulfed. He expressed the wishful assurance that Stalin's triumph would be shortlived and that the collapse of the Stalinist régime would come suddenly, with a crash. The victors of the moment, he said, relied on violence too much. True, the Bolsheviks had achieved 'gigantic results' when they had used violence against the old ruling classes and the Mensheviks and Social Revolutionaries who all stood for lost or reactionary causes. But they could not destroy in this way an Opposition which stood for historic progress. 'Expel us—you will not prevent our victory', these were the last words the party's supreme council heard from Trotsky's mouth.

.   .   .   .

Weeks of intensive activity followed. The Opposition still collected signatures for the *Platform* in the hope of impressing party opinion by the number of its adherents. Zinoviev was confident that 20,000 or 30,000 signatures would be obtained, that Stalin, confronted with evidence of such massive support, would have to stop short of further reprisals, and that the Opposition might even stage a come-back. The chiefs of the Opposition decided to make, on the anniversary of the revolution, that 'appeal to the masses' which had tempted them since the demonstration in Leningrad. It was not easy to determine the form of the appeal. Its purpose was to make the masses aware

---

[1] In a letter written to the Secretariat of the Central Committee on the next day Trotsky protested against the incomplete account of his speech in the official record and the omission of any reference to these scenes. *The Archives.*

[2] Stalin, *Sochinenya*, vol. x, pp. 172–205.

of the Opposition's demands and to stir them against the official leaders, without, however, giving the latter ground for charging the Opposition with a breach of discipline. The two things could hardly be reconciled; and the members of the Opposition spent days and nights deliberating and preparing for the trial of strength.

Trotsky, like his comrades, now spent most of his time in the suburban homes of humble workers, as he used to do when he was a young and unknown revolutionary, arguing, explaining principles and viewpoints, and instructing small groups of ardent and anxious adherents. At this moment he little resembled the Robespierre of the eve of Thermidor with whom he had compared himself. Two different characters seemed to have blended in him—Danton's and Babeuf's; but at this moment he looked more like the latter, the hunted leader of the Conspiracy of Equals, raising the cry for the regeneration of the revolution and defying the implacable builders of the new Leviathan state; and the tide of history ran against him as powerfully as it had run against Babeuf.

About fifty people filled a poor dining room [thus Victor Serge describes a typical meeting], listening to Zinoviev, who had grown fat, pale, dishevelled, and spoke with a low voice; there was about him something flabby yet also something very appealing. . . . At the other end of the table sat Trotsky. Ageing before our eyes, greying, great, stooping, his features drawn, he was friendly and always found the right answer. A woman worker, sitting cross-legged on the floor, asks him suddenly: 'And what if we are expelled from the party?' 'Nothing can prevent communist proletarians from being communists', Trotsky answers. 'Nothing can really cut us off from our party.' Half-smilingly, Zinoviev explains that we are entering an epoch when around the party there will be many expelled and semi-expelled people more worthy than the party secretaries of the name of a Bolshevik. It was simple and moving to see the men of proletarian dictatorship, yesterday still powerful, returning thus to the quarters of the poor and speaking here as men to men and looking for support and for comrades. On the staircase outside volunteers stood guard, watching passages and approaches; the G.P.U. might raid us any moment.

Once I accompanied Trotsky when he left such a meeting held in a dilapidated and poverty stricken dwelling. In the street Lev Davidovich put up the collar of his coat and pulled his cap over his

PLATE IX

Leaders of the Opposition after their expulsion from the party in 1927
Sitting from left to right: Serebryakov, Radek, Trotsky, Boguslavsky, Preobrazhensky
Standing: Rakovsky, Drobnis, Beloborodov, Sosnovsky

PLATE X

Radio Times Hulton Picture Library

(b) Rykov as Premier of the Soviet Union

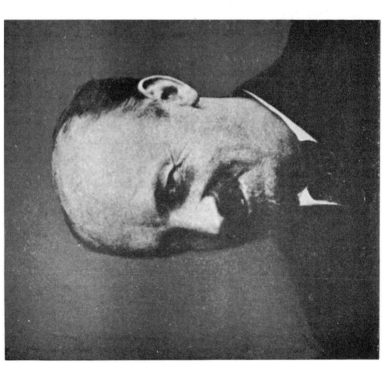

(a) Bukharin as chief of the Communist International

eyes so that he should not be recognized. He now looked an old intellectual still erect after twenty years of wear and tear. We approached a cabman. 'Please do the bargaining over the fare', Lev Davidovich said to me, 'I have very little money on me.' The cabman, a bearded peasant of the old type, leaned towards him and said: 'There is no fare for you to pay. Come on, comrade. You're Trotsky, aren't you?' The cap had not concealed sufficiently the man of the battles of Svyazhsk, Kazan, Pulkovo, and Tsaritsyn. A faint cheerful smile lit Trotsky's face: 'Don't tell anyone about this. Everyone knows that cabmen belong to the petty bourgeoisie whose favour can only discredit us.'[1]

When he told the woman worker sitting cross-legged on the floor: 'Nothing can really cut us off from our party', Trotsky did not offer a mere half-hearted consolation. He reckoned, as Zinoviev did, with mass expulsion; but he hoped against hope that this would come as a salutary shock; that the party's conscience would be aroused; that people would wish to see the *Platform* in order to find out for themselves what the Opposition stood for; and that then the great debate for which the Opposition had so many times asked in vain would begin for good. He imagined that Stalin would overreach himself: if thousands of party members were to be expelled as counter-revolutionaries, they would also have to be imprisoned. This could not but 'dismay the party' and make it realize that such an act of suppression might well mean 'the end of the proletarian dictatorship'. For the moment, indeed, many Stalinists and Bukharinists were uneasy at the thought of becoming the persecutors and jailers of their own comrades and comrades-in-arms. Stalin and Molotov had to assure them that things would not come to such a pass and that there would be no need for wholesale expulsion, because the Politbureau would so handle the Opposition as to induce it to halt before it was too late and to surrender. On 2 November Trotsky, quoting these assurances, appealed to the Opposition to remain as aggressive as ever—only then would the mass of Stalinists and Bukharinists, seeing that their leaders' boasts were deceptive, exert themselves to stop the persecution, and force the persecutors themselves to falter and surrender.[2] However, Stalin's and Molotov's boasts were not at all groundless: they sized up the weakness of the Opposition and foresaw

---

[1] Victor Serge, *Le Tournant obscur*, pp. 113–14. [2] *The Archives*.

that at the critical moment at least the Zinovievists would
stumble. Meantime, the assurances that there would be no need
for wholesale expulsion lulled disquiet and alarm and induced
the party to await events passively and thus to reconcile itself to
what was coming.

On the other hand, the torrents of abuse and threats let loose
on the Opposition hampered the Opposition's efforts. Few were
those who dared to put their signatures to the *Platform* daily
denounced as a subversive document. Instead of the 20,000 or
30,000 signatures for which Zinoviev had hoped, the Opposition
managed to collect only 5,000–6,000 at the most.[1] And the dread
of the consequences for those who signed was such that, in order
to protect their followers, the chiefs of the Opposition disclosed
only several hundred names. The campaign over the *Platform* was
thus yet another demonstration of the Opposition's weakness.

Trotsky was at this time, to quote Sedova, 'overworked, tense,
and suffering from ill-health, fever, and sleeplessness'. To enemies he presented an unyielding front, and to followers he gave
an example of self-control and heroic strength. But in domestic
privacy human frailty came back into its own. In vain did he
struggle with insomnia; drugs brought no relief. He complained
more and more often of headaches and dizziness. He was depressed and disgusted. At moments his sensitivity was almost
deadened by the stupendous venom and malignity that flooded
in from all sides. 'At breakfast we would see him opening the
newspapers', his wife writes, '. . . he glanced at them and disheartened scattered them over the table. All they contained was
stupid lies, distortions of the plainest facts, the most vulgar
abuse, hideous threats, and telegrams, from all over the world,
repeating zealously and with boundless servility the same in-
famies. . . . "What have they made of the revolution, of the party,
of Marxism, of the International!"'[2]

With Trotsky his next of kin drained the cup of defeat. Tense
and expecting the worst, the whole family suffered from in-

---

[1] This was the figure given by the Opposition. V. Serge, *Mémoires d'un Révolutionnaire*, p. 243. Stalinist sources claimed that the Opposition collected 4,000
signatures. According to N. Popov, the Stalinist historian, the Opposition received
6,000 votes out of a total of 725,000 in the elections to the congress. (*Outline
History of the CPSU*, vol. ii, p. 323.)

[2] *Vie et mort de Trotsky*, pp. 180–1.

somnia and through many a waking night awaited next day's blow—when day broke and friends came in, all put on a brave face and struggled on. Sedova, herself not very politically-minded, more at ease within the walls of museums and art galleries than amid arguing, scheming, and fighting party men, but moved by a woman's love and loyalty, was drawn completely into the cruel drama. Having given up her independent interests and placed herself in her husband's shadow, she lived with all her fibres his life, tried to think his thoughts, trembled with his anger, and shrank with worry and anxiety.

Their elder son, Lyova, now twenty-one, had spent his childhood and early youth, as he was to spend the rest of his short life, under the spell of his father's greatness. To be Trotsky's son, to share his ideas, and to follow in his footsteps had been for the adolescent and was for the young man a source of the greatest happiness. He had joined the Comsomol by subterfuge, before he had reached the age-limit, pretending to be older than he was, and he had tried to enlist in the Red Army too; he had left his parental home in the Kremlin to live in a commune-hostel, amid starving and tattered worker-students; and he had joined the Opposition the moment it had been formed. It was a galling experience for him to watch how the Comsomol, to whose members his father had so recently been a living legend and inspiration, was incited and turned against Trotskyism. With filial and revolutionary fervour he hated the men whom his father branded as bureaucrats corrupted by power. He spent years arguing and organizing Opposition groups, stumping party cells, and, side by side with such recognized leaders of the Opposition as Pyatakov and Preobrazhensky, addressing meetings in the provinces as far as the Urals. Youthful energy sustained his optimism and confidence; but in these weeks, amid rising bitterness and violence, he was seized with fear for his father's life, and became inseparable from him as assistant and body-guard, ready at any moment to jump at the throat of an assailant.

Unlike Lyova, Sergei, who was two years younger, had throughout adolescence rebelled against paternal authority and refused to be thrown in the shade by paternal greatness. The rebellion took the form of a revulsion against politics. He did not join the Comsomol; he would not hear of party affairs; and he would have nothing to do with the Opposition. Strong, brave, and adventurous, or, as his father and brother thought— light-minded, he revelled in games, sport, and the arts. Attracted by the circus (which in Russia then aspired to the dignity of an art in its own right), and, it seems, by a circus girl, he left his home in the Kremlin and spent a year or two with a troupe of performers. Having sown his wild oats, the prodigal was now back at home, still insisting on his independence and being sceptical about politics, but taking eagerly to mathematics and science for which he showed the outstanding ability that the father had shown at the same age. Yet a new sentiment began to break through the antagonism towards father and politics. The youth was moved by parental courage and sacrifice, out- raged by the things done to father and his co-thinkers, and anxiously preoccupied with the uncertainties and dangers of the day.

The other branch of the family, the one which had sprung from Trotsky's first marriage, was also deeply involved. Alexandra Sokolovskaya, ageing, yet firm in conviction and still as unafraid of voicing it against all and sundry as she had been as a solitary Marxist at Nikolayev in the 1890's, continued to be the rallying centre of the Trotskyists in Leningrad. Her two daughters, Zina and Nina, both in their middle twenties, lived in Moscow and were ardent Oppositionists. Both were as thrilled to be their father's daughters as they were when they watched him during his rise in 1917, and both were heartbroken. Both had married; each had two children; and the husbands of both, active Trotskyists, had lost jobs and livelihood and were already either expelled or on the point of being expelled from the party and deported to Siberia. Sunk into poverty, helpless, and tor- mented by anxiety about children, husbands, and parents, both women were ill with consumption, and were marked to be the first victims of a fate which was to destroy all of Trotsky's children.

.   .   .   .

As the tenth anniversary of the revolution approached, the Opposition got ready to make the 'appeal to the masses'. It in- structed its adherents to take part in the official celebrations of 7 November, but in such a way as to bring the Opposition's ideas and demands to the notice of the millions of people who would then fill the streets and squares of Soviet cities and towns. There was to be not a hint of incitement to insurrection or even disobedience. All that the members of the Opposition were to do was to march in closed ranks and as distinct groups within the official processions and to carry their own banners and slogans. These were so inoffensive outwardly, being directed against the ruling group merely by implication, that only the most politically minded of onlookers could distinguish them from the official slogans.

'Strike against the kulak, the N.E.P.-man, and the bureaucrat!', 'Down with opportunism!', 'Carry out Lenin's testament!', 'Beware of a split in the party!', 'Preserve Bolshevik unity!'—such were the Opposition's watchwords. They were designed to impress only party men and those outsiders who were intimately and sympathetically concerned with the trend of Bolshevik policy. One cannot therefore seriously describe the Opposition's action as any genuine 'appeal to the masses'—it was essentially an appeal to the party. But, driven from the party and denied access to its rank and file, the Opposition made the appeal from the outside, before the eyes of the nation and the world. Therein lay the weakness of the action. The Opposition tried to come into the open with a protest against the official conduct of party affairs and at the same time to demonstrate its own self-discipline and loyalty to the party. The protest, as it was planned, could therefore hardly become audible; and the show of self-discipline was to remain ineffective. On the dogmatically strictest interpretation of the rules—and no other interpretation could be expected from Stalin—a public demonstration against the party leaders did constitute a breach of discipline. In a word, the Opposition went either too far or not far enough. Yet such was its attitude and such were its circumstances that it had to go as far as it went and that it could go no farther.

The 7 November brought the Opposition a crushing defeat. Stalin had not been caught by surprise. He had issued strict orders for the swift suppression of any attempt at demonstration no matter how innocuous. No such attempt could from his viewpoint be innocuous, for if his adversaries succeeded this time, there was no saying whether they might not after all arouse now or later the brooding and disgruntled but cowed masses. Stalin knew that even while he was nearing the pinnacle, he might still slip and lose everything; and that, despite the crippling blows he had inflicted on his adversaries, they could still overwhelm him if he left them the slightest freedom of action. And so on 7 November squads of activists and police threw themselves on any group of Oppositionists who tried to unfurl a banner, display a picture of Trotsky or Zinoviev, or shout an unauthorized slogan. The Oppositionists were dispersed, abused, and beaten. Bare-handed, they tried to defend themselves, to regroup, and to demonstrate anew. The streets and squares became agitated with scuffles and police charges, and with dispersing and assembling crowds until even the least politically-minded person in the festive mass of onlookers became aware that he or she was witnessing a grave and critical event, that the inner-party struggle had moved from the cells into the street, and that in a way the contestants now appealed to all for support. It was indeed suppression that turned the Opposition's action into something like an appeal to the masses, and that surrounded it with an air of scandal and made it appear a semi-insurrection.

Victor Serge has left a vivid description of the day in Leningrad.[1] Since 15 October the Opposition had pinned much hope on the Leningraders, and Zinoviev had arrived there confident in their response. But the party machine on the spot, forewarned by the events of 15 October, was ready. At first Opposition groups, together with all other demonstrators, marched past the stands from which the official leaders took the parade; and they raised their banners and slogans. These attracted little attention. Then the police quietly surrounded the Oppositionists and isolated them. Serge describes how he himself, prevented by police barriers from joining the main demonstration, stopped to watch a procession of workers advancing with their red flags towards the centre of the city. Every now and then activists turned towards the marching men and women and shouted slogans. The men and women took the cues apathetically. Then Serge himself made a few steps towards the column and exclaimed: 'Long live Trotsky and Zinoviev!' or something to this effect. An astonished silence was the demonstrators' only answer. Then an activist, recovering his wits, shouted back in a voice charged with threat and fury: 'To the dustbin with them!' The marching

[1] V. Serge, *Mémoires d'un Révolutionnaire*, pp. 246–7.

workers remained silent. Serge felt that he had exposed himself and 'was going to be torn to pieces'. Suddenly a void surrounded him—he found himself facing the column alone, with only a woman and a child a few steps behind him. Through the void a student rushed to him and whispered into his ear: 'Let's go away. This may turn out badly. I am coming with you so that nobody should hit you in the back.'

In another quarter of the city, outside the Ermitage, 'a few hundred Oppositionists struggled good-humouredly with the militia'. A tall man in military uniform—this was Bakaev, former chief of the G.P.U. in Leningrad—led a 'human wave' against mounted policemen, who tried to halt it. Every time it was pushed back, the 'wave' surged forward and returned. At another place a group of workers followed a small stocky man in an attack on mounted police. The stocky man dragged a policeman off the saddle, knocked him down, then helped him to get up, and in a loud confident voice 'accustomed to command', shouted: 'You should be ashamed of yourself. You should be ashamed of charging on Leningrad workers.' The man thus venting his comradely anger was Lashevich, the ex-deputy Commissar of War, who 'had once been in command of great armies'. Similar scuffles occurred all over the city and lasted many hours. Groups of spectators watched in 'dumbfounded silence'. In the evening, at meetings of Oppositionists, Serge saw again Bakaev and Lashevich—they had come, their uniforms torn, to discuss the events of the day.

In Moscow the disturbances and fights had a far less 'good-humoured' and 'comradely' aspect. Commandos of activists and police struck with cold and swift brutality. The city was tensely aware of a crisis; and it was alarmed. 'There were rumours on the eve of the anniversary', notes an eye-witness who, however, all too eagerly picked up rumours from official quarters, 'that the army massed in the Red Square for the annual parade would demonstrate against Stalin. Some courageous soldier or officer would cry out "Down with Stalin!" and others would echo the slogan.'[1] Nothing of the sort happened, the writer remarks. At first here and there bands of Oppositionists, marching towards the Lenin Mausoleum, managed to unfold a few banners; but before they reached the Red Square they were

[1] L. Fischer, *Men and Politics*, p. 92.

surrounded by commandos who tore the banners to shreds and made the Oppositionists move on with the official demonstration. Thus, hemmed in by their adversaries and awkwardly silent, keeping in step with the rest of the procession, the Oppositionists marched past the leaders and foreign guests assembled at the Red Square. Only 'the Chinese students of Moscow's Sun Yat-sen University . . . formed a long, sinuous dragon. In the middle of the Square they threw Trotsky's proclamations in the air.' Beyond the Square the Oppositionists were kicked out of the common ranks, attacked with truncheons, and dispersed or arrested. In various places Oppositionists had hung out of be-flagged windows portraits of Lenin and Trotsky. Everywhere these were torn down and those who had hung them out were maltreated. At the House of Soviets, Smilga, who had returned from Khabarovsk, decorated his balcony with such portraits and put out the slogan 'Carry out Lenin's testament!' A gang of toughs broke into his home, tore up the pictures and the banner, demolished the dwelling, and mauled the man who had ten years ago brought the Baltic Fleet into the Neva at Petrograd to assist in the October insurrection—his crime was to display the picture of the leader of that insurrection. Among others, Sedova, finding herself in a group of demonstrators, was beaten up.

Trotsky accompanied by Kamenev and Muralov spent the day touring the city by car. At Revolution Square he stopped and attempted to harangue a column of workers marching towards the Lenin Mausoleum. At once policemen and activists assailed him. Shots were fired. There were shouts: 'Down with Trotsky, the Jew, the traitor!' The windscreen of his car was smashed. The marching column watched the scene uneasily, but moved on.

What was going on in the minds of the multitudes which crowded the festive streets? Nobody knew, nobody could even guess. They marched obediently along the prescribed routes, shouted the prescribed slogans, and observed mechanically the prescribed discipline, without betraying their thoughts or venting their feelings in a single flash of spontaneity. What a contrast they formed with the hungry, rough, warm-hearted, generous, enthusiastic, and drunken crowds of 1917! What a contrast there was between the present landscape of the cities and that of the revolution now commemorated! And what a contrast in the fortunes of the leaders! Ten years ago the workers of the two capitals were ready to give their lives at Trotsky's word of command. Now they would not even turn their heads to listen to him. Ten years ago Trotsky, as he watched Martov leading the Mensheviks in their exodus from the Soviets, shouted at him triumphantly: 'Go, go to the dustbin of history!'; and a thunder of Bolshevik applause covered his voice. 'To the dustbin with him!'—these words resounded now like a mocking echo through a square of Leningrad, when an Oppositionist tried to honour Trotsky's name. Had the wheel of history, so Oppositionists wondered, turned back or broken to pieces? Was this perhaps the Russian Thermidor?

These questions engaged Trotsky's thoughts also. He saw so many of the men who had led the Bolshevik revolution now aligned on his side. It seemed preposterous to assume that his and their defeat and humiliation had no deeper historic meaning, and that it did not mark that 'downward movement' of the revolution, that 'second chapter' of which he had spoken at the Central Committee a few months earlier. And yet he also saw that, changed as was the landscape of the revolution—its climate and colour—its broad and bare outline stood out as sharp as ever, unshaken and unaltered. It was still the Bolshevik party that ruled the republic, the party to which the Opposition still swore undying allegiance. He still regarded the republic, for all its 'bureaucratic degeneration', as a proletarian dictatorship; and he still resolutely dissociated himself and the Opposition from all those who branded it as a new police state, ruled by a 'new class' which had severed all ties with the working class and socialism. He refused to consider the bureaucracy as a new exploiting class—he viewed it as a 'morbid outgrowth on the body of the working class'. Public ownership, wherever Bolshevism had established it, was still intact. The kulak and the N.E.P.-man had not yet won. The antagonism between the workers' first state and world capitalism was unabated, even though it did not show itself in any clash of arms. So much had changed; and yet—so little. It was as if a hurricane had descended upon the scene, had flung the actors in opposite directions, had shifted all it could shift, had swayed the scene this way and that, but had left its frame solid and unimpaired. It seemed impossible that this should be the end—surely the hurricane portended an earthquake? Trotsky concluded that the 7 November was 'not yet the Soviet Thermidor', but that it was certainly 'the eve of Thermidor'.[1]

Serge relates that on the evening of 7 November, when the Oppositionists of Leningrad met, two voices could be heard: 'Nothing doing, we shall go on fighting', one voice repeated grimly. 'Against whom shall we fight?', asked another in anguish, 'Against our own people?' The same voices could be heard wherever Oppositionists met. As a rule it was the Trotskyists who asserted that they would go on fighting, and it was the Zinovievists who asked the awkward question. Zinoviev himself had returned from Leningrad in utter dejection; and he and Kamenev began to regret the luckless attempt to 'appeal to the masses' which they had undertaken with so much soaring confidence. Trotsky knew no regrets. The Opposition had done what it had to do; it could not undo what it had done: *Advienne que pourra*, he repeated. On the morrow of the fateful day he asked the Politbureau and the Presidium of the Central Control Commission for an official inquiry into the events; and he still took a rather sanguine view. He told his adherents that the up-shot of the demonstrations was not so bad: the Opposition had put on its banners the watchword 'Preserve Bolshevik unity' and had thereby shown where it stood and had at last wrested from Stalin a slogan from which he had sought to profit. Zinoviev and Kamenev replied that 7 November had brought them to the very brink of schism and that if the Opposition wished to preserve Bolshevik unity it had to retrace its steps.

For a few days they disputed what to do next. Trotsky soon abandoned his view of the consequences of the 7 November. Only five days after he had written how contented he was that the Opposition had 'wrested the slogan of unity' from Stalin, he argued that it was 'too late to talk about unity', because the party machine had become a will-less 'tool of Thermidorian forces' and was bent on routing the Opposition in the interest of kulak and N.E.P.-man.[2] Zinoviev and Kamenev were not sure of this: they noticed shifts of emphasis in Stalin's policy and

---

[1] See Trotsky's 'Balance of the Anniversary', written on 8 November, in *The Archives*.

[2] See his 'Zapiska' (Note) of 13 November in *The Archives*.

said that he was turning against the kulak and the N.E.P.-man. In any case they did not agree that it was 'too late to talk of unity'.

On 14 November the Central Committee and the Central Control Commission, convened for an extraordinary session, expelled from the party Trotsky and Zinoviev as guilty of incitement to counter-revolutionary demonstrations and virtually to insurrection.[1] Rakovsky, Kamenev, Smilga, and Evdokimov were expelled from the Central Committee; Bakaev, Muralov and others, from the Central Control Commission. From the party cells hundreds of members were ejected. Thus, after months and years in the course of which all factions hesitated and manœuvred, advanced, retreated, and went on contending, the schism was accomplished.

.   .   .   .

On the evening of 7 November Trotsky returned home and told his family that they must vacate their lodgings in the Kremlin. He himself moved out at once: he felt safer outside the Kremlin, and more than ever out of place in the residence of the ruling group. He took, provisionally, a tiny room at Granovsky Street 3, at the home of Beloborodov, an Oppositionist who was still Commissar of Home Affairs in the Russian Federal Republic, the man who, in 1918, had ordered the execution of Nicolas II at Ekaterinburg. For a few days Trotsky's whereabouts were unknown. The ruling group, somewhat alarmed, wondered what he might be up to and whether he had not 'gone underground'. He had no such intention; and for so well known a man it was impossible to go underground. On the day after his expulsion he gave his new address to the secretary of the Central Executive of the Soviets of which he was nominally still a member.[2] Leaving the Kremlin he saved himself a humiliation to which the other Opposition leaders were subjected: on 16 November they were all evicted. A friend describes their strange exodus from the Kremlin. Zinoviev left carrying only Lenin's death mask under his arm, a mask which was so depressing that the censorship had never allowed any reproduction

---

[1] *The Archives*; *KPSS v Rezolutsyakh*, vol. ii, pp. 368–70.
[2] He also notified the Executive that his wife and one of the sons were ill and unable to move, but that they would vacate their dwelling within a few days. *The Archives*.

of it to be published and so it remained Zinoviev's property. Then Kamenev came out, who in his early forties had grown suddenly white-haired and looked 'a handsome old man with very clear eyes'. Radek packed his books, with the intention of selling them; and handing out to those around him volumes of German poetry as souvenirs, muttered sarcastically: 'Haven't we been idiots! We are left penniless when we could have prepared a nice war chest. Lack of money is killing us. With our famous revolutionary probity we have been but feckless intellectuals full of scruples. . . .'[1]

Simultaneously another man made his exit in a different manner. In the evening of 16 November a revolver shot suddenly broke the stillness of the Kremlin. Adolf Abramovich Yoffe had committed suicide. He left a letter to Trotsky explaining that this was the only way in which he could protest against Trotsky's and Zinoviev's expulsion and express his horror at the indifference with which the party had received it. He had been Trotsky's disciple and friend since before 1910 when, a neurotic student, he helped Trotsky to edit the Viennese Pravda. With Trotsky he had joined the Bolshevik party in 1917, and was a member of the Central Committee at the time of the October insurrection. Soft-hearted, soft-smiling, and soft-spoken, he was one of the most determined advocates and organizers of the rising. He soon became one of the great Bolshevik diplomats: he led the first Soviet delegation to Brest Litovsk and was the first Soviet Ambassador in Berlin; he negotiated the peace treaty with Poland in 1921, and the pact of friendship between Lenin's and Sun Yat-sen's governments a year later; and he was Ambassador in Vienna and Tokyo. At the beginning of 1927 he returned from Tokyo, gravely ill with tuberculosis and polyneuritis, and was appointed Trotsky's deputy at the Concessions' Committee. In Moscow doctors held out no hope for him and urged him to take a cure abroad. Trotsky intervened on his behalf with the Commissar of Health and the Politbureau;[2] but the Politbureau refused to send him abroad on the ground that the cure would cost too much—1,000 dollars. An American publisher had just offered to pay Yoffe 20,000 dollars for

---

[1] V. Serge, Le Tournant obscur, p. 140.

[2] Trotsky's letters to Semashko, the Commissar of Health (20 January 1927), and to the Politbureau are in The Archives.

memoirs; and Yoffe asked to be allowed to leave at his own expense. Stalin then forbade him to publish memoirs, refused him an exit permit, deprived him of medical assistance, and harassed him with every kind of vexation. Bedridden, pain-stricken, penniless, and depressed by the savageness of the on-slaught on the Opposition, he blew out his brains.[1]

Yoffe's farewell letter is important not only for the light it throws on his attitude to Trotsky—it is also unique as a human and political document and a statement of revolutionary morality.

The letter begins with Yoffe's justification of his suicide, an act which revolutionary ethics normally condemned. In his youth, he recalled, he had stood up against Bebel in defence of Paul and Laura Lafargue, Marx's son-in-law and daughter, who had committed suicide when old age and infirmity had made them useless as fighters.

All my life I have been convinced that the revolutionary politician should know when to make his exit and that he should make it in time . . . when he becomes aware that he can no longer be useful to the cause he has served. It is more than thirty years since I embraced the view that human life has sense only in so far as it is spent in the service of the infinite—and for us mankind is the infinite. To work for any finite purpose—and everything else is finite—is meaningless. Even if mankind's life were to come to a close this would in any case happen at a time so remote that we may consider humanity as the absolute infinite. If one believes, as I do, in progress, one may assume that when the time comes for our planet to vanish, mankind will long before that have found the means to migrate and settle on other younger planets. . . . Thus anything accomplished in our time for mankind's benefit will in some way survive into future ages; and through this our existence acquires the only significance it can possess.

Having thus expressed in the Marxist idiom and in an atheistic spirit the ancient human longing for immortality, the immortality of mankind and of its genius, Yoffe went on to say that for twenty-seven years his life had had its full significance: he had lived for socialism; he had not wasted a single day, for

[1] Even while Yoffe was writing the letter to Trotsky his wife came to tell him that the Politbureau had declined his latest request for permission to go abroad for a month or two.

even while in prison he used every day to study and prepare
himself for future struggles. But now his life had become purposeless; and it was his duty to depart. Trotsky's expulsion and
the silence in which the party had witnessed it were the last
blows. Had he been in good health, he would have fought on
in the ranks of the Opposition. But perhaps his suicide, 'a small
event compared with your expulsion' (and 'a gesture of protest
against those who have reduced the party to such a condition
that it is unable to react in any way against this monstrosity')—
perhaps his suicide would contribute to arouse the party to the
Thermidorian danger. He feared that the hour of the party's
awakening had not yet come—all the same his death would be
more useful than his life.

With the utmost modesty, evoking their long friendship and
common work, Yoffe excused himself for 'using this tragic
opportunity' to tell Trotsky where, in his view, Trotsky's weakness lay. He had wanted to tell him this earlier, but could not
prevail upon himself to do so. He had never had any doubt that
Trotsky had been politically in the right ever since 1905. He had
heard Lenin himself saying this and admitting that not he but
Trotsky had been right in the old controversies over permanent
revolution. 'One does not tell lies before one's death, and once
again I am repeating this to you now.'[1] 'But I have always
thought that you have not enough in yourself of Lenin's unbending and unyielding character, not enough of that ability
which Lenin had to stand alone and remain alone on the road
which he considered to be the right road. . . . You have often
renounced your own correct attitude for the sake of an agreement or a compromise, the value of which you have overrated.'
In this, his last word, therefore, he wished that Trotsky should
find in himself that 'unyielding strength' which would help
their common cause to eventual even if delayed triumph.

The criticism, coming from the depth of a dying friend's devotion and love, could not but move and impress Trotsky: he
was to stand almost alone, 'unbending and unyielding', for the
rest of his life. Politically, however, Yoffe's suicide made no

---

[1] In his autobiography Trotsky relates that Yoffe had intended several times to
publish this conversation with Lenin and Lenin's admission but that Trotsky dis-
suaded him, because he feared that Yoffe would expose himself to attacks which
would ruin his health to the end. Yoffe's letter confirms this. The full text of the
letter is in *The Trotsky Archives*.

impact at all. His letter was not published—the G.P.U. had attempted to conceal it even from Trotsky who had as it were to wrest it from their hands. In the ranks of the Opposition the event spread depression; it was received as an act of despair. Trotsky feared that the example might be infectious. After the defeat of the 1923 Opposition several of its adherents—Eugene Bosch, a legendary heroine of the civil war in the Ukraine, Lutovinov, a prominent trade unionist and veteran of the Workers' Opposition, and Glazman, one of Trotsky's secretaries—took their lives. Now, when the Opposition was under an incomparably more brutal attack and saw no clear road ahead, there was even more ground for an outbreak of panic. Only after Yoffe's letter had been circulated in Opposition groups did the sense he had intended to give to his suicide become better known; and the deed came to be seen as an act of faith rather than of despair.[1]

On 19 November a long procession, headed by Trotsky, Rakovsky, and Ivan Smirnov, followed Yoffe's coffin through the streets and squares of Moscow to the cemetery of the Novodevichyi Monastery on the outskirts. It was the early afternoon of an ordinary working day—the authorities had arranged the funeral at this time in order to make it inconspicuous; but many thousands of people joined in the cortège and marched singing mournful tunes and revolutionary hymns. Representatives of the Central Committee and of the Commissariat for Foreign Affairs mingled with the Oppositionists—anxious to hush up the scandal, they had come to pay official homage to their dead adversary. When the cortège reached the Monastery —where Peter the Great once held his sister Sophia behind bars, and ordered several hundreds of her adherents to be slain under the window of her cell—police and G.P.U. tried to stop the procession outside the cemetery. The crowd forced its way into the alleys and gathered around the open grave. It received with an angry murmur an official spokesman who rose to make an oration. Then Trotsky and Rakovsky spoke. 'Yoffe left us', Trotsky said, 'not because he did not wish to fight, but because he lacked the physical strength for fighting. He feared to be-

---

[1] From the text which was circulated Trotsky omitted, as Yoffe had authorized him to do, those passages which expressed a certain pessimism about the immediate prospects of the Opposition.

come a burden on those engaged in the struggle. His life, not his
suicide, should serve as a model to those who are left behind.
The struggle goes on. Everyone remains at his post. Let nobody
leave.'

This gathering at a cemetery haunted by Russia's fearful past
was the Opposition's last public meeting and demonstration.
This was also Trotsky's last public appearance—and this his
cry for courage resounding amid graves his last public speech—
in Russia.[1]

.   .   .   .

'Everyone remains at his post! Let nobody leave!'—how
often had not these words appeared in Trotsky's Orders of the
Day at the worst moments of the civil war; and how many times
had they not led back into battle routed and disheartened
divisions, making them fight till victory! Now, however, the
words had lost their power. Zinoviev, Kamenev, and their
followers were already 'leaving their posts' and casting round
in desperation for an avenue of retreat. On the eve of Yoffe's
funeral Moscow was already astir with rumours about their surrender to Stalin. In a note, dated 18 November, Trotsky dismissed the rumours, declaring that Stalin had put them out in
order to confound the Opposition. Once again Trotsky maintained that repression worked in the Opposition's favour; and
he warned his adherents that they must continue to consider
themselves as belonging to the party, and that even expulsion and
imprisonment would not justify them in forming another party.
But if the Opposition accepted expulsion, Zinoviev and Kamenev replied, it would inevitably, even against its will, constitute
itself into another party. They were therefore obliged to do their
utmost in order to obtain an annulment of the expulsion. 'Lev
Davidovich,' they said, 'the time has come when we must have
the courage to surrender.' 'If this kind of courage, the courage

[1] The speech as well as an obituary on Yoffe are in *The Archives.* L. Fischer, who
witnessed the scene, writes that after the ceremonies 'everybody crowded towards
Trotsky to give him an ovation. Appeals were made to the people to go home. They
stayed, and Trotsky for a long time could not get out of the cemetery. Finally
young men linked elbows and formed two human chains facing one another with
a narrow corridor in between through which Trotsky could pass to the exit.' But
the crowd surged into that corridor and meantime Trotsky waited alone in a shed
on the cemetery; '. . . he never stood still. He walked like a pacing tiger. . . . I was
nearby and had the definite impression that he feared assassination.' L. Fischer,
op. cit., p. 94.

to surrender, were all that was needed,' Trotsky rejoined, 'the revolution should have been victorious by now the world over.'[1] They still agreed, however, to address a joint statement to the congress which was now convened for the beginning of December. In that statement, signed by 121 Oppositionists, they declared that they could not renounce their views, but that they recognized that the schism, leading to a struggle between two parties, was 'the gravest menace to Lenin's cause'; that the Opposition bore its share of responsibility, but not the major one, for what had happened; that the forms of inner-party contention must change; and that the Opposition, ready to disband its organization once again, appealed to the congress to reinstate the expelled and imprisoned Oppositionists.

It was clear that the congress would reject this appeal out of hand and that it would not agree to annul the expulsions. At this point the Joint Opposition was bound to dissolve and each of its two constituent groups to take its own road.

The congress was in session for three weeks; and it was wholly preoccupied with the schism. The Opposition had not a single delegate with voting rights. Trotsky did not attend; he had not even asked to be admitted in order to make a personal appeal against his expulsion. Unanimously the congress declared that expression of the Opposition's views was incompatible with membership in the party. Rakovsky tried to plead the Opposition's case; but he was driven from the rostrum. Then the assembly listened with amused astonishment to Kamenev as he gave a pathetic description of the Opposition's plight. He and his comrades, he said, were in this dilemma: either they must constitute themselves a second party—but this would be 'ruinous for the revolution' and would lead to 'political degeneration'; or else they must, 'after a fierce and stubborn struggle', declare their 'complete and thorough surrender to the party'. They had chosen surrender—they agreed, that is, to refrain from expressing any views critical of the official policy—because they were 'deeply convinced that the triumph of the correct Leninist policy could be secured only within and through our party, not outside it and despite of it'. They were therefore ready to submit to all decisions of the congress and to 'carry them out, no matter how hard they might be'.[2]

---

[1] Serge, *Le Tournant obscur*, p. 149.   [2] *15 Syezd VKP (b)*, pp. 245–6.

Having placed himself and his comrades at the mercy of the congress and going down on his knees, Kamenev then tried to stop half-way. The Oppositionists who capitulated, he said, acted as Bolsheviks; but they would not act as Bolsheviks if they were also to renounce their views. Never before had anyone in the party been asked to do this, he asserted, forgetting that he and Zinoviev had demanded it of Trotsky in 1924. 'If we should disclaim the views which we advocated a week or a fortnight ago, this would be hypocrisy on our part and you would not believe us.' He made another desperate attempt to save the capitulators' dignity: he pleaded for the release of the imprisoned Trotskyists: 'A situation in which people like Mrachkovsky are imprisoned while we are free is intolerable. We have been fighting together with those comrades. We are responsible for all their deeds.' He therefore implored the congress to give all Oppositionists the chance to undo what had happened. 'We beg you, if you wish this assembly to go down in history . . . as a congress of conciliation: give us a helping hand.'[1]

A week later the disintegration of the Joint Opposition was complete. On 10 December Zinovievists and Trotskyists separated and spoke in different voices. On behalf of the former, Kamenev, Bakaev, and Evdokimov announced their final acceptance of all decisions taken by the congress. On the same day Rakovsky, Radek, and Muralov declared that although they agreed with the Zinovievists about the 'absolute necessity' to maintain the single-party system, they nevertheless refused to submit to the decisions of the congress. 'For us to refrain from advocating our views within the party would amount to a renunciation of those views'; and in agreeing to this 'we would fail in our most elementary duty towards the party and the working class'.[2]

Zinoviev and his followers had in effect repeated what Trotsky had said in 1924—that the party was the only force capable of 'securing the conquests of October', the 'only instrument of historic progress', and that 'no one could be right against it'. It was this belief that led them to surrender. Trotsky and his adherents, on the other hand, were now convinced that they were 'right as against the party'; yet they resolved to fight on, thinking that they were fighting not against the party but for it—to save it from itself or rather from its bureaucracy. Both

---

[1] *15 Syezd VKP (b)*, p. 248. [2] Ibid., pp. 1286–7.

Trotsky and Zinoviev tried in fact to square the same circle, only that each attempted to do it in a different way. The Zinovievists hoped that by remaining within the party they might be able, circumstances permitting, to 'regenerate' it; the Trotskyists were convinced that this could be done only from without. Both repeated in the same words that any attempt to set up another party would be disastrous to the revolution; and both thereby implicitly admitted that the working class was, in their view, politically immature, that it could not be relied upon to support two Communist parties, that as yet it was therefore futile to appeal to the workers against the party bureaucracy, which, despite all its faults and vices, still acted as guardian of the proletarian interest, as trustee of the revolution and agent of socialism. If they had not thought so, the horror with which both Trotsky and Zinoviev spoke about 'another party' would have been inexplicable and ridiculous. In that case they should have, on the contrary, considered it their duty to set up another party. Recognizing their adversaries, if only implicitly and with grave reservations, as the guardians and trustees of the proletarian dictatorship, and being in conflict with them, the Oppositionists were caught in a contradiction. Zinoviev in his conscience sought to resolve the contradiction by accepting the dictates of the ruling factions. Trotsky, convinced that the ruling factions could not remain the guardians of the revolution for long, obeyed the dictates of his conscience which told him that nothing could be gained by self-renunciation.

While around him the Joint Opposition crumbled, expulsions multiplied and thousands of Oppositionists capitulated, Trotsky remained undaunted and contemptuous of the 'dead souls'— Zinoviev and Kamenev—predicting that they would be driven from surrender to surrender and from disgrace to disgrace, each worse than the other. The ruling factions were now in a rage of triumph. They were all the more boisterous because they had not been sure up to the last moment whether Stalin would indeed be able to manœuvre the Opposition into surrender. No sooner had Zinoviev and Kamenev announced their capitulation than the ruling factions declared that they did not accept it, and that the capitulators must fully repudiate their ideas and recant. At first Zinoviev and Kamenev had been given to understand that they would be reinstated if they agreed merely to refrain from voicing their views. Now that they had agreed to this, they were told that their silence would be an insult and a challenge to the party. 'Comrades', Kalinin said at the congress, 'what will the working class think . . . of people who declare that they will not advocate views which they still hold to be correct? . . . this is either deliberate deception . . . or these Oppositionists have become Philistines, keeping their views to themselves and not defending them.'[1] The ruling factions feared, in truth, that if they accepted Zinoviev's and Kamenev's first surrender, they would compromise themselves. What sort of a party is it, people would wonder, that allows its members to hold certain views but not to express them? The victors could not stop half-way. To hold the ground they had just gained, they had to gain more and drive their defeated opponents even farther. Having forbidden them to express heresy, the congress had to forbid them to profess it even by silence. Having deprived them of their voice, it had to rob them of their thought; and it had to give them back a voice so that they should use it to abjure their ideas.

Another week was filled with haggling over terms, a week during which the Zinovievists strained and struggled in the trap. They could not go back on their first capitulation; and in order to save its sense and to achieve what they had hoped to achieve through it, they stumbled into another capitulation. On 18 December Zinoviev and Kamenev returned and knocked at the doors of the congress to say that they condemned their own views as 'wrong and anti-Leninist'. Bukharin, it is related, received them with these words: 'You have done well to make up your mind—this was the last minute—the iron curtain of history is just coming down'—the iron curtain which, we may add, was to crush Bukharin as well. Bukharin was undoubtedly relieved to see Zinoviev and Kamenev return and submit, for he, like some other members of the ruling factions, had wondered anxiously what would happen if Zinoviev and Kamenev refused to recant and rejoined Trotsky. Even Ordjonikidze, who on behalf of the Central Control Commission presented the report and moved the motion for expulsion, showed his uneasiness when he said that the repressive measures hit men 'who have brought a good deal of benefit to our party and have

[1] Ibid., p. 1211.

fought in our ranks for many years'. But Stalin and the majority, drunk with jubilation, went on to kick the prostrated. They refused to reinstate them even after the recantation. By a strange quirk it was Rykov, one day to share Zinoviev's and Kamenev's fate, who went out to see them as they waited at the door and to slam the door on them. He told them that they were not readmitted into the party; that they were to remain on probation for at least six months; and that only afterwards would the Central Committee decide whether to reinstate them.

The defection of the Zinovievists left Trotsky and his adherents isolated. It soothed the none too sensitive consciences of many Stalinists and Bukharinists who saw in it the final vindication of Stalin's action. Surely Trotsky must be absolutely in the wrong, they reflected, if even his erstwhile allies turned their backs on him. The party and the nation had their eyes glued to the congress and the astounding spectacle of capitulation enacted there; they were not greatly concerned with that section of the Opposition which was not involved in the spectacle. The Trotskyists themselves were stunned. They were overcome by a sense of the finality of their break with the party. They viewed with incredulity the gulf which had opened between themselves and the Zinovievists. They wondered whether they themselves had not acted foolhardily: should they have conducted their semi-clandestine propaganda? should they have 'appealed to the masses' on 7 November? should they have precipitated the schism? Such qualms induced them to receive the verdicts of expulsion with unending and exalted declarations of their undiminished fidelity to the party. A few went in the footsteps of the Zinovievists; others wavered. The majority remained determined to fight on and face persecution. Yet no one knew who was and who was not a 'capitulator'. Immediately after the congress 1,500 Oppositionists were expelled and 2,500 signed statements of recantation.[1] But among those who signed a few withdrew, when they saw that one act of surrender entailed another; and among those who had refused to sign some weakened in their resolve when they were subjected to further intimidation, temptation, and persuasion. Those of one set viewed those of another as blacklegs or traitors. As it was not known

[1] Popov, op. cit., vol. ii, p. 327.

where one set ended and the other began, confusion and suspicion spread all through the whole of the former Joint Opposition.

Trotsky, seeing the futility of Zinoviev's surrender, was confirmed in his conviction that he had chosen the right road. He worked feverishly to impart his belief to his disheartened followers. He told them that no prudence or procrastination would have helped them, because Stalin would in any case have found the excuses needed for driving them out. What mattered was to rally those who stood fast, to draw a sharp line of division between them and the renegades, to avoid ambiguous attitudes, and to make the causes of the breach clear for contemporaries and posterity alike. Moreover, the Opposition could no longer work as it had worked hitherto—it must 'go underground' for good, find new forms of contacts between its groups and new methods of work, and establish links with its co-thinkers abroad.

Very little time was left for all this. Even before the year was out Stalin arranged to deport the Oppositionists. Yet the remorseless master of the bloody purges to come was still curiously concerned with his alibi and with appearances. He wished to avoid the scandal of an undisguised and forcible deportation and he tried so to stage the banishment of his enemies that it should look like a voluntary departure. Through the Central Committee he offered the leading Trotskyists minor administrative posts in the far corners of the immense country: Trotsky himself was to leave of his 'own free will' for Astrakhan, on the Caspian Sea. Early in January 1928 Rakovsky and Radek, delegated by the Opposition, and Ordjonikidze were engaged in fantastic bargaining over these proposals. Radek and Rakovsky protested against Trotsky's assignment to Astrakhan, saying that his health, sapped by malaria, would not stand the vaporous and hot climate of the Caspian port. The game was cut short when Trotsky and his friends declared that they were ready to accept any provincial appointments provided that these were no mere excuses for deportation, that the Opposition's consent was obtained for every appointment, and that the assignments were fixed with an eye to the health and safety of those concerned and of their families.[1]

---

[1] An account of the 'negotiations' is given in a letter written by Trotsky himself or by one of his friends to the Central Control Commission and the Politbureau in the first days of 1928. *The Archives.*

On 3 January, while the bargaining was still on, the G.P.U. summoned Trotsky to appear before it. He ignored the summons. The farce then came to an end; and a few days later, on the 12th, the G.P.U. informed Trotsky that under article 58 of the criminal code, i.e. under the charge of counter-revolutionary activity, he would be deported to Alma Ata, in Turkestan, near the Chinese Frontier. The date of the deportation was set for 16 January.

Two writers, one a complete outsider, the other a Trotskyist, have given their impression of Trotsky during his last days in Moscow. Paul Scheffer, correspondent of the *Berliner Tageblatt*, interviewed him on 15 January. At a 'superficial glance' he could see nothing indicating that a police watch was kept over Trotsky. (It may be assumed that the eye of the German journalist was not very skilled in detecting such indications.) He noticed excitement at Trotsky's home, the comings, goings, and leave-takings of men who were all about to be exiled, and the packing off for a long journey. 'In all corridors and passages there were piles of books, and once again books—the nourishment of revolutionaries, as ox blood used to be the nourishment of the Spartans.' Against this background he describes the man himself, 'somewhat less than middle-size, with a very delicate skin, a yellowish complexion, and blue, not large eyes, which at times can be very friendly, and at times are full of flashes and very powerful.' A large animated face 'reflecting both strength and loftiness of mind', and a mouth strikingly small in proportion to the face. A delicate, soft, feminine hand. 'This man, who has improvised armies and filled primitive workers and peasants with his own enthusiasm, lifting them high above their understanding . . . is at first shy, slightly embarrassed . . . that is perhaps why he is so captivating.'

Throughout the conversation Trotsky, though courteous, was on his guard, content to express himself *pro foro externo*, but extremely reticent *vis-à-vis* the bourgeois journalist about domestic issues. Not a single mention of his adversaries, no complaint, no polemics. Only once did the talk skirt inner-party affairs, when the interviewer remarked that Lloyd George had prophesied 'a Napoleonic future for Trotsky'. This was the nearest that Scheffer came to alluding to the deportation, to Trotsky's plans for the future, and so on. Trotsky, however, seized a different facet of the comparison: 'It is a strange idea', he answered somewhat amused, 'that I should be the man to put an end to a revolution. This is not the first of Lloyd George's blunders.' Characteristically, the comparison with Napoleon brought to Trotsky's mind not the obvious and superficial parallel between their personal fortunes as exiles, but the political idea, so abhorrent to him, of Bonapartism, the successor to Thermidor. With him the general problem took precedence over the personal. ('One is constantly reminded', Scheffer remarks, 'that this man is first and foremost a fighter.') He spoke mainly about the decay of capitalism and the prospects of revolution in Europe, prospects with which as always he linked the future of Bolshevik Russia. 'Trotsky's talk quickly loses its conversational tone, becomes oratorical, and soars high', and he illustrates the ups and downs of the curve of world revolution with 'beautifully melodious gestures'. The argument was interrupted by a comrade due to go into exile this very evening—he had come to ask whether there was anything he could still do for Trotsky. 'Trotsky's face, with the small upturned moustache, folds itself into many gay wrinkles: "You are going for a journey tonight, aren't you?" The man of controversy and irony will miss no opportunity. . . . The humour of the unshaken man is undimmed.' At the parting he invited Scheffer to come and visit him at Alma Ata.[1]

Unlike Scheffer, Serge described Trotsky's surroundings as 'watched day and night by comrades who were themselves watched by stool-pigeons'. In the street G.P.U. men on motorcycles noted every car coming and going.

I went up by the back stairs. . . . He whom among ourselves we called with affectionate respect The Old Man, as we used to call Lenin, worked in a small room facing the courtyard and furnished only with a field bed and a table. . . . Dressed in a well-worn jacket, active and majestic, his high crop of hair almost white, his complexion sickly, he displayed in this cage a stubborn energy. In the next room messages which he had just dictated were being typed out. In the dining room comrades arriving from every corner of the country were received—he talked with them hurriedly between telephone calls. All might be arrested any moment—what next? Nobody

[1] Paul Scheffer, *Sieben Jahre Sowjet Union*, pp. 158–61.

knew . . . but all were in a hurry to benefit from these last hours, for surely they were the last. . . .[1]

The day of 16 January, taken up by conferences, instructions, further leave-takings, and last preparations for the journey, passed as if in a fever. The hour of departure was fixed for 10 p.m. In the evening the whole family, exhausted and tense, sat waiting for the G.P.U. agents to appear. The appointed time had passed, but they did not come. The family was lost in guesses until the G.P.U. informed Trotsky by telephone, without offering any explanation, that his departure was postponed by two days. Further guesses, interrupted by the arrival of Rakovsky and other friends, all greatly excited. They had come from the station where thousands had gathered to give Trotsky a farewell. There was a stormy demonstration at the train by which he was expected to travel. Many lay down on the tracks and swore not to allow the train to leave. The police tried to remove them and to disperse the crowd; but the powers that be, seeing what turn the demonstration had taken, ordered the deportation to be postponed. The Opposition congratulated itself on the effect and planned to repeat the manifestation in two days' time. The G.P.U., however, decided to catch the Opposition by surprise and to abduct its leader surreptitiously. The plan was to take him to another station; to bring him to a little stop outside Moscow, and only there to put him on the Central Asian train. They had told him to be ready to depart on 18 January; but already on the 17th they called to apprehend him. Curiously, his followers failed to keep watch at his home; and so when the G.P.U. men arrived, they found on the spot only Trotsky and his wife, their two sons, and two women, one of whom was Yoffe's widow.[2]

A scene of rare tragi-comedy followed. Trotsky locked himself up and refused to let the G.P.U. in. This was a token of passive resistance, with which in the old days he had invariably met any police trying to lay hands on him. Through the locked door the prisoner and the officer in charge conducted a parley. Finally, the officer ordered his men to smash the door; and they broke into the room. By a weird freak the officer who had come to arrest Trotsky had served on Trotsky's military train

---

[1] V. Serge, *Le Tournant obscur*, p. 155.   [2] *Moya Zhizn*, vol. ii, p. 287.

during the civil war, as one of the body-guards. Face to face with his former chief, he lost countenance, broke down, and muttered distraught: 'Shoot me, comrade Trotsky, shoot me.' Trotsky did his best to comfort his jailer and even urged and persuaded him to carry out orders. Then he resumed the posture of disobedience and refused to dress. The armed men took off his slippers, dressed him, and, as he declined to walk out, carried him down the staircase, amid the shouts and the booing of Trotsky's family and of Yoffe's widow, who followed them. There were no other witnesses, apart from a few neighbours, high officials and their wives, who, startled by the commotion, peeped out and quickly hid their frightened faces.

The deportee and his family were bundled into a police car which then, in broad daylight, rushed through the streets of Moscow carrying away unnoticed the leader of the October Revolution and the founder of the Red Army. At the Kazan station—it was there that the escort took him—he refused to walk to the train; and armed men dragged him to a solitary carriage waiting for him at a shunting yard. The station was cordoned off and emptied of passengers, only a few busy railway men moved around. Behind the escort there followed the deportee's family. His younger son, Sergei, exchanged blows with a G.P.U. man, and the elder, Lyova, tried to arouse the railway workers: 'Look, comrades', he shouted, 'look how they are carrying off comrade Trotsky.' The workers stared with dry eyes—not a cry or even a murmur of protest came from them.

. . . .

Nearly thirty years had passed from the moment when the young Trotsky saw the towers and the walls of Moscow for the first time. He was then being transported from a jail in Odessa to a place of exile in Siberia; and it was from behind the bars of a prison van that he had his first glimpse of the 'village of the Tsars', the future 'capital of the Communist International'. It was from behind such bars also that he now had his last glimpse of Moscow, for he was never to return to the city of his triumphs and defeats. He entered it a persecuted revolutionary; and so he left it.

CHAPTER VI

A Year at Alma Ata

At a deserted little station, about 50 kilometres from Moscow, the single railway carriage in which Trotsky and his family had been taken away from the capital stopped; and it was hitched to a train bound for central Asia. Sergei, anxious to continue his academic curriculum, alighted and returned to Moscow. Sedova, ill with fever, and Lyova accompanied Trotsky into exile. A guard of about a dozen men escorted them. From the corridor through a half-open door the guardsmen kept watch on the prisoner and his wife slumped on the wooden benches inside a dark compartment dimly lit by a candle. The officer who had come to arrest Trotsky was still in command; his presence on this train was a grotesque reminder of that other and famous train, the *Predrevvoyen's*[1] field headquarters, on which he had served as Trotsky's body-guard. 'We were tired out', Sedova recalls, 'by the surprises, uncertainties, and tension of these last days; and we were resting.' As he lay in the dark or watched the endless white plain through which the train moved eastwards, Trotsky began to adjust his mind to his new circumstances. There he was, wrenched from the world with its tumult and fascination, cut off from his work and struggle, and isolated from adherents and friends. What was to follow now? And what was he to do? He bestirred himself to enter a few notes in his diary or to draft a protest; but he found—and this was something of a minor shock—that he had set out 'without writing utensils'—this had never happened to him before, not even during his perilous flight from the Far North in 1907. All was full of hazard now—he did not even know whether it was still to Alma Ata that he was being deported. Insecurity roused his defiant and contrary temper. He remarked to his wife that it was at least a consolation to know that he would not die a Philistine's death in a comfortable Kremlin bed.

On the next day the train stopped at Samara; and Trotsky

[1] *Predrevvoyen*—President of the Revolutionary Military Council.

telegraphed a protest to Kalinin and Menzhinsky saying that never during his long revolutionary career had any capitalist police treated him with such trickery and mendacity as the G.P.U. who had kidnapped him without telling him where they were taking him and made him travel without a change of linen and elementary amenities and without medicine for his sick wife.[1] The men of the escort were considerate and even friendly, however, as had been the Tsarist soldiers who had escorted him in 1907, as the convicted leader of the Petersburg Soviet. On the way they bought linen, towels, soap, &c., for the family; and they brought meals from railway stations. Their prisoner still inspired them with the awe with which a Grand Duke, deported under the old régime, might have inspired his guards: there was, after all, no knowing whether he might not be back in power soon. And so when the train arrived in Turkestan, the commander of the escort asked his prisoner to give him a certificate of good behaviour.[2] On the way Sermuks and Posnansky, Trotsky's two devoted secretaries, had joined the train, hoping to outwit the G.P.U. Such incidents broke up the monotony of the journey.

At Pishpek-Frunze[3] the railway journey came to an end. The stretch of road from there to Alma Ata, about 250 kilometres, had to be traversed by bus, lorry, sleigh, and on foot, across ice-covered and wind-swept mountain and through deep snow-drifts, with a night stop in an abandoned hut in the desert. At last, after a week's journey, on 25 January at 3 a.m., the party reached Alma Ata. The deportee and his family were put up at an inn called 'The Seven Rivers' in Gogol Street. The inn 'dated from Gogol's time'; and the spirit of the great satirist hovering over it seems to have suggested to Trotsky many of his observations on Alma Ata and the style of the frequent protests which he was to send from there to Moscow.

Towards the end of the 1920's Alma Ata was still a little town wholly Oriental in character. Although famous for gorgeous orchards and gardens, it was a slummy and sleepy Kirghiz backwater barely touched by civilization, and exposed to earth-quakes, floods, icy blizzards and scorching heat waves. The

---

[1] *The Archives.*　　　　　　　　　　[2] The text of the certificate ibid.
[3] The town Pishpek had just been renamed in honour of Frunze, Trotsky's successor as Commissar of War.

heat waves brought dense dust-clouds, malaria, and plagues of vermin. The town was to be developed as the administrative centre of Kazakhstan, but the republican administration was only begining to form itself; in the meantime officials requisitioned all available accommodation, and the local slums were more than usually overcrowded. 'In the bazaar at the centre of the town the Kirghizes sat in the mud at the doorsteps of their shops, warmed themselves in the sun, and searched their bodies for lice.'[1] Leprosy was not unknown; and during the summer of Trotsky's stay animals were struck by pestilence, and howling mad dogs swarmed in the streets.

In the same year life at Alma Ata was made even more miserable by continuous scarcity of bread. Within the first few months of Trotsky's arrival the price of bread trebled. Long queues waited outside the few bakers' shops. Foodstuffs other than bread were even scarcer. There was no regular transport. Mail was erratic; the local Soviet tried to regularize it with the help of private contractors. The gloom of the place and the helplessness and feeble-mindedness of the local caciques are well illustrated by this extract from Trotsky's correspondence: 'The other day the local newspaper wrote: "In town rumours are *functioning* to the effect that there is going to be no bread, whereas there are numerous carts coming with loads of bread." The carts are really coming, as they say; but in the meantime the rumours are functioning, malaria is functioning, but bread is not functioning.'

Here then Trotsky was to stay. Stalin was eager to keep him as far from Moscow as possible and to reduce him to his own resources. Trotsky's two secretaries were arrested, one *en route* from Moscow, the other at Alma Ata, and deported elsewhere. For the moment, however, Stalin appeared to have no further designs on his enemy; and the G.P.U. still treated Trotsky with consideration that would have been unthinkable later. It took care that his enormous library and archives, containing important state and party documents, should reach him—a lorryful of these presently arrived at Alma Ata. Trotsky protested to Kalinin, Ordjonikidze, and Menzhinsky against the conditions in which he was placed, demanding better accommodation, the right to go on hunting trips, and even to have his pet dog sent to

[1] *Moya Zhizn*, vol. ii, p. 296.

him from Moscow. He complained that he was kept at the inn at Gogol Street only to suit the G.P.U.'s convenience and that his banishment was virtual imprisonment. 'You could just as well have jailed me in Moscow—there was no need to deport me four thousand versts away.'[1] The protest was effective. Three weeks after his arrival he was given a four-roomed flat in the centre of the town, at 75 Krasin Street—the street was so named after his deceased friend. He was allowed to go on hunting trips. He showered further sarcastic telegrams on Moscow, making demands, some serious, others trivial, and mixing little quarrels with great controversy. 'Maya, my darling [Maya was his pet dog]', he wrote to a friend, 'does not even suspect that she is now at the centre of a great political struggle.' He refused, as it were, to consider himself a captive; and his persecutors made a show of leniency.

He appeared almost relaxed after so many years of ceaseless toil and tension. Thus, unexpectedly and oddly, there was a quasi-idyllic flavour about the first few months of his stay at Alma Ata. Steppe and mountain, river and lake lured him as never since his childhood. He relished hunting; and in his voluminous correspondence political argument and advice are often interspersed with poetic descriptions of landscape and humorous reports on hunting ventures. He was at first refused permission to go out of Alma Ata. Then he was allowed to go hunting but no farther than twenty-five versts away. He telegraphed to Menzhinsky that he would disregard the restriction because there were no suitable hunting grounds within that distance and he was not going to be bothered with small game—he *must* be allowed to go at least seventy versts away; and let Moscow inform the local G.P.U. about this so as to avoid trouble. He went; and there was no trouble. Then he protested to the chief of the local G.P.U. against being pursued rudely and conspicuously by sleuths and declared that because of this he would 'go on strike' and cease to hunt—unless this form of police supervision was prescribed directly by Moscow, in which case he understood the position of the local G.P.U. and waived his objections. The supervision became milder and less conspicuous.

He had begun to hunt soon after his arrival and went on as long as the spring migration of beasts along the river Ili lasted.

[1] From a protest sent out early in February. *The Archives.*

Some of the trips took as many as ten days and were strenuous and refreshing. In letters to friends he proudly described his hunting triumphs. At first he spent nights in Kirghiz mud huts or bug-ridden *yourtas*, sleeping alongside a dozen natives on the floor, boiling dirty water for tea, and barely containing nausea. 'Next time', he announced, 'I shall sleep in the open air and compel all my companions to do the same.'[1] Next time indeed —this was still before the end of March—the hunting party stayed in the open nine frosty days and nights. Crossing a river on horseback, Trotsky once slipped into the water. The booty was not large: 'about forty ducks in all'. True, he wrote to friends, bigger game could be found farther afield on the Balkhash Lake, even snow leopards and tigers; but 'I decided to make a pact of non-aggression with the tigers'. 'I enjoyed enormously ... this temporary relapse into barbarity. One does not often have such experiences as spending nine days and nights in the open, without having to wash, dress, and undress, eating venison cooked in a pail, falling from a horse into the river (this was the only time I had to undress), and staying days and nights on a small log in the midst of water, stone and reed.'[2] The hunting season over, fishing began; and then even Natalya Ivanovna joined in, although the fishing was no townsman's lazy week-end playing with tackle, for every trip was a long and arduous job with large boats, heavy loads, and elaborate tacking about.

Early in June, when the heat waves hit Alma Ata, the family moved to a *dacha* in the foothills of the mountains just outside the town, where they had rented a reed-thatched farm-house surrounded by a large apple orchard. From the house they could see the town below and the steppe beyond on one side and snow-capped mountain ranges on the other. During heavy downpours the thatched roof leaked and everyone rushed to the loft with pails and pots and pans. In the orchard a wooden hut was put up—Trotsky's study and workroom. It was soon crammed with books, newspapers, and manuscripts; and it shook with the pounding of a well-worn typewriter, which resounded through the orchard. At his writing table Trotsky watched a shrub as it pushed through a chink in the floor and

---

[1] *The Archives.*
[2] From a letter dated 1 April 1928 (no addressee) in *The Archives.*

in no time shot up to his knees. All this underlined the 'ephemeral character' of the abode; but it was a relief to have escaped from the town where through the dust-clouds people now chased and shot mad dogs in the street. Throughout the earlier months both Trotsky and Sedova had suffered from malaria and lived on a 'quinine diet'; now the attacks of fever had almost ceased.[1]

The deportee had to earn a living. True, he received an official allowance but this was a mere pittance, and, although the household was small and its needs very modest, the allowance was not enough to meet the rising cost of food. *Gosizdat*, the State Publishers, had just ceased to publish Trotsky's *Works*, of which thirteen volumes had so far appeared. These were already banished from bookshops and public libraries. Trotsky's head was full of new literary projects. He thought of writing a study of revolution in Asia and assembled a sizeable collection of reference works on China and India. In another book he planned to sum up Russian and world developments since the October Revolution. Immediately after his arrival at Alma Ata he set to work on a full-scale statement of the Opposition's principles which was to be addressed to the sixth congress of the Communist International convened for the summer. His friends, especially Preobrazhensky, urged him to write his memoirs. In April he was already at work on these, recapturing, with the help of old southern newspapers and maps of Nikolayev and Odessa, the image of his childhood and youth with which he was to open *My Life*.

None of these writings, however, could bring him any earnings, for there was no chance of their being published. Yet even a man deported under article 58, for 'counter-revolutionary activities', could still try to earn a living as translator, sub-editor, and proof-reader. When it turned out that the authors whom Trotsky would be allowed to translate, or the translations of whose works he was to supervise, were Marx and Engels, he took to the job eagerly. Ryazanov, his old friend, at present Director of the Marx-Engels Institute in Moscow, was preparing the complete edition in Russian of the *Works* of Marx and Engels; and he asked Trotsky to translate *Herr Vogt*. In this long and little-known pamphlet Marx had replied to slanders thrown at him by Karl Vogt who, as it turned out later, was an agent of

[1] See Trotsky's letter to Rakovsky of 14 July. Ibid.

Napoleon III. Reading this broadside for the first time, Trotsky remarked that it took Marx several hundred pages to refute Vogt's charges, whereas it would take his translator 'a whole encyclopaedia' to refute Stalin's slanders. Ryazanov then asked Trotsky to edit the translations and read the proofs of the other volumes of Marx and Engels, which he did.[1]

Trotsky's correspondence with Ryazanov shows the modesty and conscientiousness with which Trotsky applied himself to the work: it contains detailed, almost pedantic, criticisms of the style of the translations and minute suggestions for improvements. The correspondence is wholly non-political and studiously business-like. There is no hint of any irony on Trotsky's part about the only gainful occupation now left open to him in the Soviet Union. The fees paid him by Ryazanov supplied the family's needs and covered the cost of Trotsky's huge correspondence.[2]

. . . .

From the moment of his arrival at Alma Ata Trotsky worked hard to establish contact with friends and followers scattered all over the country and reduced to isolation and silence. At the beginning this could be done only by normal mail; and it had to be done in the most primitive conditions when it was sometimes a feat to obtain on the spot a pen, a pencil, a few sheets of rough paper, or a few candles. Lyova became his 'minister of foreign affairs and minister of post and telegraphs', body-guard, research assistant, secretary, and organizer of hunting trips. With his help a constant flow of letters and circulars began to pour out of Alma Ata in all directions. Twice or three times a week an invalid postman on horseback brought the mail bag bulging with letters, press clippings, and later even with books and newspapers from abroad. No doubt the censorship and the G.P.U. kept a watchful eye on the correspondence. Most of it was with Rakovsky, who had been deported to Astrakhan, Radek, who was at Tobolsk, Preobrazhensky, exiled to Uralsk, Smilga who was at Narym, Beloborodov, banished as far north as Ust-Kylom in the

---

[1] In one of his letters Trotsky mentioned that he was also translating the writings of Thomas Hodgkin 'the English Utopian Socialist'.

[2] Between April and October 1928 Trotsky mailed 800 political letters, many of essay length, and 550 telegrams; and he received 1,000 letters and 700 telegrams, apart from private mail.

Komi Republic, Serebriakov, who was at Semi-Palatinsk in Central Asia, Muralov at Tara, Ivan Smirnov at Novo-Bayazet in Armenia, and Mrachkovsky at Voronezh. Less systematically Trotsky corresponded with a host of other Oppositionists. Later in the year he related to Sosnovsky[1] that he was in more or less regular contact with all the major colonies of exiles in Siberia and Soviet Asia at large, with Barnaul, Kaminsk, Minussinsk, Tomsk, Kolpashevo, Yenisseisk, Novosibirsk, Kansk, Achinsk, Aktiubinsk, Tashkent, Samarkand, &c. With colonies in European Russia he communicated through Rakovsky, who from Astrakhan was in charge of Opposition centres along the southern Volga and in the Crimea, and through Mrachkovsky, who from Voronezh kept in touch with colonies in the north. In places where there were large centres of exiles, the correspondence and the circulars were duplicated and forwarded to minor colonies. Since April a secret postal service operated between Alma Ata and Moscow; it delivered and collected mail once every fortnight or three weeks.

In this way the groups of exiles, constantly growing in numbers and size, formed a community of their own with its own intense political life. Trotsky was the inspirer, organizer, and symbol of the Opposition in exile. The state of mind of the deportees was anything but settled. Some were stunned by what had happened. Others viewed the persecution to which they were subjected as little more than a bad joke. The majority at first appeared to be convinced that Stalin's triumph would not last and that soon events would vindicate the Opposition so that its adherents would return from exile to be hailed for their foresight, courage, and dedication to Marxism and Leninism.

As the conditions in which they found themselves, though painful and humiliating, were not yet crushingly oppressive, the Oppositionists reverted to a manner of existence which had been familiar to them before the revolution. The job of the political prisoners and exiles was to use enforced idleness in order to clear their thoughts, learn, and prepare for the day when they would once again have to shoulder the burdens of direct struggle or the responsibilities of government. For this kind of work the conditions seemed propitious. In many colonies there were educated men, brilliant theorists, and gifted writers whom their comrades

---

[1] Letter of 7 November in *The Archives*.

provided with a choice audience. Intensive exchange of ideas helped to keep up self-discipline and self-respect. From Alma Ata Trotsky eagerly followed this exchange and encouraged it, quoting, in letters to friends, Goethe's maxim that in intellectual and moral matters it is necessary in order to guard what one possesses to conquer it ever anew. Thus the colonies became centres of important intellectual and literary-political activity. Apart from memoranda and 'theses' on current affairs, which proliferated freely, major works were undertaken. Radek began to write a large-scale biography and study of Lenin; Rakovsky worked on a Life of Saint Simon and on the origins of Utopian socialism; Preobrazhensky wrote and completed books on the Soviet economy and on the economy of medieval Europe; Smilga was busy on a book about Bukharin and his school of thought; Dingelstedt produced essays on the social structure of India; and so on. However, these intellectual pursuits, valuable though they were, could not give any direct answer to the question which was uppermost in the thoughts of the deportees and which events were about to pose anew—the question: what next?

. . . .

Even in the remoteness of Siberia and Central Asia the shock of a new social crisis could be felt before the end of the winter. The crisis had long been in the making; and it had reached danger point as early as the autumn, just before the deportation of the Oppositionists. The state granaries were half empty; hunger threatened the urban population; and it was not even certain whether the armed forces would be supplied with provisions. The interminable queues outside the bakers and the repeated rises in the price of bread, witnessed by Trotsky at Alma Ata, could be observed all over the Soviet Union.

Yet on the face of it the agricultural situation was not bad. Almost as much land had been sown as in the best of times; and there had been a succession of three excellent harvests. But once again the 'link' between town and country was broken. The peasants refused to deliver bread and to sell it at fixed prices. Grain collections were accompanied by riots: the official collectors were driven from the villages and returned empty-handed to town. The peasantry had little or no incentive to deliver or sell its produce when now as before it could not obtain in return clothing, footwear, agricultural tools, or other industrial goods.

It demanded a steep rise in the price of grain; and clamouring for this, it followed even more clearly than before the lead of the wealthy farmers.

At the Politbureau the Bukharinists and the Stalinists came to blows over this at the very moment when they were jointly expelling the Trotskyists and crushing the Zinovievists. The Bukharinists wished to calm the peasantry with concessions, while the Stalinists were inclined, although not yet determined, to resort to force. In the first week of January, ten days before Trotsky's banishment, the Politbureau had to come to a decision about the further course of the grain collection; and undoubtedly nervousness over the situation in the country caused it to speed up Trotsky's banishment. On 6 January the Politbureau secretly instructed party organizations to proceed with greater severity against peasants who obstructed the grain collection, to levy forced 'bread loans', to resist firmly pressure for higher food prices, and to keep a sharp watch on the kulaks. The instructions failed to yield results; and five weeks later the Politbureau had to repeat them with greater emphasis and less secrecy.

In the middle of February *Pravda* sounded the alarm: 'The kulak has raised his head!' Finally, in April, the Central Committee declared bluntly, as if it had borrowed its terms from the Trotskyists and Zinovievists, that the nation was threatened by a grave crisis, and that the threat had been created by the 'growth of the kulaks' economic power' which the government's fiscal policy had failed to keep in check. 'In connexion with the further differentiation of the peasantry, the kulaks, their economic weight growing ... have acquired the power to exert considerable influence on the entire state of the market.'[1] Yet, the party, so the Central Committee said, had been and still was slack in curbing them. Emergency measures were decreed, under which compulsory loans were to be levied on kulaks to reduce their purchasing power; grain stocks were to be requisitioned; the fixed bread price was to be enforced; and finally, officials and party members inclined to treat the kulak indulgently were to be removed from their posts. These decisions were presented not as a departure from accepted policy but as *ad hoc* moves designed to deal with unexpected difficulties. The resolutions of the Central Committee contained no hint of 'wholesale collectivization'

[1] *KPSS v Rezolutsyakh*, vol. ii, p. 373.

—indeed, the idea was emphatically rejected. However, the manner in which the Central Committee explained the emergency and its insistence on the danger from the kulak and the party's failure to take counteraction already pointed to a fundamental change of policy. Within the Central Committee the Stalinists were gaining the upper hand. By obtaining powers to strengthen the party's hands against the kulak Stalin had strengthened his own hands against the Bukharinists; he was free to remove them from many posts at the lower and middle rungs of the administration and of the party machine.

The first reaction of the deported Trotskyists to these events was one of amusement, irony, even exultation. Had the predictions of the Opposition not come true? they asked. Was Stalin not being forced to adopt a 'left course', the course the Opposition had advocated? How could the party not realize now who had been right and who had been wrong in the great controversy of the last years? Most Oppositionists congratulated themselves, expecting even more confidently to be called back to play their part in overcoming the emergency and steering Bolshevik policy in the new direction. Trotsky, too, in his correspondence dwelt on the Opposition's foresight and appeared to be in a hopeful mood, although he did not share the most sanguine expectations of his followers.[1]

As the weeks passed and the 'left course' developed, while nothing changed in the official attitude towards the Opposition, the self-congratulatory mood in the colonies gave place to uneasiness and heart-searching. The turn that events had taken appeared to call into question some of the Opposition's major assumptions and predictions, especially its appraisal of the political trends within the party. Had we been right, some Trotskyists began to wonder, in denouncing Stalin as the protector of the kulak? Had we been justified in saying that once the left Opposition was defeated, the inner-party balance would be so upset that the Bukharinist right would assert itself and sweep away the Stalinist centre? Had we not overrated the strength of the

[1] See, for instance, his letter to Sosnovsky of 5 March 1928 in *The Archives*. *Inter alia* he recalls there the accusations of defeatism levelled against him after he had said that a good harvest, no less than a bad one, might, under Stalin's and Bukharin's policy, strengthen the kulak. Now *Pravda*, suddenly discovering the kulak's strength, wrote about the last three abundant harvests 'as if they had been three earthquakes'.

conservative elements in the party? The Stalinist faction, far from being overwhelmed, was beginning to overwhelm the right —had we then not exaggerated with our Cassandra cries about the danger of Thermidor? And had we, generally speaking, not gone too far in our struggle against Stalin? The great majority of deportees would not even admit such doubts to their thoughts. But a minority posed these questions ever more persistently; and every question it asked entailed other questions bringing under scrutiny ever more points of the Opposition's programme and activity. The answers turned on what view the Opposition took of the seriousness of Stalin's left course. It was still possible to regard Stalin's action against the kulak as an incidental tactical manœuvre which need not prevent him from resuming the pro-kulak policy. This was indeed what most Oppositionists thought of it. But a few were already convinced of the seriousness of the left course, saw it as the beginning of a momentous upheaval, and reflected uneasily on the Opposition's prospects. How could the Opposition remain a passive onlooker, they asked, while the party was embarking upon a dangerous struggle against the capitalist and quasi-capitalist elements in the nation, the struggle to which the Opposition had summoned it?

The Opposition had to such an extent based its own action on the idea that in all vital matters the right wing played the leading part and that the Stalinist faction, weak and vacillating, merely followed it like a shadow that Stalin's first or preliminary attack on the kulak shook the ground under its feet. Even in December, during the fifteenth congress, Zinoviev and Kamenev had excused their capitulation with the argument that Stalin was about to embark upon a left course. Soon thereafter, two eminent Trotskyists, Pyatakov and Antonov-Ovseenko, followed this example and announced their break with Trotsky. They had been the boldest and most energetic leaders of the 1923 Opposition; they had participated only half-heartedly in the struggle of the later years; and they justified their capitulation on the ground that Stalin was carrying out the Opposition's programme. The deportees at first received Pyatakov's and Antonov-Ovseenko's defection with the contempt and derision reserved for renegades; but their arguments nevertheless made an impression and stimulated self-interrogation.

Early in May Trotsky still knew little or nothing about the new ferment among the exiles; and he sent a letter to them in which he set out his views.[1] He declared that Stalin's left course marked the beginning of an important change. The Opposition, he said, had every right to regard itself with pride as the inspirer and prompter of the new policy. True, pride must be mingled with sadness when Oppositionists reflected on the price they had had to pay for their vicarious success. However, it has been the fate of revolutionaries more than once that, at the price of heavy or tragic sacrifice, they have compelled others, even enemies, to carry out parts of a revolutionary programme. Thus, the Commune of Paris had been drowned in blood but it triumphed over its hangmen, for its very hangmen had to carry out part of its programme: although the Commune had failed as a proletarian revolution, it made the restoration of the monarchy impossible in France and secured the establishment at least of a parliamentary republic. Such might be, *mutatis mutandis*, the Opposition's relation to Stalin's left course: the Opposition might be defeated; it might not see its full programme carried into effect; but at least its struggle had made it impossible for the ruling group to continue the retreat before the capitalist elements and to inaugurate a neo-N.E.P.

What was the Opposition to do? We were in duty bound, Trotsky replied, to lend critical support to Stalin's left course. Under no circumstances must we make common cause with Bukharin and Rykov against it. We should, on the contrary, encourage the vacillating Stalinist centre to break definitely with the right and make common cause with the left. An alliance between the Opposition and its Stalinist persecutors against the defenders of the kulak should not be ruled out, even though the possibility was remote. More than ever must the Opposition press for freedom within the party; and 'the left course facilitates the struggle for proletarian democracy'. In reasoning thus Trotsky was logically consistent with himself: he had, ever since 1923, maintained that the main 'function' of the Stalinist régime was to shield from the workers a party bureaucracy which protected the kulak and the N.E.P.-man. It was natural for him to conclude that once that bureaucracy had ceased to protect the kulak and the N.E.P.-man, it would draw closer to the working class, seek reconcilia-

[1] See his circular letter of 9 May in *The Archives*.

tion with its spokesmen, and restore to them freedom of expression. All the more resolutely should the Opposition, even while supporting the left course, resist Stalinist oppression and warn the party that as long as this persisted there was no guarantee that Stalin would pursue the new policy and would not once again yield to the kulak. Trotsky admitted that his was a 'dual attitude', difficult to adopt; but he claimed that it was the only attitude justified in the circumstances. Pyatakov had already described Trotsky's views as 'self-contradictory'. 'But all contradictions', Trotsky retorted, 'disappear in a man who [like Pyatakov] makes a suicide jump into a river.'

Trotsky's view had all the dialectical suppleness which the ambiguous situation demanded from him. He treated Stalin's drive against the kulak as a great and hopeful development; and he insisted all the more firmly on the need for freedom of criticism and discussion as the main guarantee of the soundness of the new policy. He offered the Opposition no axe to grind—only principles to defend. When his enemy took another leaf from his book, he acknowledged the leaf as his own and urged his followers to back his enemy in an undertaking which they had considered necessary. But there were many other leaves left in his book; and he was not going to throw them away. As to the Opposition's prospects, he shunned the extremes of optimism and pessimism: it was possible that events might compel the Stalinists to seek reconciliation with the Opposition, and in that case the Opposition would regain moral and political leadership; but the Opposition must also be ready to share the fate of the Commune of Paris and through its martyrdom to further the cause of socialism and progress.

The fact that Trotsky took a relatively favourable view of Stalin's left course and acknowledged its positive significance made a great and even bewildering impression on his followers. It added strength to the arguments of those among them who had begun to criticize the Opposition's record. If Trotsky was right now, they said, then had he not been wrong earlier in raising alarms about the Thermidorian peril? Had he not erroneously evaluated Stalin's policy? And would it be right for the Opposition to console itself with the thought that history would vindicate it, as it had vindicated the Commune of Paris? Should Trotskyists not lend a hand to the momentous struggle against private property that was going on in the country and thus help to make history rather than rely passively on history's anticipated verdict? Posterity may well extol the martyrdom of Communards; but Communards fight not for the glory of martyrdom, but for purposes which they believe to be practical and within their reach.

Such reasonings reflected a dilemma inherent in the Trotskyist attitude; and frustration injected bitterness. Exile, enforced idleness, and nagging doubts weighed down vigorous and strong-minded men who had made a revolution, fought civil wars, and built a new state. To be cast out of the party to which they had dedicated their lives, for which they had languished in Tsarist prisons, and in which they still saw humanity's highest hope— was a burden heavy enough in itself. The burden became unbearable when they realized that some of the crucial differences which had separated them from the Stalinists were vanishing and that the party was beginning to do what they had so ardently desired it to do. It is not so difficult for a political fighter to suffer defeat, privation, and humiliation as long as he knows clearly what he stands for and feels that his cause depends exclusively on what he and his comrades do for it. But even the most hardened fighter loses heart in a paradoxical situation when he sees his cause, or an important part of it, embraced by his persecutor. His cause no longer seems to depend on whether he fights for it or not. The fight itself suddenly appears to be purposeless and the persecution to which he has exposed himself—senseless. He begins to doubt whether he is justified in regarding his persecutor as an enemy.

Stalin had a cold and acute insight into the troubled mind of the Opposition; but he too had his dilemmas. Any Trotskyist commendation of his left course assisted him; but he was afraid of Trotskyist assistance. Hesitantly, falteringly, driven by circumstances, he was embarking upon an unknown and dangerous road. He risked grave conflict with the peasantry. He did not and could not gauge beforehand the scope and the violence of the resistance with which he would meet. He had cautiously turned against his erstwhile allies, the Bukharinists, whose popularity and influence he did not underrate. He could not know how far this new struggle might carry him and what dangers it might create for him. No more than Trotsky could he rule out the possibility that in an extremely critical situation he might have to seek an alliance with the left Opposition. But he too realized that this would be Trotsky's triumph; and he was determined to do all that lay in his power to defeat the Bukharinists without having recourse to reconciliation with Trotsky. He had reason to fear that the strength of his own faction would be inadequate for this and that his followers alone would not be able to manage the state machine and to cope with nationalized industry and finance in the new and difficult phase of rapid expansion. The Stalinists were primarily men of the party machine. Theorists, policy-makers, economists, industrial administrators, financial and agricultural experts, and men of political talent were found among the Trotskyists, the Bukharinists, and the Zinovievists. Stalin needed the assistance of men of ability who would be eager to carry out an anti-kulak policy and would carry it out with conviction and zeal. He could find such men in the Left Opposition. He was therefore anxious to win to his side as much Trotskyist and Zinovievist talent as he could without yielding ground to Trotsky and Zinoviev. He appealed to the Trotskyists behind Trotsky's back. Through his agents he lured them with the left course and sought to persuade them that their opposition to him had become pointless. At first the deportees rejected these appeals almost unanimously; but the appeals fell on fertile ground. In some of Trotsky's followers they intensified the doubts and the inclination to review the Opposition's record with disillusioned eyes.

Trotsky became aware of these developments only about the middle of May. Beloborodov had sent him a report on the discussions in the colonies. Another Trotskyist, who was still in Stalin's diplomatic service, informed him from Berlin about Stalin's presumed plan of action. According to this correspondent, Stalin hoped to improve his difficult position by inducing influential banished Oppositionists to recant—with their assistance he expected to put the left course into effect and to give Trotsky the *coup de grâce*. He even delayed embarking full steam upon the left course until he had secured the capitulation of many important Trotskyists. Everything now depended on whether he would succeed in this. If the Opposition could frustrate him, if it were not weakened by defections, and if it held out at least until the autumn, by which time Stalin would find his own faction

unable to cope with the difficulties, then the Opposition would
have every chance to regain the initiative and return to power.
But should Stalin succeed in sapping the Opposition's morale
and should Trotskyist capitulators come to his succour, then he
would maintain himself in power, crush the Bukharinists, and
carry on with the left course without having to conciliate Trotsky
and Trotsky's impenitent adherents. The correspondent feared
that Stalin was about to succeed: the Opposition's morale was
dangerously sagging, and all too many Oppositionists were ready
to give up the fight.[1]

Trotsky, it appears, did not believe that the Opposition's
morale was so low. There had been very few capitulations
among the deportees. One notorious case was that of Safarov,
the former leader of the Comsomol who had signed a formula of
recantation and was recalled to Moscow. However, Safarov's
case was exceptional in that he was not a Trotskyist. He had
belonged to Zinoviev's faction but had at first refused to capitulate with his leader, had gone with the Trotskyists into exile, and
only then, on second thoughts, capitulated. His behaviour, it
seemed, was not related to the mood among the Trotskyists. Yet
Safarov, as he tried to justify himself, expressed something that
struck a chord in them as well: 'Everything is now going to be
carried out without us!', he exclaimed. 'Everything' meant the
drive against the kulak and the N.E.P.-man, the expansion of
the socialist sector of the economy, accelerated industrialization,
and possibly collectivization of farming, for all these aspects of
the left course hung together. It was a galling thought for the
Trotskyists also that the great change, this 'second revolution',
might be carried out without them. The more disinterestedly
Trotsky stressed the desirability and the progressive character of
Stalin's latest moves and the more he insisted on the Opposition's
duty to support them, the greater was the frustration among his
followers, the more anxiously did they reflect over the rights and
wrongs of the Opposition's policy, and the more poignantly did
they feel that cast out of the party, in the wilderness where they

[1] This remarkable letter dated 8 May 1928 was written anonymously from
Berlin. Trotsky had known the correspondent but towards the end of his life, when
he sorted out the Archives, he could not remember who he was. In 1928 the correspondent was about to be recalled from his post and asked Trotsky whether he
should not refuse to return to Moscow. Trotsky, it seems, had already advised him
to return.

were, they could not possibly lend any practical support to the left course.

Before the end of May Trotsky again in several statements addressed his followers.[1] He defended the record of the Opposition and sought to outline new prospects. His argument may be summed up in these three points:

Firstly, it was not true that he had overrated the strength of the Bukharinist right. This was still formidable. Nor had the Opposition been wrong in seeking to arouse the party against the danger of Thermidor. In doing so it had helped to keep the Thermidorian forces at bay. The Opposition's action and pressure from the working class had compelled the Stalinists to break with the Bukharinists—otherwise the present bread crisis might have induced them to make far-reaching concessions to capitalist farming, and might thus have provoked, instead of the left course, a powerful shift to the right. He feared that those who maintained that the Opposition had magnified the danger from the right would in the end surrender to Stalin.

Secondly, the Opposition had no reason to reproach itself with having gone too far in its struggle. On the contrary, because of Zinoviev's and Kamenev's timidity, it had not gone far enough: 'All our activities had a propagandist and only a propagandist character.' The Opposition had hardly ever appealed to the rank and file strongly and boldly enough. When at last it sought to do so, on 7 November, Stalin tried to provoke it into civil war; and then it had to retreat.

Finally, the fact that Stalin was stealing the Opposition's thunder should not dishearten the Opposition. The Stalinist faction had initiated a left policy when it could do nothing else, but it would not be able to see it through to the end. Consequently, Trotsky assured his followers, 'the party is still going to need us'.

These arguments and assurances did not satisfy many of Trotsky's adherents. He offered them no clear perspective. They continued to ask whether Stalin had turned against the kulak for good or whether his left course was mere pretence; and they expected a plain answer. Trotsky did not possess it; and probably Stalin himself did not yet know definitely where he stood. Nor did Trotsky tell his followers how, in the position in which they

---

[1] See his letters to Beloborodov (23 May) and to Yudin (25 May) in *The Archives.*

were placed, they could act on his advice and how, in what manner, they could support and oppose Stalin at the same time.

Already in the spring of 1928 two distinct currents of opinion had formed themselves in the Trotskyist colonies. There were, on the one hand, those who took to heart, most of all, their obligation to support Stalin's left course, the obligation which Trotsky again and again impressed on them; and there were those, on the other hand, who were above all inclined to go on opposing Stalin, as Trotsky also urged them to do. Thus the differences that had existed within the Joint Opposition, between Trotskyists and Zinovievists, were now reproduced within the ranks of the Trotskyists themselves, splitting them into 'conciliators' and 'irreconcilables'. The conciliators were still far from the thought of surrendering to Stalin; but they wished the Opposition to mitigate its hostility towards his faction and to prepare for a dignified reconciliation with it on the basis of the left course. They held that integrity and the Opposition's own interest demanded from them that they should review critically, and modify in the light of events, the Opposition's accepted views. To this attitude rallied Oppositionists of the older generation, men of the reflective and sedate type, and those in whom the nostalgic feeling for their old party was extremely strong; and also the 'enlightened bureaucrats', economists and administrators, who had been interested more in the Opposition's programme of industrialization and economic planning than in its demands for inner-party freedom and proletarian democracy; and finally, people whose will to go on resisting the ruling group was already weakened by the ordeal they had undergone. As individuals were often moved by mixed impulses, it was in many cases wellnigh impossible to disentangle their motives.

The irreconcilable Trotskyists were mostly young men, to whom expulsion from the party had been less of a break in life than it had been to their elders; people whom the Opposition attracted by its call for proletarian democracy rather than by its economic and social *desiderata*; and the zealots of the Opposition, the doctrinaire enemies of bureaucracy, and the fanatics of anti-Stalinism. In this group too the motives of individuals could not be easily distinguished. Most often the young, those for whom the break with the party was no great moral upset, were also relatively indifferent to the complex economic and social issues, but responded ardently to the Opposition's call for freedom of expression, and viewed all bureaucracy with a fierce hostility, rendered even fiercer by persecution and exile.

Both wings of the Trotskyist Opposition tended to overlap with other groups outside it. The conciliators moved closer and closer towards the Zinovievists whom they had hitherto despised. They began to see them in a new light, and even if they were not ready to follow them, they began to appreciate the reasons for their surrender, to listen eagerly to their arguments, and to watch sympathetically their doings. The most extreme irreconcilables, on the other hand, found that they had much in common with the unrepentant Mohicans of the Workers' Opposition and of the Decemists, who were led by Sapronov and Vladimir Smirnov, and had been exiled together with the Trotskyists. In their enmity towards the bureaucracy they had been far less inhibited than the Trotskyists. More or less openly they had renounced all allegiance to the existing state and party. They proclaimed that the revolution and Bolshevism were dead, and that the working class had to begin from the beginning, that is to start a new revolutionary struggle in order to free itself from exploitation by the new 'state capitalism', the N.E.P. bourgeoisie, and the kulaks. To many a young Trotskyist this plain and single-minded message sounded more convincing than did Trotsky's carefully balanced analyses and 'dual policy'. It was easier to digest it, for in it yes was yes and no was no, without any dialectical complication. To denounce Stalin as the grave-digger of the revolution, the Decemists said, and to dwell, as Trotsky did, on the progressive implications of the left course was absurd; to fight Stalin meant to fight him and not to support him.

Both sets of Trotskyists looked to Trotsky for guidance, although each was inclined to accept only that part of his advice that suited it. Both sets evoked the Opposition's first principles and common interests. But as the differences widened, the sense of comradeship wore thin and mutual suspicions grew until the two sets had little more for each other than black looks and harsh words. To the irreconcilables their more moderate comrades were men of small faith, if not yet deserters. The moderates looked down on the irreconcilables as ultra-lefts or crude *anarchisants*, devoid of Marxist intellectual discipline and responsibility for the fortunes of the revolution. The irreconcilables suspected that knowingly or unknowingly the conciliators worked for Stalin, while the conciliators held that nothing compromised the Opposition and assisted Stalin more effectively than the exaggerations and the excesses of the doctrinaires and the zealots of Trotskyism.

The spokesmen for each of the two sets were Oppositionists of long standing and Trotsky's trusted and respected friends. Preobrazhensky was the first to speak of the need for a more conciliatory attitude towards Stalinism. He had never faltered as an Oppositionist, and not the slightest taint of self-regard or opportunism attached to his character. His weakness, if weakness it be, lay rather in his utter disregard of expediency and popularity and in the theoretical consistency of his views. He began to preach conciliation from deep conviction, which can be traced back to his writings of 1924–5. He had been, we know, the chief theoretical exponent of primitive socialist accumulation. 'The period of primitive socialist accumulation', he had written in *The New Economics,* 'constitutes the most critical era in the life of the socialist state after the conclusion of civil war. . . . To go through this period as rapidly as possible and to reach as soon as possible the stage at which the socialist system develops all its advantages *vis-à-vis* capitalism is for the socialist economy a matter of life and death.' During that period the socialist state was bound to get the worst of both worlds: it would benefit neither from the advantages of capitalism nor from those of socialism. It would have to 'exploit' the peasantry in order to finance accumulation in the socialist sector. On this point, it will be remembered, Preobrazhensky had clashed with Bukharin and the neo-Populist school, 'our Soviet Manchester school of thought', as he dubbed it. 'The pressure of [foreign, mainly American] capitalist monopolism', he argued then, 'can find a barrier only in socialist monopolism.' This must subordinate to itself, by means of fiscal policy and through a state-regulated price mechanism, the private sector of the economy, especially farming. To Bukharin's outcry of indignation Preobrazhensky replied: 'But can it be otherwise? To put it in the simplest terms: can the burden of the development of the state-owned industry . . . be thrown on to the shoulders of our three million industrial workers only—or should the twenty-two million of our peasant smallholders also contribute their share?'

Even he had not advocated the expropriation and forcible collectivization of the smallholders; but more than anyone else he
had been aware of the inherent violence of the conflict between
State and peasantry under 'the iron heel of the law of primitive
socialist accumulation'.[1]

No wonder that Preobrazhensky responded eagerly to Stalin's
left course. He received it as a confirmation of his own theory.
He saw it as an inevitable and wholly desirable development.
From the outset he was convinced of its momentous significance;
and he was convinced of it more firmly than Trotsky was. The
differences between him and Trotsky, hitherto only implicit in
their writings but of no practical consequence, now began to
affect their attitudes. Trotsky had never committed himself to
the view that the workers' state must as a rule 'exploit' the
peasantry—at any rate, he had never expounded this view as
bluntly as Preobrazhensky had done it. Nor had he advocated
a pace of industrialization as forcible as that which Preobrazhensky had anticipated. Preobrazhensky's theorem of *The New
Economics* had not been incompatible with socialism in one
country—it had implied that primitive accumulation, the most
difficult part of the transition from capitalism to socialism, might
be accomplished within a single and industrially underdeveloped
nation-state. Finally, unlike Trotsky, Preobrazhensky had dwelt
on the 'objective force of the laws' of the transition to socialism,
a force which would assert itself and compel the party leaders to
act *malgré eux-mêmes* as the agents of socialism. Nationalization
of all large-scale industry, he held, led ineluctably to planned
economy and rapid industrialization. In opposing these the
Stalinists and Bukharinists opposed a historic necessity—a necessity which the Opposition alone saw in time and of which it
sought to make the Bolsheviks aware. Stalin and Bukharin might
defeat the Opposition; but 'they could not outwit the laws of
history'. 'The structure of our state economy [which] often
proves itself to be more progressive than is the whole system of
our economic leadership' would eventually force them to carry
out the Opposition's programme.

These ideas, which were little more than asides and hints in
Preobrazhensky's earlier writings, now came to govern all his
thoughts. Stalin declaring war on the kulak was in his eyes but

[1] See Chapter V, pp. 234-40.

the unconscious and reluctant agent of necessity. While Trotsky still viewed the left course somewhat incredulously and wondered whether it might not be only a temporary shift, Preobrazhensky had no doubt whatsoever that Stalin was not trifling, that he could not retreat from the left course, that he would be compelled to wage war on the kulak ever more ruthlessly, and that this created a completely new situation for the country at large and for the Opposition in particular. He insisted that the country was on the brink of a tremendous revolutionary upheaval. The kulaks, he said, would go on refusing to sell grain and would threaten the town with famine. The middle and poor peasants would not be able to supply enough food; and the official attack on the kulak would antagonize them too and would lead to a colossal clash between the government and the bulk of the peasantry. In a survey written in the spring of 1928 Preobra- zhensky maintained that Stalin's threats and emergency mea- sures had already aroused in the country a storm so violent that to calm it the government would have to make such vast and dangerous concessions to capitalism that not only Stalin but even Bukharin and Rykov would recoil and refuse to make them.[1] Only a drastically right or drastically left policy could avert a calamity; and everything indicated that Stalin would move farther to the left.

What was to be the Opposition's role in this upheaval? The Opposition, Preobrazhensky replied, had acted as the conscious interpreter of a historic necessity. It had displayed superior foresight: its ideas were 'reflected in Stalin's new policy as in a distorting mirror'. The present crisis would not have been as grave if the party had acted on the Opposition's advice earlier. The Opposition must still go on advocating accelerated indus- trialization; and it must call as insistently as ever for proletarian democracy. However, although the Opposition had correctly interpreted the needs of the time, it was not given to it to meet those needs in practice. Stalin and his adherents were taking charge of the practical task. They were the agents of the historic necessity, although they had not understood it and had long resisted it. Somewhere then the Opposition had gone wrong. It had exaggerated the danger from the right and the Stalinist con- nivance with the kulak. It had misjudged the trends within the

[1] See Preobrazhensky's 'Levyi Kurs v Derevnie i Perspektivy' in The Trotsky Archives.

party and their relationship to the social classes without, a grave error for Marxists to commit. It was therefore incumbent on the Opposition to modify its attitude and to contribute to a *rapprochement* with the Stalinist faction.

With this aim in view Preobrazhensky proposed that the Opposition should ask for official permission to call a conference of its members, at which all exiled colonies should be represented, in order to discuss the new situation and the Opposition's conduct. Trotsky had spoken of the possibility and desirability of an alliance between left and centre against the right; but he had not proposed any move designed to bring it about. Preobrazhensky was not satisfied with this. If there was to be such an alliance, he argued, the time for it was now, when the Stalinists struck against the right; and the Opposition's duty was to act instead of waiting until events produced the alliance ready-made—they might never produce it.

Trotsky was dead against Preobrazhensky's proposal. He held that, desirable though a centre-left coalition was in theory, the Opposition could do nothing to bring it about. The jailer and the jailed were not allies. He feared that Preobrazhensky took too favourable a view of the left course; but, even if this were not so, the gulf between Stalinism and the Opposition remained fixed. Persecution continued. The party was still robbed of its freedom, and its régime was getting worse and worse. The dogma of the Leader's infallibility was established; and it was applied to past as well as present. The entire history of the party was falsified to suit the requirements of that dogma. Under such conditions the Opposition could take no step to meet the ruling faction half-way. It would be a disgrace for it to ask its persecutors for permission to hold a conference—the request alone would flavour of capitulation.[1]

In May the colonies discussed Preobrazhensky's proposal— this was the first test of the exiles' reaction to the left course. The proposal was rejected out of hand. The great majority was in an irreconcilable mood, sceptical about the left course, inclined as before to see in Stalin the defender of the kulak and the accomplice of the Thermidorians, confident in the Opposition's cause, and reluctant to contemplate any revision of its attitude.

Despite this rebuff to Preobrazhensky, his ideas began to

[1] See Trotsky's *'Pismo Drugu'* (24 June 1928) in *The Archives*.

germinate in many minds. Radek, it seems, was the first of the leaders of the Opposition to come under their influence. He had not so far belonged to those inclined to pull punches. Throughout 1927 he had urged the Opposition to attack the ruling group more boldly, to appeal to the factory workers who stood outside the party, and to voice aggressively their grievances, instead of contenting itself with 'honour-saving gestures' and highbrow theory. He had not recoiled from the idea of a new party and had favoured the admission to the Opposition of the Decemists who stood for it. He was still in this militant mood after his deportation, when he wrote scornfully of Zinoviev's and Pyatakov's recantations and the morbid odour of *Dostoevshchyna* they exuded. 'They have denied their own convictions and have lied to the working class—you cannot help the working class with lies.'[1] Even in May, when Preobrazhensky called for the conference, Radek still appeared to be opposed to the idea—at any rate, he criticized Preobrazhensky's conciliatory attitude.

Barely a month later, the man seemed to have completely changed; he preached conciliation himself, with all the ingenuity, verve, and wit peculiar to him. His accession enormously strengthened the 'moderate' wing, for he and Preobrazhensky were, next to Trotsky and Rakovsky, the most authoritative leaders in exile. Subsequently, as his prolific correspondence shows, his will to resist Stalinism crumbled almost from week to week, although nearly a year was to elapse before his actual surrender.

It would be too simple to attribute the change merely to Radek's volatility or lack of courage. His motives were tangled. No doubt he did not possess all the 'Bolshevik hardihood' that others had gained in underground politics, in Tsarist prisons, and through years of Siberian exile. His spells of underground work had been brief: up to 1917 he had spent his political life mainly in the open socialist movements of Austro-Hungary and Germany. He was essentially a Western European and a Bohemian, sociable, accustomed to breathe the air and the excitement of great cities and to be at the centre of public affairs. In the course of more than twenty-five years he had held famous Central Committees and great editorial offices spell-

---

[1] See Radek's letter to Zhenya written from Tobolsk, on 10 May 1928, and his letter to Preobrazhensky of 25 May in *The Trotsky Archives*.

bound by his views and witticisms. For ten years he had been one of the leading lights of the Bolshevik party and the Communist International. As long as he was surrounded by the bustle of political life, his confidence and mettle had not abandoned him—he had remained bold and active and had still been at the centre of affairs even in the Moabit prison of Berlin in 1919. But cast out suddenly into the empty, bleak, and severe wilderness of northern Siberia, his spirit began to sink. Solitude oppressed him. He felt as if he had been exiled from life itself. His sense of reality was shaken. Had all the years he had spent at Lenin's side, as valued comrade and adviser, helping to direct the affairs of a world-wide movement, been but a dream? Men of far greater resilience were beset by similar feelings. This, for instance, is what Ivan Smirnov, hero of the civil war, wrote to Radek, from southern Armenia to northern Siberia:

> You, dear Karlyusha,[1] are pained that we have found ourselves outside the party. To me, too, and to all the others this is agony indeed. At the beginning I was haunted by nightmares. I would suddenly awake at night and could not believe that I was a deportee —I, who had worked for the party ever since 1899, without a day's break, not like some of those scoundrels of the Society of Old Bolsheviks who after 1906 deserted the party for full ten years.[2]

But it was not only this predicament that troubled Radek and his friends. They brooded over the fate of the revolution. They were accustomed to regard themselves as the true guardians of the 'conquests of October' and as the sole depositaries of Marxism and Leninism which the Stalinists and Bukharinists had diluted and falsified. They were accustomed to think that whatever was beneficial to Marxism and the revolution was beneficial to the Opposition too, and that the Opposition's defeats were the revolution's defeats. They now saw the Opposition reduced to a small group, almost a sect, utterly impotent and estranged from the great state and party with which they had identified themselves. Was it possible, they wondered, that a movement claiming for itself so high a mission should be reduced to so low a state? They were confronted with this dilemma: if they really were the sole reliable and legitimate guardians of October, then their cruel defeat could not but bring irretrievable disaster to the

---

[1] Diminutive from Karl, Charlie.
[2] The letter, written in 1928 (no more precise date), is in *The Trotsky Archives*.

revolution, and the heritage of October was lost. But if this was not so, if the 'conquests of October' were more or less intact and the Soviet Union, despite all that had happened, was still a workers' state, then had the Opposition not been wrong and guilty of arrogance in regarding itself as the sole depositary of Marxism-Leninism and in denying its adversaries all revolutionary virtue? Were the few thousand Oppositionists all that was left of the great and world-shaking Bolshevik movement? Had the mountain of revolution brought forth a mouse? 'I cannot believe', Radek wrote to Sosnovsky, 'that Lenin's entire work and the entire work of the revolution should have left behind only 5,000 Communists in the whole of Russia.'[1] Yet, if one were to take literally some of the Opposition's claims and if one were to believe that the other Bolshevik factions merely paved the way for counter-revolution, then one could not escape this conclusion, from which both realism and the Marxist sense of history recoiled. Surely, the Bolshevik epic with all its heroism, sacrifice, hope, blood, and toil, could not have been mere sound and fury, signifying nothing. As long as Stalinists and Bukharinists had jointly protected the kulaks and the N.E.P.-men there was substance in the Opposition's claims and charges. But the left course, bringing the Stalinist faction into mortal conflict with private property, demonstrated that the work of Lenin and the October Revolution had left behind something more than a handful of righteous men, something more than 'five thousand communists in the whole of Russia'. The volcano of the revolution, far from having given birth to a mouse and becoming extinguished, was still active.

Preobrazhensky argued that it was the 'objective force' of social ownership that provided the impulse for Russia's further revolutionary and socialist transformation. The 'objective force' asserted itself through men, its subjective representatives. The Stalinist faction was the agent of historic necessity; and despite confusion, mistakes, and even crimes committed, it acted as guardian of the heritage of October and champion of socialism. The Stalinists, Radek discovered, had proved themselves worthier than the Opposition had thought them to be. The Opposition should and could admit this without any self-depreciation. In the new advance towards socialism the Opposition had acted as the

[1] The letter (dated Tomsk, 14 July 1928) is in *The Archives*.

vanguard while the Stalinist faction had formed the rearguard. The conflict between them had not been a clash of hostile class interests, but a breach between two sections of the same class, for both vanguard and rearguard belonged to the same camp. It was time to heal the breach. Many Oppositionists were startled by the idea of a reconciliation between Stalinists and Trotskyists; but, Radek remarked, such a regrouping would not be stranger than had been earlier reversals of inner-party alliances. 'There was a time when we thought that Stalin was a good revolutionary and that Zinoviev was hopeless. Then things changed—they may change once again.'

There was an unmistakable note of despair in these pleadings —but it was a despair which sought to escape from itself and to change into hope. The conciliators' mood was nurtured in the deepening isolationism of Bolshevik Russia. It was within the Soviet Union, not without, that Radek and Preobrazhensky— and many others—looked for a great and promising change in the fortunes of communism. And this fact accounts for much that was to follow.

This was the aftermath of the Chinese Revolution. In December 1927 the communist rising of Canton had been suppressed. The rising had been the last act or rather the epilogue of the drama of 1925–7. The shock of the defeat was making itself felt in all Bolshevik thinking: it sapped even further and submerged the internationalist tradition of Leninism; and it enhanced Russian self-centredness. More than ever socialism in one country appeared to offer the only way out and the only consolation. This time, however, the tide of isolationism affected the Opposition as well; it reached the remote colonies of deportees and worked on the thoughts of the conciliators. Like Stalin's turn to the left this latest defeat gave Preobrazhensky and Radek a fresh motive for disillusionment with the Opposition's record. The Opposition, they argued, had been partly mistaken in its estimate of internal Russian developments—had it not been mistaken in its view of the international prospects as well? Trotsky had erred over the Soviet Thermidor—was not his Permanent Revolution, too, a fallacy?

Barely a few weeks after their deportation Trotsky and Preobrazhensky were already in correspondence over the Canton rising. Knowing little about the actual circumstances of the event and trying to form an opinion from *Pravda*'s belated and scanty reports, Trotsky resumed an exchange of views which he had had with Preobrazhensky in Moscow. Like many an old Bolshevik in Opposition, Preobrazhensky had not accepted the idea of Permanent Revolution and its corollary that the Chinese Revolution could conquer only as a proletarian dictatorship. Like Zinoviev and Kamenev he held that China could not go beyond a bourgeois revolution. From their places of exile Trotsky and Preobrazhensky discussed the bearing of the Canton rising on this difference. *Pravda* had reported that the insurgents of Canton had formed a Council of Workers' Deputies and had set out to socialize industry. Although the rising had been crushed —so Trotsky wrote to Preobrazhensky on 2 March—it left a message and a significant pointer to the course of the next Chinese Revolution which would not be arrested in its bourgeois phase but would establish Soviets and aim at socialism. Preobrazhensky replied that the rising had been staged by Stalin merely in order to save his face after all his surrenders to the Kuomintang, that this had been a reckless venture, and that the 'Soviet' of Canton and its 'socialist' slogans, not having resulted organically from any mass movement, had not reflected the inherent logic of any genuine revolutionary process.[1] Preobrazhensky was, of course, closer to the facts than Trotsky, who in this case relied on dubious evidence for his conclusion about the character of the next Chinese Revolution. His conclusion was nevertheless correct: the revolution of 1948–9 was to transcend its bourgeois limits; and to this extent it was to be a 'permanent revolution', even if its course and the alignment of social classes in it were to prove very different from what the Trotskyist and indeed the Marxist and Leninist theories of revolution had envisaged.

'We, the old Bolsheviks in opposition, must dissociate ourselves from Trotsky on the point of permanent revolution', Preobrazhensky declared. The statement itself could not have been surprising to Trotsky, but its emphatic tone certainly was. Trotsky had been accustomed to hear such reminders of his non-Bolshevik past from his adversaries and lately again from Zinoviev and Kamenev, but hardly from Preobrazhensky, his close

[1] Preobrazhensky's (undated) answer is in *The Archives*.

co-thinker since 1922. He knew that such reminders never cropped up fortuitously. What surprised him even more was that Radek, too, produced a *critique* of Permanent Revolution —Radek, no old Bolshevik himself, had hitherto whole-heartedly defended that theory. Even now he acknowledged that Trotsky had in 1906 anticipated the course of the Russian Revolution more correctly than Lenin; but he added that it did not follow that the scheme of permanent revolution was valid in other countries. In China, Radek maintained, Lenin's 'democratic dictatorship of the proletariat and the peasantry' was preferable because it allowed for a possible hiatus between the bourgeois revolution and the socialist one.

Apparently this controversy had no direct bearing on the issues of the day; and Trotsky was drawn into it reluctantly. He replied that China had freshly demonstrated that any contemporary revolution which did not find its consummation in a socialist upheaval was bound to suffer defeat even as a bourgeois revolution. Whatever the pros and the cons, the fact that the two conciliators attacked Permanent Revolution was all the more symptomatic because Trotsky had not attempted to make of his theory the Opposition's canon. This was not the first time that frustration with defeats of communism abroad and isolationist propensities induced Bolsheviks to turn against the theory which in its very title challenged their isolationism. The result of all the dogmatic battles fought over Permanent Revolution since 1924 had been to make of it in the party's eyes the symbol of Trotskyism, Trotsky's master heresy, and the intellectual fount of all his political vices. To Stalin's and Bukharin's followers Permanent Revolution had become a horror-inspiring taboo. An Oppositionist beset by doubts and second thoughts and looking for a way back to the party—his lost paradise—sought instinctively to rid himself of any association with that taboo. It will be remembered that Trotsky, anxious to render it easier for Zinoviev and Kamenev to make common cause with him, had declared that his old writings on Permanent Revolution had their place in the historical archives and that he would not defend them on every point, even though for himself he was convinced that his idea had stood the test of time. Yet he did not succeed in relegating his theory to the archives. Not only his enemies dragged it out and forced him to defend it. Again and again his allies did likewise; and whenever they did so this was a sure sign that one of his political alliances or associations was about to break down.

Presently dissension came into the open over a more topical and less theoretical issue. The sixth congress of the Communist International was convened in Moscow for the summer of 1928. The Opposition had the statutory right to appeal to the congress against its expulsion from the Russian party; and it intended to do so. There was no chance that the appeal would obtain a proper hearing, or that the leaders of the Opposition would be allowed to appear before the congress to state their case. '. . . the Congress will probably make an attempt to cover us in a most authoritative manner with the heaviest of tombstones . . .', Trotsky wrote. 'Fortunately, Marxism will rise from this *papier-mâché* grave and like an irrepressible drummer sound the alarm!'[1] He intended to produce a brief and blunt criticism of the Comintern's policy and a concise declaration of the Opposition's purpose to be addressed to the congress. But the work grew in his hands into a massive treatise the writing of which kept him occupied throughout the spring and summer.[2] The congress was expected to adopt a new programme of which a draft, written largely by Bukharin and centred on socialism in one country, had been published. Trotsky gave his statement the form of a *Critique* of the new programme. He completed this in June; and in July he followed it up by a message to the congress under the title *What Next?* He summed up 'five years of the International's failures' and five years of the Opposition's work in a manner 'free from all traces of reticence, duplicity, and diplomacy' and calculated to mark clearly the gulf between the Opposition and its adversaries. He sent copies to the colonies just before the opening of the congress; and he asked all Oppositionists to endorse his statements in their collective and individual messages addressed to the congress.

Meantime Radek and Preobrazhensky had prepared statements of their own which were more conciliatory in content and tone. True, Preobrazhensky drew a devastating balance-sheet of the Comintern's policy of recent years; and he was outspoken about the differences which opposed Trotskyists of all shades to

[1] Trotsky's circular letter of 17 July 1928. *The Archives.*
[2] The work is known in English under the title *The Third International After Lenin.*

Stalinism and the Comintern. But in his conclusion he declared that 'many of these differences have disappeared as a result of the change that has occurred in the International's policy', because the International, following the Russian party, had also 'veered to the left'.[1] Radek expressed the same opinion and at once dispatched his statement to Moscow. 'If history shows', he wrote, 'that some of the party leaders with whom yesterday we crossed swords are better than the viewpoints they defended, nobody will find greater satisfaction in this than we shall.'[2]

The fact that Trotsky and Radek had addressed different and partly conflicting messages to the congress could only damage the Opposition's cause. Instead of demonstrating its unity, the Opposition spoke with two voices. When Trotsky learned what had happened, he telegraphed to the major centres of the Opposition asking all the exiles to dissociate themselves publicly from Radek. The colonies were agog with indignation, disavowed Radek, and sent appropriate statements to Moscow. In the end Radek himself informed the Congress that he was withdrawing his message and that he was in complete agreement with Trotsky. To his comrades he apologized for his *faux pas*, saying that it had been due to difficulties of communication with Trotsky, whose *Critique* of the Comintern had reached him too late. Trotsky accepted the apology; and there the matter rested for the moment. The Opposition, Trotsky said, had 'straightened out its front'. However, the incipient split was not mended—it was only thinly covered over.

.   .   .   .

An important event had helped Trotsky to rally the exiles. In July the Central Committee held a session at which Bukharin's faction appeared to have gained the upper hand over Stalin's. The critical issue was still the same: the bread crisis and the threat of famine suspended over urban Russia. The emergency measures earlier in the year had not averted the threat; and the situation was aggravated by a partial failure of winter crops in the Ukraine and in the northern Caucasus. The peasantry was in uproar. It delivered and sold only 50 per cent. of the grain it

[1] Preobrazhensky, '*Chto nado skazat Kongresu Kominterna*' in *The Archives.*

[2] Radek's memorandum to the congress, written in Tomsk in June 1928, is in *The Trotsky Archives.* Trotsky must have read 'psychoanalytically' the passage quoted here: he underlined with red pencil the word 'yesterday' in Radek's phrase about the party leaders 'with whom yesterday we crossed swords'.

used to sell before the revolution. All exports of grain had to be stopped.[1] The strong-arm methods in grain collection had been just enough to enrage the farmers but not enough to intimidate them. The Central Committee noted 'the discontent among . . . the peasantry which showed itself in demonstrations of protest against arbitary administrative proceedings'; and it declared that such proceedings 'had helped capitalist elements to exploit the discontent and turn it against the Soviet Government . . . and had given rise to talk about the [forthcoming] abolition of N.E.P.'[2]

At the session of the Committee, after Mikoyan had made a report, the Bukharinist faction called for an end to the left course. Rykov demanded the cancellation of the antikulak policy; Frumkin, the Commissar of Finance, went even farther and asked for a revision of the entire peasant policy enunciated at the fifteenth congress (at which Stalin, to confound the Trotskyists and the Zinovievists, had adopted some of their ideas) and for a return to the predominantly Bukharinist policy of the previous congress. The Central Committee declared that it stood by the decisions of the fifteenth congress; but it cancelled its own emergency measures 'against the kulak'. It proclaimed that henceforth the 'rule of law' must prevail. It forbade searches and raids on barns and farmsteads. It stopped the requisitioning of food and the forcible levying of grain loans. Last but not least, it authorized a 20 per cent. rise in the price of bread, the rise it had so categorically prohibited three months earlier.[3] Seen in retrospect, this was the Central Committee's last attempt to appease the peasants, the last before it proceeded to suppress private farming. At the time, however, it looked as if the kulak had won a round, as if Stalin had abandoned the left course, and as if Bukharin and Rykov had dictated policy.

It may be imagined how the Trotskyist deportees received this. They were back on familiar ground. The old scheme of things, within which they had been accustomed to think and to argue, seemed re-established. They saw 'the defenders of the kulak' reasserting themselves. They saw Stalin's 'vacillating centre' yielding as always. Authorizing the higher bread price, the Central Committee had hit the industrial workers and had acted in the interests of the wealthy farmers. This was surely not

---

[1] *KPSS v Rezolutsyakh*, vol. ii, p. 392.   [2] Ibid., p. 395.   [3] Ibid., p. 396.

yet the end. The struggle was on: the right wing would resume its offensive; and the Stalinists would continue to retreat. The Thermidorian danger was closer than ever—the Thermidorians were on the move. Trotsky thought likewise: 'In Rykov's speech...', he declared, 'the right wing has thrown its challenge at the October Revolution. . . . The challenge must be taken up. . . .' The rise of the price of bread was only the beginning of a neo-N.E.P. To appease the kulak, the right wing would soon make a determined attempt to undermine the state monopoly of foreign trade. He saw Rykov and Bukharin as the victors who would soon 'hunt down Stalin as a Trotskyist, just as Stalin had hunted down Zinoviev'. Rykov had said at the Central Committee that 'the Trotskyists regarded it as their main task to prevent a victory of the right wing'. Trotsky replied that this was indeed the Opposition's main task.[1]

Among the Trotskyists the conciliators were completely isolated for the moment. 'Where is Stalin's left course?' the deportees exultantly demanded of Radek and Preobrazhensky. 'It was all a flash in the pan. But this was enough for you to try and throw overboard our old and well-tested ideas and views and to urge us to conciliate the Stalinists!' Once again they saw Stalin's ascendancy as a mere incident in the fundamental struggle between themselves and the Bukharinists; and they believed even more ardently than before that all Bolsheviks who had remained faithful to the revolution would soon see the issues in this light, as a conflict essentially between right and left, and would opt for the left. Stalin's apparent discomfiture raised their hopes sky-high. 'The day is not far off', wrote as eminent a Trotskyist as Sosnovsky, 'when the call for Trotsky's return will resound throughout the world.'[2]

In the middle of all this political excitement tragedy visited Trotsky's family. Both his daughters, Zina and Nina, had been ill with consumption. The health of Nina, the younger of the two—she was twenty-six—broke down after the imprisonment and deportation of Nevelson, her husband. The news reached Trotsky in the spring, during a fishing expedition. He was not yet fully aware of the gravity of Nina's illness; but he spent the

[1] '*Yulskiye Plenum i Prava Opasnost*' in *The Archives*.
[2] See Sosnovsky's letter to Rafail of 24 August in *The Archives*.

next weeks in anxiety and worry. He knew that both his daughters and their children lived in utter poverty, that they could not count on the help of friends, and that Zina, herself wasted by consumptive fever, spent her days and nights at Nina's bed. 'Am aggrieved', he telegraphed her, 'that cannot be with Ninushka to help her. Communicate her condition. Kisses for both of you. Papa.' Again and again he asked for news but received no answer. He wrote to Rakovsky begging him to make inquiries in Moscow. Finally, he learned that Nina had died on 9 June. Much later he received the last letter she had written him—it had been on the way and held up by the censors for over ten weeks. It was a painful thought for Trotsky that on her death-bed she had waited for his answer in vain. He mourned her as 'an ardent revolutionary and member of the Opposition' as well as a daughter; and to her memory he devoted the *Critique* of the Comintern's programme on which he had been working at the time of her death.

Messages of condolence from many deportees were still arriving at Alma Ata when another blow caused Trotsky much sorrow and grief. After Nina's death Zina had intended to go to Alma Ata. Her husband too had been deported and she had strained her health in nursing her sister. From week to week she delayed the journey until news reached Alma Ata that she was dangerously ill and unable to travel. Her malady was aggravated by a severe and protracted nervous disorder; and she was not to rejoin her father before his banishment from Russia.

A family reunion, nevertheless, occurred at the *dacha* outside Alma Ata, when Sergei arrived to spend his holiday there. With him came Lyova's wife and child. They stayed only a few weeks; and this was an anxious and mournful reunion.

. . . .

After the 'rightward turn' of official policy, the extreme and irreconcilable Trotskyists had the upper hand in nearly all the centres of the Opposition. The mass of deportees would not even hear of any attempt at narrowing the gulf between themselves and the Stalinists. However, the extreme irreconcilables had no spokesmen of Preobrazhensky's and Radek's authority and ability. Their views were formulated by men like Sosnovsky, Dingelstedt, Elzin, and a few others who expressed a mood rather than any definite political ideas.

Of this group Sosnovsky was the most gifted and articulate; and when he confidently asserted that 'the cry for Trotsky's return will soon resound throughout the world' he expressed the fervid hope of many of his comrades. He was Trotsky's trusted friend and one of the most effective Bolshevik journalists, very popular far outside the ranks of the Opposition. But he was no political leader or theorist. He was distinguished as a chronicler of Bolshevik Russia and a sharp-eyed critic of morals and manners. A rebel by temperament, animated by an intense hatred of inequality and injustice, he indignantly watched the rise, in the workers' state, of a privileged bureaucracy. He pungently exposed its greed and corruption (the 'harem-cum-motor-car-factor'), its snobbery and its upstart ambition to assimilate itself to the old bureaucracy and aristocracy and to intermarry with them. He felt only distaste for those who could even think of any conciliation with the ruling group. He was in this respect poles apart from Radek. It was to Sosnovsky that Radek had written that he could not believe that all that was left of Lenin's party was a mere handful of righteous Opposition-ists—to Sosnovsky the Opposition was indeed the sole guardian of the heritage of October. Nothing characterizes him more strikingly than a letter he wrote to Vardin, his old comrade, who had together with Safarov deserted the Opposition and 'capitula-ted'. Merciless in contempt, Sosnovsky recalled an old Jewish funeral custom which required that when a dead man was brought to the cemetery, his synagogue associates should shout into his ears: 'So-and-So, Son of So-and-So, know that thou art dead!' He, Sosnovsky, now shouted this into the ear of his old comrade, and he would shout it into the ear of every capitulator. Distrustfully he watched Radek's evolution, wondering whether he ought not to shout these words into Radek's ear as well.[1]

The other spokesmen of this wing of the Opposition were younger men of lesser stature. Dingelstedt was a promising scholar, a sociologist and economist; a Bolshevik since 1910, distinguished as an agitator in the Baltic Fleet in 1917, he was

[1] About the same time Radek also wrote to Vardin; and his and Sosnovsky's letters make a curious contrast. This was in May when Radek had only just begun to develop the conciliatory mood. He chided Vardin, but he did this gently and sympathetically and was far from treating the capitulator as 'morally dead'. Radek's and Sosnovsky's letters are in *The Archives*.

still in his early thirties. Elzin had been one of Trotsky's gifted secretaries. These men were not sure whether Trotsky himself was not showing signs of vacillation. Thus Dingelstedt wrote to him that 'some comrades were gravely disturbed' by his opinion that Stalin's left course was 'an indubitable step in our direction' and that the Opposition should 'support it unconditionally'.[1] They also reproached Trotsky for the 'indulgence' with which he treated Radek and Preobrazhensky. Nor did they share Trotsky's hope for a reform in the party and the revival in it of proletarian democracy.

Thus, while at one extreme the Opposition included those who were more and more anxious to come to terms with their persecutors, its other extreme became almost indistinguishable from the followers of V. Smirnov and Sapronov, the Decemists, and the remnants of the Workers' Opposition. These 'ultra-left' groups, we remember, had acceded to the Joint Opposition in 1926; but then they left it or were expelled. In the places of deportation their adherents mingled with the Trotskyists and endlessly argued with them. They carried the ideas of the Trotskyists to extreme conclusions, which were sometimes logical, sometimes absurd, and sometimes absurd in their very logicality. In an exaggerated manner they expressed all the emotions that stirred in Trotskyist hearts, even if many of Trotsky's reasonings were above their heads. They therefore occasionally said things which Trotsky at first indignantly rejected only to take them up and say them at a later stage. They criticized Trotsky for indecision and pointed out that it was hopeless to count on a democratic reform in the party. (It was to take Trotsky another five or six years to reach the same conclusion.) The party led by Stalin was 'a stinking corpse', V. Smirnov wrote in 1928. He and his adherents maintained that Stalin was the victorious chief of the Russian Thermidor, which had occurred as far back as 1923, and the authentic leader of the kulaks and of the men of property at large. They denounced the Stalinist régime as a 'bourgeois democracy', or 'a peasant democracy', which only a new proletarian revolution could overthrow. 'The liquidation in 1923 of inner-party democracy and proletarian democracy at large', Smirnov wrote, 'has proved to be a mere prologue to the

---

[1] See Dingelstedt's letters to Trotsky of 8 July and 24 August 1928 in *The Archives*. Also his letter to Radek of 22 August.

development of a peasant-kulak democracy.'[1] Sapronov held that 'bourgeois parties are already legally organizing themselves in this country'—and this in 1928![2] They thus accused Stalin of restoring capitalism just when he was about to destroy private farming, the main potential breeding ground of capitalism in Russia; and of favouring a bourgeois multi-party régime just when he was driving the single-party system to its ultimate conclusion and establishing himself as the single leader. This was Quixotry indeed. An element of it might be found in Trotsky too, but his realism and self-discipline kept it in check. V. Smirnov, Sapronov, and their followers were held back by no such inhibitions when they threw themselves on the windmills of Stalin's 'kulak-democracy'; and some of Trotsky's younger and unreasoning followers were tempted to join them, especially after the 'liquidation of the left course' in July had momentarily given the windmills the slightest semblance of a foe on the move.[3]

Amid all these cross-currents Trotsky did what he could to prevent the Opposition from falling to pieces. He saw its dissensions as a conflict between two generations of Oppositionists, a clash between 'fathers and sons', the former over-ripe and weary with knowledge and experience, the latter full of innocent ardour and audacity. He himself felt with both, understood both, and was apprehensive of both. He had forebodings about Radek and Preobrazhensky: in their mood and reasonings he discerned the impulses that were to lead them to capitulation. But he was wary of alienating them; he gave them the benefit of the doubt; and he defended them from abuse by over-zealous Trotskyists. Patiently but firmly he argued with the two men: he granted them that there was truth in what they said about the left course and the changing outlook of the country; but he begged them not to draw rash conclusions and not to exaggerate the chances of any genuine conciliation with Stalinism. At the same time he tried to curb the extremists on the other side, telling them

---

[1] The quotation is from a Decemist essay 'Pod Znamya Lenina', the authorship of which Trotsky ascribes to V. Smirnov. *The Archives.*

[2] See Sapronov's statement of 18 June, addressed to an unknown friend, in *The Archives.*

[3] Trotsky described those who shared V. Smirnov's and Sapronov's views as the lunatic fringe of anti-Stalinism, but he favoured co-operation with the more moderate Decemists, such as Rafail, V. Kossior, Drobnis, and Boguslavsky. See his circular letter on the Decemists of 22 September 1928 in *The Archives.*

that they were taking an over-confident view of the Opposition's prospects and were courting disillusionment: they should not imagine that the latest attempt at appeasing the kulaks was 'Stalin's last word' which could be followed only by the 'inevitable collapse' of the Stalinist régime. The outlook, as he saw it, was far more complex: it was impossible to be sure what might emerge from the cauldron. At any rate, although he had said that 'the party will still need us', he was far less confident than Sosnovsky that 'the call for Trotsky's return will soon resound throughout the world'.[1]

He endeavoured to maintain the Opposition's unity on the ground of a 'sustained and uncompromising struggle for inner party reform'. His adamant rejection of 'illusions about a *rapprochement* with Stalinism' commended his attitude to the young irreconcilables, while his emphasis on inner party reform provided the link between himself and the conciliators. He repudiated the 'wholly negative and sterile' Decemist attitude towards the party; and he sought to counteract the nostalgic hankering after the party, the creeping feeling of isolation, and the sense of their own uselessness to which the older Oppositionists tended to succumb. He tried to rekindle the sense of mission—the conviction that even in exile they still spoke for the mute working class, that what they said still mattered, and that sooner or later it would reach the working class and the party. This conviction, he added, should not induce self-righteousness or arrogance in the Opposition: although it alone stood consistently for the Marxist and Leninist tradition, it must not dismiss all its adversaries as worthless—indeed, it must not assume that all that had been left of Lenin's party was a few thousand Oppositionists. The Opposition was right in exposing the party's 'bureaucratic degeneration'; but even in this a sense of proportion was necessary, because there were 'various degrees of degeneration'; and there were still many uncorrupted and sound elements in the party. 'Stalin has owed his position not only to the terror exercised by the machine, but also to the confidence or semi-confidence of a section of Bolshevik workers.' With those workers the Opposition must not lose touch—to them it must appeal.[2]

---

[1] See Trotsky's letter to 'V. D.' (Elzin?) of 30 August 1928.

[2] See his circular letter on the Opposition's differences with the Decemists of 11 November 1928 and also his letters of 15 July, 20 August, 2 October, and 10 November, dealing with the same subject.

Trotsky's finely balanced interventions were not always well received. The ultra-radicals went on carping at his leniency towards the conciliators, while Preobrazhensky and Radek reproached him with countenancing the 'Decemist attitude' of those Trotskyists who behaved as if the Opposition were a new party, not a faction of the old. The estrangement between the groups grew continually. But as long as Trotsky remained at Alma Ata and from there exercised his influence, and as long as Stalin's policy, being in a state of suspense, did not further accentuate the Opposition's dilemmas, Trotsky succeeded in preventing the various sets of his followers from straying too far apart and scuttling the Opposition.

In these trying circumstances he found the strongest moral support in Rakovsky. Their old and close friendship had now acquired a new depth of affection, intimacy, and intellectual concord. After his great career as head of the Bolshevik government in the Ukraine and diplomat, Rakovsky worked at Astrakhan, the place of his exile, as a minor official of the local Gosplan. His correspondence with Trotsky and eye-witness accounts give impressive evidence of the stoical calm with which he bore his fortune and of the intensity and range of his intellectual work in exile.[1] In the wanderer's bag he had brought to Astrakhan the works of Saint-Simon and Enfantin, of many French historians of the revolution, and of Marx and Engels, Dickens's novels, and classics of Russian literature. In the first weeks of deportation Cervantes was his favourite reading. 'In a situation like this', he wrote to Trotsky, 'I go back to *Don Quixote* and find enormous satisfaction.' Longing for his native Dobrudja, he reread Ovid. Concerned with economic planning in the Astrakhan area, he assiduously studied the 'geological profiles' of the Caspian steppes; and describing this work to Trotsky, he interspersed his remarks with references to Dante and Aristotle. Above all, he was eagerly restudying the French Revolution;[2]

---

[1] Louis Fischer, who visited Rakovsky at Astrakhan, relates that he once saw him employed by local authority to act as an interpreter to a group of American tourists. Rakovsky looked worn and haggard and, when he had finished interpreting, the American visitors tried to tip him. With a polite gesture, half-sad and half-amused, Rakovsky withdrew.

[2] As Ambassador in Paris Rakovsky had done much to promote the study by Soviet historians of the archives of the French Revolution, in which he himself took a close interest. Among the books he took with him into exile and cherished

and he wrote a *Life of Saint-Simon*. He reported to Trotsky on the progress of his work and quoted to him Saint-Simon's predictions about Russia and the United States as the two antagonistic giants of the future (predictions which are less known but more original than those made by Tocqueville later). Grumbling about the effect of age on his memory and imagination—he was fifty-five at the time of deportation—he nevertheless worked 'with enormous zest—*avec ardeur*!' With a hint of paternal tenderness, he urged Trotsky not to spend his energy and talents on current affairs only: 'It is extremely important that you should also choose a large subject, something like my Saint-Simon, which would compel you to take a fresh look at many issues and re-read many things from a definite angle.'[1] He procured for Trotsky books and periodicals which could not be obtained at Alma Ata. He kept in touch with Trotsky's children in Moscow, and shared in the family's sorrows. Politically, he supported Trotsky against both conciliators and ultra-radicals; and to no one of the leaders of the Opposition was Trotsky attached as strongly as he was to Christian Georgevich.[2]

Rakovsky's political temperament was in many respects different from Trotsky's. He did not, of course, possess Trotsky's powers of thought, passion, and expression, nor his tempestuous energy. But he had a very clear and penetrating mind; and perhaps also a greater capacity for philosophical detachment. For all his devotion to the Opposition, he was less of a partisan, at least in the sense that his views transcended in their largeness the Opposition's immediate aims and tactics. Convinced of the Opposition's rightness and its ultimate vindication, he was far less confident about its chances of political success. He stood back and took in the immense picture of the revolution and grasped clearly the tragic motif that ran through it and affected all the warring factions. That motif was 'the inevitable disintegration of the party of revolution after its victory'.

He developed this idea in his 'Letter to Valentinov,' an essay which caused a stir in the Trotskyist colonies in the summer of

greatly was a copy of Aulard's *Histoire politique de la Révolution Française* dedicated to him by the author.

[1] See Rakovsky's letter to Trotsky of 17 February 1928 in *Bulleten Oppozitsii*, nr. 35.

[2] 'To Christian Georgevich Rakovsky, Fighter, Man, and Friend' Trotsky had dedicated his *Literature and Revolution*.

1928.[1] How is one to account, Rakovsky asked, for the abysmal wickedness and moral depravity which had revealed itself in the Bolshevik party, a party that had consisted of honest, dedicated, and courageous revolutionaries? It was not enough to blame the ruling group or the bureaucracy. The deeper cause was 'the apathy of the masses and the indifference of the victorious working class after the revolution'. Trotsky had pointed to Russia's backwardness, the numerical weakness of the working class, isolation, and capitalist encirclement as the factors responsible for the 'bureaucratic degeneration' of state and party. To Rakovsky this explanation was valid but insufficient. He argued that even in a most advanced and thoroughly industrialized nation, even in a nation consisting almost entirely of workers and surrounded only by socialist states, the masses might, after the revolution, succumb to apathy, abdicate their right to shape their own life, and enable an arbitrary bureaucracy to usurp power. This, he said, was the danger inherent in any victorious revolution—it was the 'professional risk' of government.

Revolution and civil war are as a rule followed by the social decomposition of the revolutionary class. The French Third Estate disintegrated after it had triumphed over the *ancien régime*. Class antagonisms in its midst, the conflicts between bourgeois and plebeian, destroyed its unity. But even socially homogeneous groups were split because of the 'functional specialization' of their members, some of whom became the new rulers while others remained among the ruled. 'The function adjusted its organ to itself and changed it.' Because of the disintegration of the Third Estate, the social basis of the revolution narrowed and power was exercised by ever fewer people. Election was replaced by nomination. This process had been well advanced even before the Thermidorian *coup*; it was Robespierre who had furthered it and then became its victim. First it was the exasperation of the people with hunger and misery that did not allow the Jacobins to trust the fate of the revolution to the popular vote; then Jacobin arbitrary and terroristic rule drove the people into political indifference; and this enabled the Thermidorians to destroy Robespierre and the Jacobin party. In Russia similar changes had occurred in the 'anatomy and physiology' of the

---

[1] The text of the letter, written on 2 August 1928, is in *The Archives*. Valentinov had been editor in chief of *Trud* and was exiled as a Trotskyist.

working class and had led to similar results: the abolition of the elective system; the concentration of power in very few hands; and the replacement of representative bodies by hierarchies of nominees. The Bolshevik party was split between rulers and ruled; it disintegrated; and it changed its character so much that 'the Bolshevik of 1917 would hardly recognize himself in the Bolshevik of 1928'.

A deep and shocking apathy still paralysed the working class. Unlike Trotsky, Rakovsky did not think that it was the workers' pressure that had forced Stalin to embark upon the 'left course'. This was a bureaucratic operation carried out exclusively from above. The rank and file had no initiative and was all too little eager to defend its freedoms. Rakovsky recalled one of Babeuf's sayings in 1794: 'To re-educate the people in the love of liberty is more difficult than to conquer liberty.' Babeuf raised the battle-cry: 'Liberty and an Elected Commune!'; but his cry fell on deaf ears. The French had 'unlearned' freedom. It was to take them thirty-seven years, from 1793 to 1830, before they re-learned it, recovered from apathy, and rose in another revolution. Rakovsky did not explicitly pose the question which suggested itself: how long would it take the Russian masses to regain their political vitality and vigour? But his argument implied that a political revival could occur in Russia only in a relatively remote future, after great changes had taken place in society, and after the working class had grown, developed, become reintegrated, and had recovered from many shocks and disillusionments. He 'confessed' that he had never expected early political triumphs for the Opposition; and he concluded that the Opposition should direct its efforts mainly towards the long-term political education of the working class. In this respect, he said, the Opposition had not done or attempted to do much, although it had done more than the ruling group; and it should keep in mind that 'political education bears fruit only very slowly'.

The unspoken conclusion was that the Opposition had little chance, if any, to influence the course of events in its time, although it could confidently look forward to its ultimate, perhaps posthumous, vindication. Rakovsky threw into relief the Opposition's basic predicament: its position between a demoralized, treacherous, and tyrannical bureaucracy on the one side, and a hopelessly apathetic and passive working class on the other. 'I think', he emphasized, 'that it would be utterly unrealistic to expect any inner-party reform based on the bureaucracy.' Yet he anticipated no regenerative movement from the masses either, for many years to come. It followed (although Rakovsky did not say it) that the bureaucracy, such as it was, would remain, perhaps for decades, the only force capable of initiative and action in the reshaping of Soviet society. The Opposition was bound by its principles to persist in irreducible hostility towards the bureaucracy; but it could not effectively appeal against it to the people. Consequently, it could not play any practical part in the evolution of party and state; and it was eliminated in advance from the great historic process by which Soviet society was presently to be transformed. It could only hope to work for the future mainly in the field of ideas.

A conclusion of this kind, implicit in Rakovsky's 'Letter to Valentinov', may in certain situations satisfy a small circle of theorists and ideologues; but it spells the death sentence for any political movement. Rakovsky viewed the course of the revolution and the Opposition's prospects with cool, profound insight and stoical equanimity. No such detachment or equanimity could be expected from the several thousand Oppositionists who read the 'Letter to Valentinov'. Whether workers or intellectuals, they were practical revolutionaries and fighters, passionately interested in the immediate outcome of their struggle and in the upheavals which were shaking and shaping their nation. They had joined the Opposition as a political movement, not as a coterie of philosophers or ideologues; and they wished it to triumph as a political movement. Even the most heroic and selfless rebels or revolutionaries struggle as a rule for objectives which they believe to be in some measure within the reach of their generation—only few and very exceptional men, thinkers, can fight for a prize which history may award them posthumously.

The mass of the Oppositionists had striven to strengthen the socialist sector of the Soviet economy, to further industrialization, to revive the spirit of internationalism, and to restore some freedom within the party. They could not bring themselves to believe that these objectives were for them unattainable. They had already found out that they could not attain them by themselves, and that they had to look either to the masses or to the bureaucracy for help. They could not accept the view that it was hopeless to look to either. To exist politically, they had to believe either that the masses would sooner or later rise against the bureaucracy, or that the bureaucracy would, for its own reasons, carry out many of the reforms for which the Opposition stood. The radical Trotskyists looked to the masses; the conciliators to the ruling group or to a section of it. Each of these hopes was illusory, but not to the same extent. There was no sign in the country of any spontaneous mass movement in favour of the Opposition's objectives. But the bureaucracy was clearly in ferment; it was divided against itself over such issues as industrialization and peasant policy. The conciliators saw that on these issues the Stalinist faction had, after all, moved closer to the Opposition; and this encouraged them to expect that it might move closer in other respects as well. The fact that the bureaucracy was the only force showing effective social initiative induced the hope that the bureaucracy might even restore freedom to the party. The alternative was too gloomy to contemplate: it was that inner-party freedom and proletarian democracy at large were bound to remain empty dreams for a long time to come.

Trotsky was greatly impressed by Rakovsky's views and commended them to the Opposition; but he missed, so it seems, some of their deeper and relatively pessimistic implications. In Trotsky the detached thinker and the active political leader were now at loggerheads. The thinker accepted an analysis from which it followed that the Opposition was virtually doomed as a political movement. The leader could not even consider such a conclusion, let alone reconcile himself to it. The theorist could admit that Russia, like France before her, had 'unlearned freedom' and might not relearn it before the rise of a new generation. The man of action had to banish this prospect from his mind and try and give his followers a practical purpose. The thinker could run ahead of his time and work for the verdict of posterity. The chief of the Opposition had to go back to his time, live in it, and believe with his followers that they had a great and constructive part to play in it. Both as thinker and political leader, Trotsky refused to contemplate his country in isolation from the world. He remained convinced that the worst predicament of Bolshevism lay in its isolation and that the spread of revolution to other

countries would help the peoples of the Soviet Union to relearn
freedom much earlier than they could otherwise do.

. . . .

Late in the summer of 1928 startling news, coming from
clandestine Trotskyist circles in Moscow, reached Alma Ata. It
brought detailed evidence that Stalin was about to resume the
left course and that the breach between his faction and Bukharin's was complete and irreparable. Moreover, the reports
from Moscow claimed that both the Bukharinists and the Stalinists mooted an alliance with the Left Opposition and that both
were already vying for Trotskyist and Zinovievist support. It
looked indeed as if the cry for Trotsky's return was about to go
up, after all.

The Moscow Trotskyists were in fairly close contact with
Kamenev, who gave them an account of talks he had had with
Sokolnikov during the July session of the Central Committee.
Sokolnikov, still a member of the Committee and himself something of a semi-Bukharinist and semi-Zinovievist, appeared to
entertain the hope of forming a coalition of right and left against
the Stalinist centre; and he himself tried to act as go-between.
He related to Kamenev that Stalin had boasted at the Central
Committee that in the struggle against the Bukharinists he would
soon have the Trotskyists and Zinovievists on his side, and that
indeed he had them 'in his pocket' already. Bukharin was dismayed. Through Sokolnikov he implored the Left Opposition to
refrain from assisting Stalin and even suggested joint action
against Stalin. However, the July session of the Central Committee ended with Bukharin's seeming success, or rather with a
compromise between him and Stalin. But shortly thereafter they
were at loggerheads again; and Bukharin secretly met Kamenev
in Sokolnikov's presence. He told Kamenev that both he and
Stalin were going to be compelled to turn to the Left Opposition
and try and make common cause with it. Bukharinists and
Stalinists still were afraid of appealing to their former enemies;
but both knew that this course would become 'inevitable within
a couple of months'. It was, in any case, certain, Bukharin said,
that the expelled and deported Oppositionists would soon be
recalled to Moscow and reinstated in the party.[1]

[1] The reports of the Moscow Trotskyists are in *The Archives*. The account of
Sokolnikov's talks with Kamenev is dated 11 July 1928; and that of the meeting

Of his meeting with Bukharin, Kamenev wrote a detailed account for Zinoviev who was still in semi-exile at Voronezh; and this account allows us to reconstruct the scene with its peculiar colour and atmosphere. The Bukharin closeted with Kamenev and Sokolnikov was a very different man from the one who had only seven months earlier, at the fifteenth congress, helped to crush the Opposition. There was no trace in him now of that earlier self-confident and bragging Bukharin who had mocked Kamenev for 'leaning on Trotsky' and whom Stalin had congratulated for 'slaughtering' the leaders of the Opposition 'instead of arguing with them'. He arrived at Kamenev's home stealthily, terrified, pale, trembling, looking over his shoulders, and talking in whispers. He began by begging Kamenev to tell no one of their meeting and to make no mention of it in writing and over the telephone because they were both spied upon by the G.P.U. Broken in spirit, he had come to 'lean' on his old adversary who was himself morally crippled. Panic made his speech partly incoherent. Without pronouncing Stalin's name he repeated obsessively: '*He* will slay us', '*He* is the new Genghiz Khan', '*He* will strangle us'. On Kamenev Bukharin already made 'the impression of a doomed man'.

Bukharin confirmed that the crisis in the leadership had been caused by the conflict between government and peasantry. In the first half of the year, he said, the G.P.U. had had to quell 150 sporadic and widely scattered peasant rebellions—to such despair had Stalin's emergency measures driven the muzhiks. In July the Central Committee was so alarmed that Stalin had to feign a retreat: he revoked temporarily the emergency measures but he did so only in order to weaken the Bukharinists and to prepare himself better for a new attack. Since then he had succeeded in winning over to his side Voroshilov and Kalinin, who had been in sympathy with the Bukharinists; and this had given him a majority in the Politbureau. Stalin, so Bukharin related, was now ready for the final offensive against private farming. He had adopted Preobrazhensky's idea and argued that only by 'exploiting' the peasantry could socialism proceed with primitive

between Bukharin and Kamenev bears the date of 11 August. A further report on a meeting between the Trotskyists and Kamenev is of 22 September. The account of the talk between Kamenev and Bukharin was clandestinely circulated by the Trotskyists in Moscow a few months later, at the time of Trotsky's deportation from Russia.

accumulation in Russia, because, unlike early capitalism, it could not develop through the exploitation of colonies and with the help of foreign loans. From this Stalin drew the conclusion (which Bukharin characterized as 'illiterate and idiotic') that the further socialism advanced, the stronger would popular resistance to it become, a resistance which only 'firm leadership' could hold down. 'This meant a police state', Bukharin commented; but 'Stalin will stop at nothing'; 'his policy is leading us to civil war; he will be compelled to drown rebellions in blood'; and 'he will denounce us as the defenders of the kulak'. The party was on the brink of an abyss: if Stalin were to win, not a shred of freedom would be left. And again: '*He* will slay us', '*He* will strangle us'. 'The root of the evil is that party and state are so completely merged.'

This was the situation in which Bukharin resolved to appeal to the left Opposition. The old divisions, as he saw it, had become largely irrelevant: 'Our disagreements with Stalin', he said to Kamenev, 'are far, far graver than those we have had with you.' What was now at stake was no longer normal differences of policy, but the preservation of party and state and the self-preservation of all of Stalin's adversaries. Though the Left Opposition stood for an anti-kulak policy, Bukharin knew that it would not wish to pursue it by the reckless and bloody methods to which Stalin would resort. In any case, it was not ideas that mattered to Stalin: 'He is an unprincipled intriguer who subordinates everything to his lust for power . . . he knows only vengeance and . . . the stab in the back. . . .' And so Stalin's opponents should not allow their old differences over ideas to prevent them from joining hands in self-defence.

Eager to encourage his would-be partners, Bukharin then enumerated the organizations and influential individuals whom he supposed to be ready to line up against Stalin. The workers' hatred of Stalin, he said, was notorious: Tomsky, when he was drunk, whispered into Stalin's ear: 'Soon our workers will start shooting at you, they will.' At party cells members were so disgusted with Stalin's lack of principles that, when the left course was initiated, they asked: 'Why is Rykov still at the head of the Council of People's Commissars, while Trotsky is exiled to Alma Ata?' The 'psychological conditions' for Stalin's dismissal were not yet ripe; but they were ripening, Bukharin claimed. True,

Stalin had won over Voroshilov and Kalinin; Ordjonikidze who had come to hate Stalin had no guts; but Andreev, the Leningrad leaders—was Kirov one of them?—and Yagoda and Trillisser, the two deputy chiefs of the G.P.U., and others, were ready to turn against Stalin. Alleging that the two effective chiefs of the G.P.U. were on his side, Bukharin nevertheless did not cease to speak in terror of the G.P.U.; his account of the forces that he could marshal against Stalin could not sound reassuring to his interlocutor.

A few weeks later the Muscovite Trotskyists reported to Alma Ata on another meeting they had had with Kamenev. 'Stalin is on the point of making overtures to the Left Opposition'—so confident was Kamenev about this that he had already warned Zinoviev not to compromise their position by responding too eagerly to Stalin's approaches. He held that a *dénouement* was imminent; and he was 'at one with Trotsky' in thinking that Stalin's policy had antagonized the entire peasantry, not merely the kulaks, and that the tension had reached explosion point. Consequently, a change in the party leadership was unavoidable: it 'was bound to occur even before the end of the year'. But Kamenev implored Trotsky to take a step which would ease his re-entry into the party. 'Lev Davidovich should make a statement now, saying: "Call us back and let us work together." But Lev Davidovich is stubborn. He will not do that; he will rather stay at Alma Ata until they send a special train to fetch him. But by the time they have made up their minds to send that train, the situation will be out of hand and Kerensky will be *ante portas*.'[1]

Stalin, however, did not make the direct approaches that Kamenev expected. Instead, he dropped many broad hints of a possible reconciliation; and he made sure that these hints reached Trotsky by roundabout ways. Thus he told a foreign, Asian, communist that he recognized that even in exile Trotsky and his followers had, unlike the Decemists, remained 'on the ground of Bolshevik ideology'; and that he, Stalin, was only wondering how to bring them back at the earliest opportunity. Stalin's entourage, Ordjonikidze in particular, talked about Trotsky's

---

[1] Kamenev resented Trotsky's attacks on the capitulators; nevertheless he and Zinoviev intervened with Bukharin and Molotov on Trotsky's behalf and protested against his being kept in exile in conditions detrimental to his health.

reinstatement openly and freely; and at the sixth Comintern congress foreign delegations had been told in confidence to reckon with the possibility or even probability of a coalition between Stalin and Trotsky.[1]

The sense of crisis had by now spread from the Russian party to the International. Despite a show of unanimity and official enthusiasm, the sixth congress was disappointed by Stalin's and Bukharin's joint conduct of the International's affairs. Trotsky's *Critique* of the new programme had, in a censured version, been circulated at the congress, where, according to Trotsky's correspondents, it made its impression.[2] Even those foreign communist leaders who passed as ardent Stalinists spoke in private with disgust about the dogmas and rites Stalin had imposed upon the Communist movement. Togliatti-Ercoli was reported to have complained about the unreality of the proceedings at the congress, the 'dull and sad parades of loyalty', and the arrogance of the Russian leaders. 'One felt like hanging oneself from sheer despair', he allegedly said. 'The tragedy is that one cannot speak the truth about the most important current issues. We dare not speak. . . .' Togliatti found Trotsky's *Critique* 'extraordinarily interesting . . . a very sensible analysis of socialism in one country'. Thorez, the French leader, characterized the mood at the congress as one of 'uneasiness, discontent, and scepticism'; and he too approved much in Trotsky's criticism of socialism in one country. 'How has it happened', he asked, 'that we have been made to swallow this theory?' Even if the Russian party had had to combat Trotskyism, it should not have accepted Stalin's dogma. He found the degradation of the International 'almost unbearable'. It was not possible to conceal from the congress the conflict between Stalin and Bukharin; and it was in this connexion that trusted foreign delegates were forewarned that, in the case of a definite breach with Bukharin, Stalin might consider it desirable or necessary to form a coalition with Trotsky.

Similar reports continued to reach Alma Ata from many sides throughout August and September. Stalin himself was

[1] See an undated letter entitled '*Podgotovka Kongresa*' and other undated correspondence from Moscow in *The Archives*.

[2] It was in this version that American Communists brought *The Critique* out of Russia and published it in the States in 1928.

undoubtedly still lending colour to the belief that he favoured Trotsky's imminent recall. In part this was deceit and *ruse de guerre*. By hinting that he was ready to make peace with Trotsky Stalin sought to intimidate Bukharin and Rykov, to confuse the Trotsky-ists and to make the conciliators among them even more impatient for conciliation than they were. But Stalin was not only bluffing. He could not yet be quite sure of the outcome of his show-down with Bukharin, Rykov, and Tomsky, and of his ability to cope simultaneously, and in the midst of a national crisis, with both oppositions, left and right. He worked tirelessly to bring both oppositions to their knees; but as long as he had not fully succeeded in this, he had to keep his door ajar to agreement with one of them. His position was already so much stronger than Bukharin's that he had no need to make direct overtures. But he threw out *ballons d'essai* and he watched how Trotsky and his associates received them.

Trotsky was well prepared to meet some of these developments; yet others caught him by surprise. The recrudescence in so dangerous a form of the conflict between town and country, the breach between Stalin and Bukharin, and the fact that the eyes of some of his adversaries and of the capitulators were once again turned on himself—all this corresponded to Trotsky's expectations. He was still inclined to think that the Stalinist faction would not be able to extricate itself and that it would be compelled to beg the Left Opposition to come to its rescue. He had repeatedly declared, in the most formal and solemn manner, that in such a situation the Opposition 'would do its duty' and would not refuse co-operation. He now reiterated this pledge. But he added that he spurned all 'bureaucratic combinations': he was not prepared to bargain behind the scenes for his seat in the Politbureau or to content himself with such share of control over the party machine as Stalin might offer him *in extremis*. He and his associates, he declared, would re-enter the party only on terms of proletarian democracy, reserving full freedom of expression and criticism—and on the condition that the party leadership would be elected by the rank and file in secret ballot instead of being picked by caucusdom through the familiar interfactional contrivances.[1]

Stalin's situation, difficult though it was, was not so desperate

[1] See, for instance, Trotsky's letter to S. A. (20 August 1928).

that he should accept Trotsky's conditions. Trotsky expected, however, that it would deteriorate further and then the bulk of the Stalinist faction, with or without its leader, might be compelled to seek agreement on his terms. As a matter of both principle and self-interest, he would not contemplate any other terms—after all his experiences he would not rely on the favours of the 'apparatus'.

Meantime, however, Trotsky was confronted by an unexpected turn of events. For years he had not ceased to speak of the 'danger from the right' and to warn the party against the defenders of the kulak and the Thermidorians. He had been prepared to form a 'united front' with Stalin against Bukharin. But it was Bukharin who implored the Left Opposition to make common cause against Stalin, their common enemy and oppressor. When Bukharin whispered in terror: '*He* will strangle us—*He* will slay us' Trotsky could not dismiss this as the imaginings of a befogged and panicky man—he himself had repeatedly spoken about the holocaust the 'grave-digger of the revolution' was preparing for the party. True, Bukharin's appeal had come very late in the day, after he had helped Stalin to crush the Opposition and to destroy the party's freedom. But he was not the first of Stalin's adversaries to behave in this manner. Zinoviev and Kamenev had done the same; yet this had not prevented Trotsky from joining hands with them. Should he then spurn Bukharin's outstretched hand? If Stalin was taking one leaf out of Trotsky's book, the left course, Bukharin took another: he appealed to the Left Opposition in the name of proletarian democracy. Trotsky was in a quandary: he could not turn a deaf ear on Bukharin's appeal without denying one of his own principles; and he could not respond to it without acting, or appearing to act, against another principle of his by which he was committed to support the left course.

Seeking a way out, he took a more reserved attitude towards Stalin's left course and became less emphatic in proclaiming the Opposition's support for it. Quite apart from Bukharin's approaches, he had his own reasons for this. From all over the Soviet Union followers wrote to him about the terror Stalin had let loose on the country in the spring and early summer and about the 'orgies of brutality' to which he had subjected the middle and even the poor peasants. Officialdom tried to disclaim

responsibility by telling the people that Trotskyist and Zinovievist pressure had provoked the drive against the peasants. Everything indicated that, if and when Stalin resumed the left course,
it would lead to a bloody cataclysm. Trotsky refused in advance
to bear any share of responsibility for this. In August 1928,
nearly a year before the 'liquidation of the kulaks' started, he
wrote to his followers that although the Opposition had pledged
itself to back the left course, it had never proposed to deal with
the peasantry in the Stalinist manner. It had stood for higher
taxes on the wealthy, for government support to the poor
farmers, for an equitable and equable treatment of the middle
ones, and for encouragement of voluntary collectivization—but
not for a 'left course' of which the chief ingredient was administrative force and brutality. In judging Stalin's policy 'it
was necessary to consider not only *what* he did but also *how* he
did it'.[1] Trotsky did not suggest that the Opposition should not
support the left course, but he stressed more than ever that it
must combine support with severe criticism. He set his face
against the conciliators who regained spirit from the latest
evidence that the breach between Stalin and Bukharin was
unhealed and that Stalin was about to resume the 'drive against
the kulak'. He rejected Kamenev's promptings with scorn and
contempt. He declared that he would do nothing to 'ease' his
own re-entry into the party and that he would not beg his
persecutors to recall him to Moscow. It was up to them to do so
if they wished, but even then he would not cease to attack them
and the capitulators as well.[2]

This was Trotsky's reply not only to Kamenev's suggestions,
but also to Stalin's vague and allusive blandishments. Conciliation between them was out of the question. He responded far
more favourably to Bukharin's appeal. He did this in 'A Frank
Talk with a Well-wishing Party-man', a circular letter of 12
September. The 'well-wishing party-man' was a Bukharinist
who had written to Trotsky inquiring about his attitude towards

---

[1] See Trotsky's letter of 30 August to Palatnikov, a 'red professor', an economist,
exiled to Aktiubinsk. In a letter to Rakovsky of 13 July Trotsky wrote that Radek
and Preobrazhensky imagined that the Stalinist faction, having moved leftwards,
had only a 'rightist tail' behind it and should be persuaded to rid itself of it. Even if
this were true, Trotsky remarked, it would help little: 'an ape freed of its tail is
not yet a human being.' *The Archives.*

[2] '*Pismo Druzyam*' of 21 October.

the right wing, now the Right Opposition. Trotsky replied that on major issues of industrial and social policy the gulf between them was as wide as ever. But he added that he was ready to co-operate with the right for one purpose, namely, the restoration of inner-party democracy. If Rykov and Bukharin were prepared to work with the left in order to prepare jointly an honestly elected and truly democratic party congress, he favoured an agreement with them.

This statement caused astonishment and even indignation in Trotskyist colonies. Many exiles, not only conciliators, protested against it and reminded Trotsky how often he himself had described coalitions of right and left directed against the centre as unprincipled, pernicious, and responsible for the wrecking of more than one revolution. Had Thermidor not been precisely such a combination of right and left Jacobins nefariously united against Robespierre's centre? Had not the whole conduct of the Opposition so far been determined by its readiness to coalesce, on conditions, with the Stalinists against the Bukharinists, not vice versa? Had Trotsky himself not solemnly restated this principle only quite recently, when he assured the Communist International that the Left Opposition would never enter into any combination with those who opposed Stalinism from the right?

Trotsky replied that he still viewed the Bukharinist right rather than the Stalinist centre as the chief antagonist. He had not proposed to Bukharin any coalition over issues of policy. But he saw no reason why they should not join hands for the one clearly defined purpose of re-establishing inner-party freedom. He was ready to 'negotiate with Bukharin in the same way that duellists parley through their seconds over the rules and regulations by which they will abide'.[1] The left could only wish to pursue its controversy with the right under rules of inner-party democracy; and if this was also what the right wanted, nothing would be more natural than that they should collaborate in order to make those rules prevail.

This plea carried little conviction with Trotsky's followers. They had been so much accustomed to see in Bukharin's faction their chief enemy that they could not contemplate any agreement with it. They had so long and so persistently assailed

[1] See '*Na Zloby Dnya*' (no precise date), Trotsky's reply to his critics, in *The Archives*.

PLATE XI

(*a*) Christian Rakovsky, as Soviet envoy
at the Genoa Conference (1922)

(*b*) Karl Radek

(*c*) Adolfe Yoffe: committed suicide to
protest against Trotsky's expulsion

(*d*) Antonov-Ovseenko, Trotskyist, as
chief political commissar of the Red
Army

PLATE XII

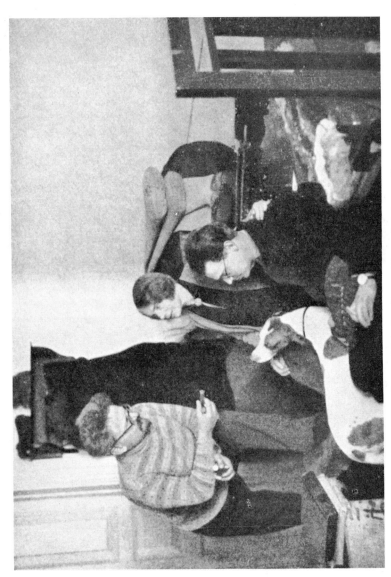

Trotsky and his family at Alma Ata in 1928

the Stalinists as the two-faced accomplices of the right that they were horrified to think that they themselves might appear as its accomplices. Nor could they accept Trotsky's explanation that he had proposed to the Bukharinists only a technical arrangement, something like the fixing of regulations for a duel. For one thing, this was not a duel but a three-cornered fight, in which any agreement between two sides was automatically directed against the third. For another, inner-party democracy was a political problem *par excellence* with a bearing on all major issues. An alliance of right and left, no matter how limited its purpose, would, if successful, result in the overthrow of the Stalinist faction, and this after the latter had initiated the left course. The left course would be brought to an immediate standstill. The sequel would hang on the uncertain outcome of the contest between right and left. If the right were to win, it would surely proclaim that neo-N.E.P., the danger of which had haunted the Trotskyists. Could they take such a risk? With the country on the brink of economic catastrophe and the peasantry in turmoil, should they expose the party to a convulsion in the course of which the Stalinists might be overthrown but the Bukharinists and the Trotskyists might be unable to resolve their differences democratically, let alone rule jointly? They might thus unwittingly ruin the party, and give anti-Bolshevik forces their opportunity. This would indeed be a classically Thermidorian situation, for it was precisely such a coalition of right and left, both exasperated by the terror, that had brought about Robespierre's downfall. Was not Trotsky now playing with Thermidorian fire—he who had all these years warned others against it?

Trotsky and the Opposition were in an *impasse*. If any chance of self-preservation was left to them, that chance lay in a broad alliance of all anti-Stalinist Bolsheviks. Yet they could hardly hope that even such an alliance would save them. They had reason to fear that its outcome might be the end of the Bolshevik party. In considering for a while the idea of a coalition, both Trotsky and Bukharin were moved by a fleeting reflex of self-defence. Neither, however, could act on this reflex any further. Both factions were more anxious to preserve the party as it was than to preserve themselves; or else they did not see clearly their inexorable dilemma. Undoubtedly, some of the leaders saw it.

Kamenev's account of his meeting with Bukharin contains these grim words: 'Sometimes I say to Yefim: "Is not our situation hopeless? If our nation is crushed we are crushed with it; and, if it extricates itself and Stalin changes course in time, we are still going to be crushed."' Radek, in a letter written to his comrades, described the choice before them as a choice 'between two forms of political suicide'—one which consisted in being cut off from the party, and the other in re-entering the party after having abjured one's convictions.[1]

Bukharin's panicky offer of an alliance and Trotsky's tentative answer had therefore no sequel. The Bukharinists could not but react to their leader's proposal with the same resistance with which the Trotskyists met their leader's reply. They had seen their chief enemies in the Trotskyists and Zinovievists; and their most recent charge against Stalin was that he had become a crypto-Trotskyist (or, as Bukharin put it, that he had adopted Preobrazhensky's ideas). How then could they themselves contemplate a partnership with the Trotskyists? They knew that the latter and the Zinovievists viewed the left course with a sneaking sympathy—Bukharin too must have inferred this from his talk with Kamenev. And if even the banished Trotskyists were frightened of the shock to which a right–left coalition might expose the party, how much more must the Bukharinists have dreaded this prospect—they who had been in and of the ruling group and were still in it. They were dismayed by Stalin's hints that if they misbehaved he would ally himself with Trotsky. They decided that they would not misbehave. They did not even try to carry the struggle against Stalin into the open as the Trotskyists and Zinovievists had done; or, if they did, they found that having robbed the Left Opposition of freedom of expression they had also robbed themselves of it. Bukharin therefore could not follow up his approaches or respond to Trotsky's idea of a 'limited agreement'.

These developments strengthened the Trotskyist conciliators. Three of the most authoritative leaders of the Opposition in exile, Smilga, Serebriakov, and Ivan Smirnov, were now with Radek and Preobrazhensky. It was clear, they argued, that Stalin had not said 'his last word' in July when he appeared to yield to the kulak—the left course was on. Trotsky had im-

---

[1] The letter dated 16 September is in *The Trotsky Archives*.

plicitly admitted that the Left Opposition could not persist in splendid isolation and that it ought to look for allies; but its natural allies were the Stalinists, not the Bukharinists. This is not to say that the conciliators were happy about the manner in which Stalin dealt with the right opposition. 'Today the régime strikes at Bukharin', Smilga wrote, 'in the same way in which it struck at the Leninist Opposition . . . [the Bukharinists] are being strangled behind the back of the party and the working class.' But 'the Leninist Opposition has no reason to express political sympathy for the right because of this': its watchword was still 'Down with the Right!'[1] This had been Trotsky's watchword in the summer, but hardly in the autumn. The relations between him and the conciliators grew tense and inimical. He was scarcely in touch with Preobrazhensky; and his correspondence with Radek became acrimonious and intermittent. Radek protested against Trotsky's scathing attacks on Zinoviev, Kamenev, and the other capitulators. 'It is ludicrous to think', Radek wrote, 'that they have surrendered only from cowardice. The fact that group after group one day speaks against capitulation and on the next day agrees to capitulate, and that this has happened repeatedly and many times shows that we are confronted here with a clash of principles and not merely with fear of repression.'[2] It was true that the capitulators committed political suicide; but so also did those who had refused to capitulate. There remained only this hope that further shifts within the party and its further evolution towards the left would clear the air and enable the Left Opposition to re-enter the party with dignity.

While he thus excused Zinoviev's and Kamenev's motives, Radek circulated among his comrades a long treatise which he had written in refutation of Trotsky's Permanent Revolution.[3] He did not, however, send it to Trotsky, who received it second-

---

[1] The quotation is from Smilga's 'Platforma Pravovo Kryla VKP (b)' (23 October 1928) which was a comment on Bukharin's 'Zametki Ekonomista' which had appeared in Pravda on 30 September. (This was Bukharin's only public statement of his objections to the left course.) Smilga was also writing a book on Bukharin and Bukharinism, but it is not known whether he completed it.

[2] See Radek's circular letter to comrades of 16 September.

[3] The text of the treatise Razvitie i Znachenie Lozunga Proletarskoi Diktatury (hitherto unpublished) is in The Trotsky Archives. In reply to this Trotsky wrote Permanentnaya Revolutsia, the most extensive historical-theoretical defence of his conception.

hand—from Moscow. With an ironic reply to Radek Trotsky enclosed Radek's own earlier writings in defence of Trotskyism, saying that there he would find the best rejoinder to his latest arguments.[1] He did not yet suspect Radek of the intention to capitulate. He trusted that Radek's sense of humour and European-Marxist habits of mind would not allow him to go through a 'Byzantine' ritual of recantation. Still fond of the man and admiring him, Trotsky ascribed Radek's behaviour to 'moodiness' and went on defending him and Preobrazhensky against the suspicions of the young irreconcilables.[2]

Even now all Oppositionists, conciliators and irreconcilables alike, still looked to Trotsky as to their undisputed leader. Their feelings for him are best expressed in a protest which no other than Radek sent to the Central Committee in October, when news of a deterioration in Trotsky's health greatly disturbed the exiles:

Trotsky's illness has brought our patience to an end [Radek wrote]. We cannot look on in silence while malaria wastes the strength of a fighter who has served the working class all his life and was the Sword of the October Revolution. If preoccupation with factional interests has extinguished in you all memories of a common revolutionary struggle, let intelligence and plain facts speak. The dangers against which the Soviet Republic is contending are piling up. . . . Only those who do not understand what is needed in order to meet those dangers can remain indifferent towards the slow death of that fighting heart that is Comrade L. D. Trotsky. But those among you —and I am convinced that they are not few—who think with dread of what the next day may bring . . . must say: Enough of this inhuman playing with Comrade Trotsky's health and life![3]

.   .   .   .

Since the summer Trotsky's health had indeed deteriorated. He suffered again from malaria, severe headaches, and the chronic stomach infection which was to trouble him till the end of his days. Reports of his illness brought forth a spate of letters

[1] See Trotsky's letter to Radek of 20 October in *The Archives*.

[2] Even many months later, towards the end of May 1929, in Prinkipo, Trotsky received the first news of Radek's capitulation with the utmost incredulity and wrote: 'Radek has behind him a quarter of a century of revolutionary Marxist work . . . it is doubtful whether he is capable of joining the Stalinists. In any case, he will not be capable of staying with them. For this he is, after all, too much a Marxist and too internationally minded.' *The Archives*.

[3] Quoted from the *Militant*, 1 January 1929.

and telegrams from exiles expressing sympathy with him and protesting to Moscow. Some of the deportees were anxious for more vigorous action in Trotsky's defence and planned a collective hunger strike. Trotsky had some difficulty in dissuading them from so desperate a decision. There was no need—he said in messages to the colonies—to worry greatly about his health which was not so bad as to prevent him working. It was advisable to circulate more widely the protests which the Opposition had already made; but it would be rash to resort to drastic action which might only worsen the lot of those involved.[1]

As the autumn advanced more clouds began to gather over Trotsky's head. In October he ceased to receive letters from friends and followers—only communications from men who were ready to desert the Opposition were still delivered to him. The censorship worked selectively. His own letters and messages failed to reach their destination. He could obtain no reply even to telegrams in which he inquired about Zina's health, which continued to cause him anxiety. He passed the days of the revolution's anniversary in loneliness and apprehension—none of the usual greetings had arrived. Thereafter the ill omens multiplied. A local official, who had secretly sympathized with the Opposition and had kept in touch with Trotsky, was suddenly imprisoned. An Oppositionist who had come all the way from Moscow, had taken a driver's job at Alma Ata, and used to meet Trotsky stealthily at the public baths, and who had, it seems, run the 'secret mail' between Moscow and Alma Ata, vanished without a trace. The household had by now moved back from the *dacha* with its orchards and flower-beds into the dreariness of the town. 'Since the end of October', Sedova wrote to a friend, 'we have received no letters from home. We get no answers to our telegrams. We are under a postal blockade. Things will not stop at this, of course. We are awaiting something worse. . . . There is a severe frost here. The cold in our rooms is agony. Houses here are not built for the cold weather. The price of wood is incredibly high.'

Finally, rumours reached Trotsky from many quarters that he

---

[1] This, for instance, is the text of a telegram to the deportees of Yenisseisk (14 October 1928): 'Categorically object to forms of protest contemplated by yourselves. . . . My illness not immediately dangerous. Please observe common line [of conduct]. Fraternal greetings. Trotsky.' *The Archives.*

would not be left at Alma Ata and that he was presently to be deported farther away and isolated far more rigorously. At first he dismissed the rumours. 'I do not expect this to happen— where on earth could they send me?', he wrote to Elzin on 2 October. He looked forward to a winter of intensive study and literary work at Alma Ata and, of course, to hunting trips in the surrounding wilds. But the rumours persisted; and the postal blockade and other signs indicated that 'something worse' was indeed about to happen.

This was a strange autumn. On the anniversary of the revolution there resounded from the Red Square in Moscow these official calls: 'The Danger is on the Right!', 'Strike Against the Kulak!', 'Curb the N.E.P.-men!', 'Speed up Industrialization!'; and the slogans reverberated throughout the country, penetrating to the remotest backwaters, even to Alma Ata. How long had not Trotsky sought to persuade the party to adopt this policy! Only the year before on the same anniversary day, his followers had descended on the streets of Moscow and Leningrad with the same slogans on their banners. They were dispersed, maltreated, and accused of counter-revolution. There could be, one might have thought, no greater vindication of the Opposition than the fact that the ruling group was compelled to appropriate its ideas. No one who took the mildest interest in public affairs could fail to notice it. The ferocious attacks on Trotsky, the 'super-industrializer' and the 'enemy of the muzhik', were still fresh in everyone's mind. Now the dishonesty and wretchedness of these attacks were exposed and cried to heaven. Was not Stalin himself turning into a super-industrializer and enemy of the peasants, many a Bolshevik wondered. Yet this year as the year before, millions of citizens marched in the official processions, followed the prescribed routes, and shouted the prescribed slogans as if nothing unusual had occurred and as if they had been unable to think, reflect, and act.

Popular apathy allowed Stalin to steal Trotsky's clothes with impunity. Trotsky still consoled himself with the thought that Stalin would not be able to wear them, because they would not fit him. He still expected that as the national crisis deepened the Stalinist faction alone would not be able to cope with it. The crisis had indeed deepened. With the peasantry in rebellion and the towns gripped by fear of hunger, the nation lived in unbearable tension. There was feverish nervousness in the air and a sense of danger and alarm. The party machine sternly marshalled its strength and called on all to be ready for a grim, though as yet undefined, emergency. But it showed no inclination to recall the exiled Oppositionists.

Towards the end of the year Stalin was in a position far stronger than that in which he had been in the summer. He was less afraid of having to contend with two Oppositions simultaneously. The right was cowed and demoralized and was already surrendering. The left was torn by dissensions and paralysed. Stalin watched the disputes between Trotsky, Radek, Preobrazhensky, the irreconcilables, and the Decemists, and concluded that time was on his side. He was still engaged in the preliminaries to his all-out drive for industrialization and collectivization; and already the Trotskyist conciliators felt that they must not stand aside. How much more would they feel this once he had passed from the preliminaries to real action? True, they were not yet in a mood to surrender; but they were steadily approaching that stage; and all that they needed to reach it was time and a little encouragement. Through his agents Stalin encouraged them by every means at his command: he invoked the revolution's supreme interest; he appealed to Bolshevik loyalty; he mixed cajolery with threat; and he intensified the terror against the irreconcilable Trotskyists and the Decemists.[1] In this way he expected to make good his boast, which was premature when he made it, that he had the left Opposition 'in his pocket'. He did, in fact, need the left's assistance for his new policy. But he was bent on getting that assistance, not by allying himself with the left, but by splitting it, taming a large section of it, and turning it against Trotsky. He hoped to inflict on Trotsky a defeat far more harrowing than all the blows he had aimed at him so far.

Yet, for all his strength, Stalin could not be sure that he would be able to achieve what he had set out to achieve. He was about to embark upon a gigantic enterprise such as no ruler had ever undertaken: he was to expropriate at a stroke twenty-odd

---

[1] In the autumn the police supervision over the deportees was suddenly tightened and many deportees were imprisoned. V. Smirnov was sent to jail because he was five minutes late in reporting to the local G.P.U. for a routine check. Butov, one of Trotsky's secretaries, died in prison after a hunger strike lasting fifty days.

million farmers and drive them and their families into collective farms; he was about to force urban Russia into an industrial drive in which the horrors of primitive capitalist accumulation were to be reproduced on an immense scale and condensed within a very short time. He could not know how the nation would take it, what despair, anger, violence, and revolt the upheaval might not breed; to what pass he himself might not be driven; and whether his adversaries might not then try and seize their opportunity. If they were to seize it, they were sure to look to Trotsky for leadership. Even from Alma Ata Trotsky's ideas and his personality, surrounded by the halo of heroic martyrdom, fascinated the Bolshevik *élite*. Despite all the confusion and dejection among the exiles, Trotskyism was gaining new followers in the party cells. The G.P.U. had to deal with so many of them that towards the end of 1928 between 6,000 and 8,000 Left Oppositionists were imprisoned and deported, whereas at the beginning of the year the strength of both Trotskyists and Zinovievists was estimated at between 4,000 and 5,000 only. Kamenev was not alone in thinking that in an emergency the party would have to 'send a special train' for Trotsky. There was much heart-searching among capitulators and even among Stalinists, some of whom wondered whether, if the left course was justified, Trotsky had not been right all along; they were therefore sick of the calumny and brutality with which he had been treated. Stalin knew that, for nearly every one of the six or eight thousand Oppositionists who rather than renounce their views had chosen prison and exile, there were one or two capitulators who at heart agreed with their less pliable comrades, and one or two doubters or 'conciliators' (*dvurushniki*, the double-faced, as he called them) in his own faction. They all lay low now; but would they not rise against him when the tide had turned?

Nor could Stalin take lightly the threat of an alliance between Trotsky and Bukharin. Although it had not materialized this time, the threat remained in being as long as Trotsky was the undisputed chief of the Left Opposition and could be brought back by a 'special train'. Stalin therefore redoubled his efforts to break the spirit of the Opposition. His agents held out every possible hope and temptation to Radek, Preobrazhensky, and their friends, promising rehabilitation, invoking common purposes, and speaking of the great, fruitful, and honourable work they could still perform for the party and for socialism. All these efforts, however, met with a formidable obstacle in the influence which Trotsky exercised from Alma Ata and which had so far prevented the Opposition in exile from disintegrating. Stalin was determined to remove this obstacle from his path.

But how was he to do it? He still shrank from sending the killer; he did not yet dare even to throw his enemy into jail. The odium would have been too heavy, because, despite all that had happened, Trotsky's part in the revolution was still too fresh and vivid in the nation's mind. He therefore planned to expel Trotsky from Russia. He knew that even this would shock; and he carefully prepared public opinion. First, he put out rumours about the new banishment; next he ordered the rumours to be denied; and, finally, he gave them fresh currency. In this way he blunted public sensitivity. Only after rumour, denial, and recurrent rumour had made the thought of Trotsky's expulsion from the U.S.S.R. familiar and therefore less shocking could Stalin carry out his intention.

. . . .

Amid all the uncertainties about his future, Trotsky posed anew the great and baffling question: 'Whither is the revolution going?' The Soviet Union was now in the grey interval between two epochs—between N.E.P. and the Stalinist 'second revolution'.[1] The shape of things to come was blurred; at best, it could be seen only through a glass, darkly. Trotsky was becoming aware that some of the ideas he had expounded in recent years were on the point of being surpassed by events. He tried to go beyond these ideas, but their pull on him was strong. He attempted to outline new prospects; but habits of mind, formed during N.E.P. and suited to its realities, and historical memories of the French Revolution kept intruding upon his vision.

He realized, for instance, that his conception of the Soviet Thermidor had become untenable. It had become absurd to maintain that Bukharin and Rykov were still *the* defenders of

---

[1] I first used the term 'second revolution' in *Stalin, a Political Biography*, pp. 294 ff., and I have been criticized for using it. Collectivization and industrialization, the critics say, do not amount to a revolution. But if a change in property relations resulting from the dispossession at a stroke of twenty-odd million smallholders is not an economic and social revolution, what is?

private property, that Stalin was their will-less helpmeet, and that they were bound to be the ultimate beneficiaries of his policy. Trotsky therefore virtually abandoned his conception of the Soviet Thermidor.[1] In a 'Letter to Friends', written in October 1928,[2] one of his most remarkable essays of the Alma Ata period (although it is written in the Opposition's peculiar phraseology), he argued that Bukharin and the Bukharinists were Thermidorians *manqués* who had lacked the courage to act on their convictions. He gave this ironical and vivid description of their behaviour: 'Bukharin has gone farther than any of the leaders of the right [in promoting the interests of the kulak and of the N.E.P.-man], while Rykov and Tomsky have watched him from a safe distance. But every time Bukharin steps into the cold water [of Thermidor] he shivers, shudders, and jumps out; and Tomsky and Rykov run to the bushes for shelter.' Consequently, the kulak, the N.E.P.-man, and the conservative bureaucrat, disillusioned with the leaders of the Bolshevik right, were inclined to look for effective leadership elsewhere, notably to the army. With the French precedents in mind, Trotsky spoke about the closeness of the 'Bonapartist danger', implying that the Russian Revolution might skip Thermidor and pass directly from the Bolshevik to the Bonapartist phase.

The Bonapartist danger, he went on, could assume two different forms: it could either materialize as a classical military *coup d'état*, a Russian 18th Brumaire; or it could take the shape of Stalin's personal rule. He considered it probable that the army, appealing directly to the property-owning peasantry and backed by it, would attempt to overthrow Stalin and to put an end to the Bolshevik régime at large. It was to him a secondary question which of the army leaders might place himself at the head of the movement: in favourable conditions even mediocrities like Voroshilov or Budienny might take the initiative and succeed. (Trotsky quoted a proverb of which, he said, Stalin was fond: *Iz gryazi delayut Knyazia*—a Prince can be made out of filth.) The conditions favouring a *coup* were there: the peasantry felt nothing but hostility towards the party led by Stalin, and the working class was disgruntled and listless. A military

---

[1] He did, however, take it up again and defend it, after his exile to Turkey, but only to 'revise it' once again a few years later.
[2] '*Pismo Druzyam*' of 21 October in *The Archives*.

dictatorship, if it came into being, would therefore be broadly based. It would be counter-revolutionary in its character and consequences. It would seek to guarantee security, stability, and expansion to the private sector of the economy. It would dismantle or cripple the socialist sector. It would bring about the restoration of capitalism. Confronted with such a danger, Trotsky concluded, all Bolsheviks eager to defend socialism would have to unite; and the Left Opposition would have to co-operate with Stalin and his faction, because Stalin spoke not for the men of property, but for the 'proletarian upstart' and had so far avoided an open breach with the working class.

It was possible, on the other hand, that Stalin himself would become the Soviet Bonaparte. This would create a different situation for the country and the Opposition. Stalin could exercise his personal rule only through the party machine, not through the army. His dictatorship would not immediately have the counter-revolutionary consequences that would follow a military *coup*. But it would be based very narrowly; and it would be utterly insecure. Stalin would find himself in chronic conflict with all classes of society; he would try to overpower now this class and now that and play them against one another. He would have to struggle continuously in order to keep in subordination the party machine, the state bureaucracy, and the army; and he would rule in incessant and unrelievable fear of defiance from either of them. He would suppress all spontaneous social and political activity and all freedom of expression. In such conditions there could hardly be room for any 'united front' between the Left Opposition and the Stalinists—there would be only struggle beyond conciliation.

In this context Trotsky analysed succinctly and with powerful foresight the social background, the mechanics, the shape and the outlook of Stalin's rule as it was to evolve over the next twenty years. He portrayed in advance the General Secretary transformed into the fully fledged totalitarian dictator. However, having done this, he himself viewed the portrait rather incredulously; he thought that, on balance, the danger of a purely military dictatorship was more real. It seemed to him far more probable that Voroshilov, Budienny, or some other general would lead the army against Stalin, and that Trotskyists and Stalinists would then fight together 'on the same side of the barricade'.

He added that on the long historical view it might matter little which of them, Stalin or Voroshilov, 'rode the white horse' and which of them was trampled underfoot. In the short run, however, the difference was important—it was the difference between an open and immediate triumph of the anti-socialist forces (under a military dictator) and a far more complex, confused, and protracted development (under Stalin). In the long run, he held, Stalin's dictatorship too would be detrimental to socialism; and he saw the triumphant kulak and N.E.P.-man at the end even of Stalin's road. 'The film of revolution is running backwards, and Stalin's part in it is that of Kerensky in reverse.' Kerenskyism epitomized Russia's transition from capitalism to Bolshevism; and victorious Stalinism could mark only the return passage.

It is all too easy to see in retrospect the fallacies of this reasoning; and it is even easier to miss the core of truth they contained. That Trotsky could imagine Voroshilov or Budienny acting the part of Bonaparte must appear almost absurd. Out of this 'filth' no Prince was to be made. Yet as a political analyst Trotsky had to keep an eye on potentialities as well as actualities; and the potentiality of a military coup was there. Although it did not become an actuality, at least in the next thirty years, the threat repeatedly haunted first Stalin and then his successors; witness Stalin's conflicts with Tukhachevsky and other generals in 1937 and with Zhukov in 1946 and Khrushchev's clash with Zhukov in 1957. Here Trotsky touched a trend latent in Soviet politics; but he evidently overrated its strength. He also overrated the force of what was, in Marxist theory, the social impulse behind that trend: the determination and the power of the peasantry to defend its property, and its capacity to assert itself, through the army, against the town. Trotsky himself had written in 1906 that 'the history of capitalism is the history of the subordination of the country to the town'; and in this context he had analysed the amorphousness and the political helplessness of the Russian peasantry under the old régime.[1] That subordination of country to town characterizes *a fortiori* the history of the Soviet Union. Stalin's hammer blows were just about to descend upon private farming with a terrific impact and to crush the peasantry. They could not prevent the peasants from resisting collectivization.

[1] See *The Prophet Armed*, pp. 156 ff.

That resistance, shapeless, scattered, and protracted, was to result in the chronic inefficiency and backwardness of collectivized farming; but it could not focus itself into any effective political action on the national scale. And in the defeat of the property-loving muzhik lay the secret of the failures of the military candidates to the post of the Soviet Bonaparte.

The helplessness and dumbness of the peasantry were part and parcel of the political lethargy of the post-revolutionary society at large; and this formed the background to the extraordinary activity and seeming omnipotence of the ruling bureaucracy. Trotsky repeatedly grappled with this aspect of the situation; and repeatedly his mind ran away from it. Krupskaya once made the remark, which in all probability she had picked up from Lenin, that Trotsky was inclined to underrate the apathy of the masses.[1] In this Trotsky was true to himself and his character as a revolutionary. The revolutionary is in his element when society is in action, when it unfolds all its energies, and when all social classes pursue their aspirations with the maximum of vigour and *élan*. Then his perception is at its most sensitive, his understanding at its acutest, and his eye at its quickest and sharpest. But let society be overcome by torpor and let its various classes fall into a coma, and the great revolutionary theorist, be he Trotsky or even Marx, loses something of his vision and penetration. This condition of society is most uncongenial to him and he cannot intellectually accommodate himself to it. Hence Trotsky's errors of judgement. Even when he made the utmost allowance for the post-revolutionary weariness of the masses he still shrank from fathoming its full depth. Thinking ahead, he still envisaged all social classes and groups —kulaks as well as workers and army leaders as well as the various Bolshevik groupings—in action and motion, in a state of self-reliance and animation, ready to jump at one another and to fight their titanic battles. His thought was baffled at the sight of Titans drowsy and indolent whom a bureaucracy could tame and tie hand and foot.

Because ultimately he identified the process of revolution with the social awareness and activity of the toiling masses, the evident absence of that awareness and activity led him to conclude that, with Stalinism victorious, 'the film of the revolution was

[1] N. Krupskaya, 'K Voprosu ob Urokakh Oktyabrya' in *Za Leninizm*, p. 155.

running backwards', and that Stalin's part in it was that of
Kerensky in reverse. Here again the fallacy is obvious; but the
core of truth in it should not be overlooked. The film was not
running as the precursors and the makers of the revolution had
expected: it was moving partly in a different direction—but not
backwards. Stalin's role in it was not that of Kerensky in reverse.
The film is still on; and it may still be too early to pass final
judgement on it. In theory it may yet be possible for it to end
in a setback for the revolution as grave as that which earlier
great revolutions, the French and the English, had suffered. But
this possibility appears to be extremely remote. When Trotsky
wrote that the film was running backwards he meant that it was
moving towards the restoration of capitalism. Actually it moved
towards planned economy, industrial expansion, and mass
education; and these, despite all bureaucratic distortion and
debasement, Trotsky himself recognized to be essential prerequisites for socialism, the *sine qua non* for the ultimate fulfilment
of the revolution's promise. The prerequisites were admittedly
not the fulfilment; and the Soviet Union of the 1950's had
enough reasons to look back upon the record of Stalinism, or
at least upon some of its facets, with sorely disillusioned eyes.
But it did not see the triumphant kulak and N.E.P.-man at the
end of Stalin's road.[1]

Was Stalin's record one of Bonapartism? Trotsky did not use
the term in the accepted meaning as signifying merely 'government by the sword' and personal rule. The wider Marxist
definition of Bonapartism is that of a dictatorship exercised by
the state machine or the bureaucracy at large, of which military
autocracy is only one particular form. What, in the Marxist
view, is essential to Bonapartism is that the State or the Executive should acquire *political* independence of all social classes and
establish its absolute supremacy over society. In this sense
Stalin's rule had, of course, much in common with Bonapartism.
Yet, the equation offers only a very general and vague clue to
the understanding of the phenomenon in all its complexity and
contradictoriness. Stalin exercised his rule not so much through
an 'independent' state machine as through the 'independent'

---

[1] Eastern Europe (Hungary, Poland, and Eastern Germany), however, found
itself almost on the brink of bourgeois restoration at the end of the Stalin era; and
only Soviet armed power (or its threat) stopped it there.

party machine through which he also controlled the state. The difference was of great consequence to the course of the revolution and the political climate of the Soviet Union. The party machine considered itself to be the only authorized guardian and interpreter of the Bolshevik idea and tradition. Its rule therefore meant that the Bolshevik idea and tradition remained, through all successive pragmatic and ecclesiastical re-formulations, the ruling idea and the dominant tradition of the Soviet Union. This was possible only because the idea and the tradition were firmly anchored in the social structure of the Soviet Union, primarily in the nationalized urban economy. If any partial parallel to this state of affairs were to be drawn from the French Revolution, it would have to be an imaginary one: we would have to imagine what revolutionary France would have looked like if the Thermidorians had never overthrown Robespierre, and if he had ruled France, in the name of a crippled and docile Jacobin party, throughout all those years that the historian now describes as the eras of the Directory, the Consulate, and the Empire—in a word, what France would have looked like if no Napoleon had ever come to the fore and if the revolution had run its full course under the banner of Jacobinism.[1]

We have seen that the rule of the party machine had in fact been initiated at the close of the Lenin era. It had been inherent in the dominance of the single party which Lenin himself saw as being essentially the dominance of the Bolshevik Old Guard. Lenin's government in his last years might therefore be described, in accordance with Trotsky's use of the term, as 'Bonapartist', although it lacked the feature which formed the true consummation of Bonapartism, namely, personal rule. Thus, when in 1928 Trotsky spoke about the danger of Bonapartism he saw a phase of development which had been largely accomplished many years earlier as still looming ahead. Since Lenin's day the despotism of the party machine had, of course, become

[1] Auguste Blanqui described Robespierre as *un Napoléon prématuré* or *un Napoléon avorté* while Madame de Staël said of the First Consul: *c'est un Robespierre à cheval.* (Daniel Guérin in *La Lutte de classes sous la Première République*, vol. ii, pp. 301–4, devotes some interesting passages to this subject.) However, the *Robespierre à cheval* had social forces behind him different from those behind the leader of the Jacobins: his mainstay was the army, not the *petite bourgeoisie*, and he was not constrained by Jacobin ideology. Of Robespierre Michelet said: 'Il eut le cœur moins roi que prêtre.' Napoleon was only King not Priest. Stalin was both Pope and Caesar.

increasingly aggressive and brutal. But the specific content of
the stormy political history of these years, from 1921 till 1929,
consisted not merely and not so much in this as in the transformation of the rule of a single party into the rule of a single faction.
This was the only form in which the political monopoly of
Bolshevism could survive and become consolidated. In the opening pages of this volume we found the single-party system to be
a contradiction in terms. The various Bolshevik factions, groups,
and schools of thought formed something like a shadowy multiparty system within the single party. The logic of the single-
party system implicitly required that they be eliminated. Stalin
spoke with the voice of that logic when he declared that the
Bolshevik party must be monolithic or it would not be Bolshevik. (Up to a point, of course, the party ceased to be Bolshevik as it was becoming monolithic.)

The logic of the single-party system might not have asserted
itself as strongly as it did, it might never have become as ruthless
as it was and its implication might never have become explicit,
or the system might even have been undone by the growth of a
workers' democracy, if the whole history of the Soviet Union,
encircled and isolated in its age-old poverty and backwardness,
had not been an almost uninterrupted sequence of calamities,
emergencies, and crises threatening the nation's very existence.
Almost every emergency and crisis posed all major issues of
national policy on the knife's edge, set the Bolshevik factions
and groupings at loggerheads, and gave to their struggles that
indescribable vehemence and intensity which led to the substitution of the rule of a single faction for that of the single party.
At the point our narrative has reached, in the show-down between
the Stalinists and the Bukharinists, this process was coming to a
close. What still lay ahead was the quasi-Bonapartist consummation: the substitution, in the early 1930's, of the rule of the
single leader for that of a single faction. It was this consummation
—Stalin's autocracy—that Trotsky clearly foreshadowed, however he erred in other respects.

Even now, however, Trotsky did not perceive the ascendancy
of Stalinism as an inevitable result of the Bolshevik monopoly of
power. On the contrary, he saw it as the virtual end of Bolshevik
government. Thus, while Stalin presented the undivided rule
of his own faction as the consequence and final affirmation of the

rule of the single party, Trotsky viewed it as a negation. In truth, the Bolshevik monopoly of power, as established by Lenin and Trotsky, found in Stalin's monopoly both its affirmation and its negation; and each of the two antagonists now dwelt on a different aspect of the problem. We have traced the transitions through which the rule of the single party had become the rule of the single faction and through which Leninism had given place to Stalinism. We have seen that the things that had been implicit in the opening phase of this evolution became explicit and found an extreme or exaggerated expression in the closing phase. To this extent Stalin dealt in realities when he claimed that in his conduct of party affairs he followed the line set by Lenin. But Trotsky's emphatic denial of this was no less strongly based on realities. The rule of the single faction was indeed an abuse as well as a consequence of the rule of the single party. Trotsky, and following him one Bolshevik leader after another, protested that when they had, under Lenin, established the Bolshevik political monopoly they had sought to combine it with a workers' democracy; and that, far from imposing any monolithic discipline on the party itself, they had taken the party's inner freedom for granted and had indeed guaranteed it. Only the blind and the deaf could be unaware of the contrast between Stalinism and Leninism. The contrast showed itself in the field of ideas and in the moral and intellectual climate of Bolshevism even more strongly than in matters of organization and discipline. Here indeed the film of revolution ran backwards, at least in the sense that Stalinism represented an amalgamation of Marxism with all that was primitive and archaically semi-Asiatic in Russia: with the illiteracy and barbarism of the muzhik on the one hand, and the absolutist traditions of the old ruling groups on the other. Against this Trotsky stood for undiluted classical Marxism, in all its intellectual and moral strength and also in all its political weakness—a weakness which resulted from its own incompatibility with Russian backwardness and from the failures of socialism in the West. In banishing Trotsky, Stalin banished classical Marxism from Russia.

Yet, such were the paradoxical fortunes of the two antagonists, that just when Trotsky was being ejected from his country Stalin set out to uproot, in his own barbaric manner, that Russian backwardness and barbarism which had as if regurgitated classical Marxism, and the Stalinist bureaucracy was about to put into effect Trotsky's programme of primitive socialist accumulation. Trotsky was the authentic inspirer and prompter of the second revolution of which Stalin was to be the practical manager in the coming decade. It would be futile to speculate how Trotsky might have directed that revolution, whether he would have succeeded in carrying out Russia's industrialization at a comparable pace and scale without condemning the mass of the Soviet people to the privation, misery, and oppression they suffered under Stalin, or whether he would have been able to bring the muzhik by persuasion to collective farming rather than to coerce him into it. These questions cannot be answered; and the historian has more than enough work in analysing events and situations as they were, without trying to ponder events and situations that might have been. As things were, the political evolution of the 1920's predetermined the manner in which Russia's social transformation was to be accomplished in the 1930's. That evolution led to autocracy and monolithic discipline and consequently to *forcible* industrialization and collectivization. The political instruments that would be needed for primitive socialist accumulation had been forged throughout the 1920's; and they were now ready for use. They had been forged not in any deliberate and conscious preparation for the task ahead but rather in the unpremeditated course of the inner-party struggles by which the Bolshevik monopoly of power had become the Stalinist monopoly. However, if autocracy and monolithic discipline formed, as the Marxist would say, the political superstructure of primitive socialist accumulation, they also derived from this a measure of self-justification. Stalin's adherents could argue that without autocracy and monolithic discipline that accumulation, on the scale on which he carried it out, could not be undertaken. To put it plainly, from the long contests of the Bolshevik factions there had emerged Stalin's 'firm leadership' for which he may have striven for its own sake. Once he wielded it, he employed it to industrialize the Soviet Union, to collectivize farming, and to transform the whole outlook of the nation; and then he pointed to the use he was making of his 'firm leadership' in order to vindicate it.

Trotsky repudiated Stalin's self-vindicatory claims. He continued to denounce his adversary as a Bonapartist usurper. He was to acknowledge the 'positive and progressive' aspects of Stalin's second revolution and to see them as the realization of parts of his own programme. He had, we remember, already compared his and the Opposition's fate with that of the Communards of Paris, who although they failed to conquer as proletarian revolutionaries in 1871 had nevertheless barred the way to a monarchist restoration. This had been their victory in defeat. The great transformation of the Soviet Union in the 1930's was Trotsky's victory in defeat. But the Communards had not been reconciled to the Third Republic, the *bourgeois* republic which might never have prevailed without them. They remained its enemies. Similarly, Trotsky was for ever to remain unreconciled to the *bureaucratic* second revolution; and against it he was to call for the self-assertion of the working classes in a workers' state and for freedom of thought in socialism. In doing this he was condemned to political solitude, because all too many of his closest associates allowed themselves, partly from frustration and weariness and partly from conviction, to be captivated or bribed by Stalin's second revolution. The Opposition in exile was on the point of virtual self-liquidation.

Was Trotsky then in conflict with his time? Was he fighting a hopeless battle 'against history'? Nietzsche tells us:

If you want a biography, do not look for one with the legend: 'Mr. So-and-So and his times', but for one in which the title page might be inscribed 'A fighter against his time'. . . . Were history nothing more than an 'all embracing system of passion and error' man would have to read it as Goethe wished *Werther* to be read—just as if its moral were 'Be a man and follow me *not*!' But, fortunately, history also keeps alive for us the memory of the great 'fighters against history', that is against the blind power of the actual . . . and it glorifies the true historical nature in men who troubled themselves very little about the 'Thus It Is', in order that they might follow a 'Thus It Must Be' with greater joy and greater pride. Not to drag their generation to the grave, but to found a new one—that is the motive that ever drives them onwards. . . .

These are excellent words despite their underlying subjectivist romanticism. Trotsky was indeed a 'fighter against his time', though not in the Nietzschean sense. As a Marxist he was greatly concerned with the 'Thus It Is' and was aware that the 'Thus

It Must Be' is the child of the 'Thus It Is'. But he refused to bow to 'the blind power of the actual' and to surrender the 'Thus It Must Be' to the 'Thus It Is'.

He fought against his time not as the Quixote or the Nietzschean Superman does but as the pioneers do— not in the name of the past but in that of the future. To be sure, as we scrutinize the face of any great pioneer we may detect in it a quixotic trait; but the pioneer is not a Quixote or a Utopian. Very few men in history have been in such triumphant harmony with their time as Trotsky was in 1917 and after; and so it was not because of any inherent estrangement from the realities of his generation that he then came into conflict with his time. The precursor's character and temperament led him into it. He had, in 1905, been the forerunner of 1917 and of the Soviets; he had been second to none as the leader of the Soviets in 1917; he had been the prompter of planned economy and industrialization since the early 1920's; and he was to remain the great, though not unerring harbinger of some future reawakening of the revolutionary peoples (—to that political reawakening the urge to transcend Stalinism which took hold of the Soviet Union in the years 1953–6 was an important pointer; still faint yet sure). He fought 'against history' in the name of history itself; and against its accomplished facts, which all too often were facts of oppression, he held out the better, the liberating accomplishments of which one day it would be capable.

.   .   .   .

At the beginning of December Trotsky protested to Kalinin and Menzhinsky against the 'postal blockade' under which he was held. He waited a fortnight for the answer. On 16 December a high official of the G.P.U. arrived from Moscow and presented him with an 'ultimatum': he must cease at once his 'counter-revolutionary activity' or else he would be 'completely isolated from political life' and 'forced to change his place of residence'. On the same day Trotsky replied with a defiant letter to the leaders of the party and of the International:

To demand from me that I renounce my political activity is to demand that I abjure the struggle which I have been conducting in the interests of the international working class, a struggle in which I have been ceaselessly engaged for thirty-two years, during the whole of my conscious life. . . . Only a bureaucracy corrupt to its roots can demand such a renunciation. Only contemptible renegades can give such a promise. I have nothing to add to these words!¹ A month of sleepless suspense followed at Alma Ata. The G.P.U. envoy did not return to Moscow, but waited on the spot for further orders. These still depended on the Politbureau's decision; and the Politbureau had not yet made up its mind. When Stalin urged it to sanction the expulsion order, Bukharin, Rykov, and Tomsky vehemently objected; and Bukharin, remorseful for what he had done to Trotsky and ever more dreading the 'new Genghiz Khan', screamed, wept, and sobbed at the session. But the majority voted as Stalin wished it to vote; and on 20 January 1929—it was a full year now since Trotsky had been deported from Moscow—armed guards surrounded and occupied the house at Alma Ata; and the G.P.U. official presented Trotsky with the new order of deportation, this time 'from the entire territory of the U.S.S.R.'. 'The decision of the G.P.U.', Trotsky wrote on the receipt, 'criminal in substance and illegal in form, was communicated to me on 20 January 1929.'²

Once again there followed tragi-comic scenes similar to those that had attended his arrest in Moscow. His jailers were embarrassed by their orders, and went in awe, as it were, of their charge; worrying because they did not know whither they were to take him, they made anxious inquiries of his family and showed him furtively their solicitude and friendliness. But their orders were harsh: they were to disarm him, to carry him off within twenty-four hours, and to tell him that only *en route* would he receive a message indicating whither he was being deported.

At dawn on 22 January the prisoner, his family, and a strong escort drove from Alma Ata towards Frunze through mountainous desert and the pass of Kurday. They had driven by the same road through a snow blizzard this time last year. The present journey was far worse. This was a winter memorable for its severity, perhaps the cruellest winter for a century. 'The powerful tractor which was to tow us over the pass got stuck and almost disappeared in the snow-drifts, together with the seven motor-cars behind it. Seven men and a good many horses were frozen to death. . . . We had to change to sleighs. It took us more than seven hours to advance about thirty kilometres.'³

¹ *The Archives.*   ² Ibid.   ³ *Moya Zhizn*, vol. ii, p. 314.

At Frunze, Trotsky and his family were put on a special train bound for European Russia. Under way the message came through that he was being deported to Constantinople. He at once protested to Moscow. The government, he declared, had no right to banish him abroad without his consent. Constantinople had been an assembly place for the remnants of Wrangel's army who had come there from the Crimea. Did the Politbureau dare expose him to the vengeance of the White Guards? Could they not at least obtain for him an entry permit to Germany or another country? He demanded to be allowed to see those members of his family who lived in Moscow. This last demand was granted: Sergei and Lyova's wife were brought from Moscow and joined the deportees on the train. Once again Trotsky refused to proceed to Constantinople. The G.P.U. envoy, who accompanied him on the journey, forwarded his protests and waited for instructions. Meantime, the train was diverted from its route and stopped on a side-line 'near a dead little station'.

There it sinks into a coma between two thin stretches of woods. Day after day passes. More and more empty tins are lying by the side of the train. Crows and magpies gather for the feast in larger and larger flocks. Wasteland . . . solitude . . . the fox has laid his stealthy tracks to the very train. The engine, one carriage hitched to it, makes daily trips to a larger station to fetch our midday meal and newspapers. Influenza has invaded our compartments. We re-read Anatole France and Klyuchevsky's *History*. . . . The cold is 53 degrees below zero. Our engine keeps rolling back and forth . . . to avoid freezing . . . we do not even know where we are.[1]

Thus twelve days and twelve nights passed during which no one was allowed to leave the train. The newspapers brought the only echoes from the world—they were full of the most violent and threatening invective against Trotskyism and of reports about the discovery of a new 'Trotskyist centre' and arrests of hundreds of Oppositionists.[2]

After twelve days the journey was resumed. The train went full steam ahead southward, through familiar Ukrainian steppe. Since the German government had, as Moscow claimed, refused

---

[1] *Moya Zhizn*, vol. ii, p. 315.

[2] Among those imprisoned was Voronsky, editor of *Krasnaya Nov*, Budu Mdivani, and several of the Georgian Bolsheviks who had opposed Stalin since 1921, and 140 Moscow Oppositionists who had circulated Trotsky's 'Letter to Friends' quoted above.

to grant Trotsky an entry permit, it was, after all, at Constantinople that he was to be cast out. Sergei, anxious to continue his academic course, and Lyova's wife, returned to Moscow hoping for the family's early reunion abroad. The parents embraced them with forebodings; but uncertain of their own future they did not dare to ask them to share exile. They were never to see them again.

From this train, through the darkness of night, Trotsky saw Russia for the last time. The train ran through the streets and the harbour of Odessa, the city of his childhood and his first ambitions and dreams of the world. In his memories there always stood out the old Tsarist governor of Odessa who had exercised 'absolute power with an uncurbed temper' and who 'standing in his carriage, fully erect, shouted curses in his hoarse voice across the street, shaking his fist'. Another cursing and hoarse voice and another shaking fist—or was it the same?—now pursued the man in his fiftieth year through the streets of his childhood. Once the sight of the satrap made him shrink, 'adjust the school bag and hurry home'. Now the prison-train hurried through the harbour where he was to embark on a boat which would take him to the unknown; and he could only reflect on the incongruity of his fate. The pier in the harbour was densely surrounded by troops which had only four years earlier been under his orders. As if to mock him, the empty ship waiting for him bore Lenin's patronymic—*Ilyich*! She left harbour precipitately in the dead of night and in a gale. Even the Black Sea was frozen that year; and an ice-breaker had to force a passage to a distance of about sixty miles. As the *Ilyich* lifted anchor and Trotsky looked back at the receding shore, he must have felt as if the whole country he was leaving behind had frozen into a desert and as if the revolution itself had become congealed.

There was no power on earth, no human ice-breaker, to cut open a return passage.

Bibliography

(See also Bibliography in *The Prophet Armed*)

BAJANOV, B., *Avec Staline dans le Kremlin*. Paris, 1930.
BALABANOFF, A., *My Life as a Rebel*. London, 1938.
BELOBORODOV, Unpublished Correspondence with Trotsky, quoted from *The Archives*.
BRANDT, SCHWARTZ, Fairbank, *A Documentary History of Chinese Communism*. London, 1952.
BRUPBACHER, F., *60 Jahre Ketzer*, 1935.
BUBNOV, A., 'Uroki Oktyabrya i Trotskizm', in *Za Leninizm*, 1925.
—— *Partiya i Oppozitsia, 1925 g*. Moscow, 1926.
—— *VKP (b)*. Moscow–Leningrad, 1931.
BUKHARIN, N., *Proletarskaya Revolutsia i Kultura*. Petrograd, 1923.
—— *Kritika Ekonomicheskoi Platformy Oppozitsii*. Leningrad, no date.
—— *K Voprosu o Trotskizme*. Moscow, 1925.
—— 'Teoriya Permanentnoi Revolutsii' in *Za Leninizm*.
—— *V Zashchitu Proletarskoi Diktatury*. Moscow, 1928.
—— (co-author with Preobrazhensky, E.) *The ABC of Communism*. London, 1922.

CHEN TU-HSIU, 'Open Letter to members of the Chinese Communist Party'. The American translation under the title 'How Stalin–Bukharin destroyed the Chinese Revolution' was published in *The Militant*, November 1930 (and not 1929, as stated on p. 317, n. 1).

DEGRAS, J. (ed.), *Soviet Documents on Foreign Policy*. London, 1952.
DINGELSTEDT, I., Unpublished essays, articles, and letters to Trotsky, Radek, and others, quoted from *The Trotsky Archives*.
DZERZHINSKY, F., *Izbrannye Stati i Rechi*. Moscow, 1947.

EASTMAN, M., *Since Lenin Died*. London, 1925.
ENGELS, F., *Dialektik der Natur*. Berlin, 1955.

FISCHER, L., *Men and Politics*. New York, 1946.
—— *The Soviets in World Affairs*, vols. i–ii. London, 1930.
FISCHER, R., *Stalin and German Communism*. London, 1948.
FOTIEVA, L. A., 'Iz Vospominanii o Lenine', in *Voprosy Istorii KPSS*, nr. 4, 1957.
FROSSARD, L-O., *De Jaurès à Lénine*. Paris, 1930.
—— *Sous le Signe de Jaurès*. Paris, 1943.

GUÉRIN, D., *La Lutte de Classes sous la Première République*, vols. i–ii. Paris, 1946.

HERRIOT, E., *La Russie Nouvelle*. Paris, 1922.
HOLITSCHER, A., *Drei Monate in Sowjet Rußland*. Berlin, 1921.

474 BIBLIOGRAPHY

ISAACS, H., *The Tragedy of the Chinese Revolution*. London, 1938.

KAMEGULOV, A. A., *Trotskizm v Literaturovedeni*. Moscow, 1932.
KAMENEV, L., 'Partiya i Trotskizm' and 'Byl-li Deistvitelno Lenin Vozhdem Proletariata i Revolutsii', in *Za Leninizm*.
—— His letter about his meeting with Bukharin in the summer of 1928 is quoted from *The Trotsky Archives*, as are also other documents of which he was co-author.
—— His speeches are quoted from the records of party congresses and conferences.
KAROLYI, M., *Memoirs*. London, 1956.
KHRUSHCHEV, N., *The Dethronement of Stalin* (this is the *Manchester Guardian* edition of the 'secret' speech at the XX Congress), 1956.
KOLLONTAI, A., *The Workers' Opposition in Russia*. London, 1923.
KPSS v Rezolutsyakh, vols. i–ii. Moscow, 1953.
KRITSMAN, L., *Geroicheskii Period Velikoi Russkoi Revolutsii*. Moscow, 1924 (?).
KRUPSKAYA, N., 'K Voprosu o Urokakh Oktyabrya', in *Za Leninizm*.
—— Speeches, quoted from party records.
KUUSINEN, O., 'Neudavsheesya Izobrazhenie "Nemetskovo Oktyabrya" ', in *Za Leninizm*.

LATSIS (SUDBARS), *Chrezvychainye Komissii po Borbe s Kontrrevolutsiei*. Moscow, 1921.
LENIN, V., *Sochinenya*, vols. i–xxxv. Moscow, 1941–50. All quotations are from this, the fourth, edition of Lenin's Works, except in one case, indicated in a footnote, where the 1928 edition (vol. xxv) is quoted.
—— *Sochinenya*, vol. xxxvi. Moscow, 1957. (The first of the additional volumes of the fourth edition, published after the XX Congress and containing Lenin's previously suppressed or unknown writings.)
—— Still unpublished parts of Lenin's correspondence with Trotsky and others are quoted from *The Trotsky Archives*.
Leninskii Sbornik, vol. xx. Moscow, 1932.

Manifest der Arbeitergruppe der Russischen Kommunistischen Partei. Berlin, 1924.
MAO TSE-TUNG, *Izbrannye Proizvedeniya*, vols. i–ii. Moscow, 1952–3.
MARX, K., *Das Kapital*.
—— *Das Kommunistische Manifest*.
—— *Der 18 Brumaire des Louis Bonaparte*.
—— *Herr Vogt*.
MOLOTOV, V., 'Ob Urokakh Trotskizma', in *Za Leninizm*.
—— Speeches quoted from records of party congresses and conferences.
MORIZET, A., *Chez Lénine et Trotski*. Paris, 1922.
MURALOV, N., Unpublished correspondence with Trotsky and others in *The Trotsky Archives*.
MURPHY, J. T., *New Horizon*. London, 1941.

POKROVSKY, M. N., *Oktyabrskaya Revolutsia*. Moscow, 1929.
—— *Ocherki po Istorii Oktyabrskoi Revolutsii*, vols. i–ii. Moscow, 1927.

BIBLIOGRAPHY 475

Popov, N., *Outline History of the C.P.S.U. (b)*, vols. i–ii (English translation from 16th Russian edition). London, no date.
Preobrazhensky, E., *Novaya Ekonomika*, vol. i, part 1. Moscow, 1926.
—— Essays, memoranda ('Levyi Kurs v Derevnie i Perspektivi' 'Chto Nado Skazat Kongresu Kominterna', &c.,) and his correspondence with Trotsky, Radek, and others, quoted from *The Trotsky Archives.*
—— (co-author with Bukharin), *The ABC of Communism.*
Pyat Let Sovietskoi Vlasti. Moscow, 1922.
Radek, K., *In den Reihen der Deutschen Revolution.* Munich, 1921.
—— 'Noyabr, iz Vospominanii', *Krasnaya Nov*, nr. 10, 1926.
—— *Pyat Let Kominterna*, vols. i–ii. Moscow, 1924.
—— *Portrety i Pamflety.* Moscow, 1927. (This edition includes the essay, omitted from later editions, 'Lev Trotski, Organizator Pobedy' (Trotsky, the Organizer of Victory), which originally appeared in *Pravda*, 14 March 1923.)
—— *Kitai v Ognie Voiny.* Moscow, 1924.
—— *Razvitie i Znachenie Lozunga Proletarskoi Diktatury.* (This is Radek's long, unpublished, treatise about Trotsky's theory of permanent revolution, written in exile in 1928 and available in *The Trotsky Archives.* It was in reply to this treatise that Trotsky wrote his book *The Permanent Revolution.*)
—— Unpublished correspondence with Trotsky, K. Zetkin, Dingelstedt, Sosnovsky, Preobrazhensky, Ter-Vaganyan, and others, *The Trotsky Archives.*
—— Speeches quoted from party and Comintern records.
Rakovsky, Ch., 'Letter to Valentinov', unpublished memoranda and correspondence with Trotsky and others, *The Trotsky Archives.* A French translation of the 'Letter to Valentinov' is in *Les Bolcheviks contre Staline.* Paris 1957.
Ransome, A., *Six Weeks in Russia in 1919.* London, 1919.
Rosmer, A., *Moscou sous Lénine.* Paris, 1953.
Roy, M. N., *Revolution und Kontrrevolution in China.* Berlin, 1930.
Rykov, A., 'Novaya Diskussiya' in *Za Leninizm.*
—— Speeches quoted from records of party congresses and conferences.
Sapronov, T., Memoranda and Correspondence in *The Trotsky Archives.*
Scheffer, P., *Sieben Jahre Sowjet Union.* Leipzig, 1930.
Sedova, N. (part author with V. Serge), *Vie et Mort de Trotsky.* Paris, 1951.
Serge, V., *Le Tournant Obscur.* Paris, 1951.
—— *Mémoires d'un Revolutionnaire.* Paris, 1951.
—— *Vie et Mort de Trotsky.* Paris, 1951.
Sheridan, C., *Russian Portraits.* London, 1921.
Smilga, I., Correspondence and essays ('Platforma Pravovo Kryla VKP (b)') are quoted from *The Trotsky Archives.*
Smirnov, I., Unpublished correspondence with Trotsky, Radek, and others. Ibid.
Smirnov, V., *Pod Znamya Lenina* (an unpublished essay expounding the Decemist viewpoint in 1928. Trotsky attributes its authorship to V. Smirnov, but is not certain of it.)

476 BIBLIOGRAPHY

SOKOLNIKOV, G., 'Teoriya tov. Trotskovo i Praktika Nashei Revolutsii' and 'Kak Podkhodit k Istorii Oktyabrya', in *Za Leninizm*.

SORIN, V., *Rabochaya Gruppa*. Moscow, 1924.

SOSNOVSKY, L., *Dela i Lyudi*, vols. i–iv. Moscow, 1924–7.

—— Unpublished correspondence with Trotsky and others quoted from *The Archives*.

STALIN, J., *Sochinenya*, vols. v–x. Moscow, 1947–9.

TANG LEANG-LI, *The Inner History of the Chinese Revolution*. London, 1930.

THALHEIMER, A., *1923, Eine Verpasste Revolution?*. Berlin, 1931.

TROTSKY, L., *Sochinenya*. Moscow, 1925–7. The following volumes of this edition of Trotsky's collected writings are quoted or referred to in this work:

Vol. III: (part 1) *Ot Fevralya do Oktyabrya*; (part 2) *Ot Oktyabrya do Bresta*. It is as a preface to this volume that the much debated 'Uroki Oktyabrya' (*The Lessons of October*) first appeared.

Vol. XII: *Osnovnye Voprosy Proletarskoi Revolutsii*.

Vol. XIII: *Kommunisticheskii Internatsional*.

Vol. XV: *Khozaistvennoe Stroitelstvo v Sovietskoi Rossii*.

Vol. XVII: *Sovietskaya Respublika i Kapitalisticheskii Mir*.

Vol. XX: *Kultura Starovo Mira*.

Vol. XXI: *Kultura Perekhodnovo Vremeni*.

—— *Kak Vooruzhalas Revolutsia*, vols. i–iii. Moscow, 1923–5.

—— *Pyat Let Kominterna*, vols. i–ii. Moscow, 1924–5. An American edition under the title *The First Five Years of the Communist International*, vols. i–ii, appeared in New York, 1945 and 1953.

—— *Moya Zhizn*, vols. i–ii. Berlin, 1930. The English edition *My Life*. London, 1930.

—— *Terrorism i Kommunism*. Petersburg, 1920.

—— *Voina i Revolutsia*. Moscow, 1922.

—— *Literatura i Revolutsia*. Moscow, 1923. An American edition *Literature and Revolution* appeared in New York, 1957.

—— *Voprosy Byta*. Moscow, 1923. The English edition *Problems of Life*. London, 1924.

—— *Mehzdu Imperializmom i Revolutsiey*. Moscow, 1922.

—— *Novyi Kurs*. Moscow, 1924. The American edition *The New Course*. New York, 1943.

—— *O Lenine*. Moscow, 1924.

—— *Zapad i Vostok*. Moscow 1924.

—— *Pokolenie Oktyabrya*. Moscow, 1924.

—— *Kuda idet Angliya?*. Moscow, 1925. The English edition *Where is Britain Going?*, with a preface by H. N. Brailsford, appeared in London, 1926.

—— *Kuda idet Angliya? (Vtoroi Vypusk)*. Moscow 1926. This is not, as the title may suggest, a second edition of the previous work, but a collection of criticisms by British authors, Bertrand Russell, Ramsay Mac-Donald, H. N. Brailsford, George Lansbury, and others, and of Trotsky's replies to his critics.

TROTSKY, L., *Europa und Amerika*. Berlin, 1926.

—— *Towards Socialism or Capitalism*. London, 1926.

—— *The Real Situation in Russia*. London, no date. The English version of the 'Platform' of the Joint Opposition of which Trotsky and Zinoviev were the co-authors.

—— *Problems of the Chinese Revolution*. New York, 1932.

—— *The Third International After Lenin*. New York, 1936. The American edition of the *Critique* of the Programme of the Third International, written in 1928.

—— *Chto i Kak Proizoshlo ?*. Paris, 1929.

—— *Permanentnaya Revolutsia*. Berlin, 1930.

—— *Stalinska Shkola Falsifikatsii*. Berlin, 1932. The American edition *The Stalin School of Falsification*. New York, 1937.

—— *The Suppressed Testament of Lenin*. New York, 1935.

—— *Écrits*, vol. i. Paris, 1955.

—— *The Revolution Betrayed*. London, 1937.

—— *Stalin*. New York, 1946.

—— *The Case of Leon Trotsky*. London, 1937. Trotsky's depositions and cross-examination before the Dewey Commission in Mexico in 1937.

This list of Trotsky's published works includes only books and pamphlets quoted or referred to in this volume.

The Trotsky Archives, Houghton Library, Harvard University. A description of *The Archives* was given in the bibliography of *The Prophet Armed*.

Speeches and statements by Trotsky are quoted either from *The Archives* or from published records of party congresses and conferences, as indicated in footnotes.

YAROSLAVSKY, E., *Rabochaya Oppozitsia*. Moscow, no date.

—— *Protiv Oppozitsii*. Moscow, 1928.

—— *Vcherashny i Zavtrashny Den Trotskistov*. Moscow, 1929.

—— *Aus der Geschichte der Kommunistischen Partei d. Sowjetunion*, vols. i–ii. Hamburg–Berlin, 1931.

—— *Ocherki po Istorii VKP (b)*. Moscow, 1936.

YOFFE, A., 'Farewell letter to Trotsky'. The full text is in *The Trotsky Archives*.

Za Leninizm, Leningrad, 1925. A collection of the most important contributions by Stalin, Zinoviev, Kamenev, Rykov, Bukharin, Sokolnikov, Krupskaya, and others to the discussion on Trotsky's *Lessons of October*. It includes also *The Lessons of October*.

Zayavlenie o Vnutripartiinom Polozheni. The statement of the 'Forty Six', of 15 October 1923, is in *The Trotsky Archives*.

ZINOVIEV, G., *Sochinenya*, vols. i, ii, v, xvi. Moscow, 1924–9.

—— *Dvenadtsat Dney v Germanii*. Petersburg, 1920.

—— 'Bolshevizm ili Trotskyizm', in *Za Leninizm*.

—— *Istoriya RKP (b)*. Moscow, 1924.

—— *Lenin*. Leningrad, 1925.

—— *Leninizm*. Leningrad, 1926.

Memoranda, essays, and other unpublished documents are quoted from *The Trotsky Archives*, and speeches from the published party records.

The following editions of protocols and verbatim reports have been quoted:

Congresses and conferences of The Communist Party of the Soviet Union:

10 Syezd RKP (b). Moscow, 1921.
11 Konferentsya RKP (b). Moscow, 1921.
11 Syezd RKP (b). Moscow, 1936.
12 Syezd RKP (b). Moscow, 1923.
13 Konferentsya RKP (b). Moscow, 1924.
13 Syezd RKP (b). Moscow, 1924.
14 Syezd VKP (b). Moscow, 1926.
15 Konferentsya VKP (b). Moscow, 1927.
15 Syezd VKP (b), vols. i–ii. Moscow, 1935.
Protokoly Tsentralnovo Komiteta RSDRP. Moscow, 1929.

Congresses of Soviet trade unions:

3 Syezd Profsoyuzov. Moscow, 1920.
4 Syezd Profsoyuzov. Moscow, 1921.
5 Syezd Profsoyuzov. Moscow, 1922.
6 Syezd Profsoyuzov. Moscow, 1925.
7 Syezd Profsoyuzov. Moscow, 1927.

Congresses of Soviets:

8 Vserosiiskii Syezd Sovetov. Moscow, 1921.
9 Vserosiiskii Syezd Sovetov. Moscow, 1922.

The Communist International:

International Congresses:

3 Vsemirnyi Kongress Kominterna. Petrograd, 1922.
4 Vsemirnyi Kongress Kominterna. Moscow, 1923.
5 Vsemirnyi Kongress Kominterna, vols. i–ii. Moscow, 1925.

Sessions of the Executive:

Rasshirenny Plenum IKKI. Moscow, 1923.
Rasshirenny Plenum IKKI. Moscow, 1925.
Shestoi Rasshirenny Plenum IKKI. Moscow, 1927.
Puti Mirovoi Revolutsii (session of November–December 1926), vols. i–ii. Moscow, 1927.
Rasshirenny Plenum IKKI, vols. i–xii. Moscow, 1930– .
The Lessons of the German Events. London (?), 1924.
   (Report on the discussion at the Praesidium of the Executive on the 'German Crisis' of 1923.)

Varia:

The Second and Third International and the Vienna Union (report of the 1922 conference of the Three Internationals in Berlin), no date.

Newspapers and periodicals:

Bolshevik, Bulleten Oppozitsii, Ekonomicheskaya Zhizn, Die Freiheit, L'Humanité, Izvestya Ts. K. RKP (b), Iskusstvo Kommuny, Internationale Presse Korrespondenz, Kommunist, Kommunisticheskii Internatsional, Krasnaya Letopis, Krasnaya Nov, Kuznitsa, Labour Weekly, Labour Monthly, The Militant, The New International, Na Postu, The Nation, The New Leader, The New York Times, Oktyabr, Pechat i Revolutsia, Pod Znamyenem Marksizma, Pravda, Proletarskaya Revolutsia, Przegląd Socjal-Demokratyczny, Revolutsionnyi Vostok, Trud, Voprosy Istorii KPSS, Znamya, Z Pola Walki.

INDEX

Andreev, A., 246, 254, 443.
Anglo-Soviet Trade Union Council, 217, 221, 223, 267; attacked by Trotsky, 269, 278; defended by Stalin, 284; breaks up, 333, 348; effects on Soviet foreign policy, 361–2.
Antonov-Ovseenko, V. A., 31, 113, 115, 203, 207, 338; as leader of 1923 Opposition, 116–17, 160–1, 265; in Joint Opposition, 267, 334 n. 1; his capitulation, 406.
Aristotle, 197.
Aulard, A., 434 n. 2.

Babel, I., 259 n. 1.
Babeuf, G., 292, 368, 437.
Bajanov (Bazhanov), B., 112 n. 1, 138 n. 1.
Bakaev, I., 261, 265, 343, 351 n. 2; on 7 November 1927, 375, 379; is expelled and capitulates, 386.
Balabanoff, A., 78 n. 1.
Bebel, A., 381.
Beloborodov, 379, 401, 410.
Belyi, A., 182.
Bernstein, E., 310 n. 3.
Blanqui, A., 463 n. 1.
Blok, A., 183–4.
Blücher, B., 319.
Bogrebinsky, 255 n. 1.
Bogushevsky, 243.
Boguslavsky, M., 432 n. 3.
Bolshevism, Bolsheviks (see also Communist Party of the Soviet Union), 1–3; and social classes after civil war, 5–9 and passim; and early N.E.P., 38 ff.; and non-Russian nationalities, 49–51, 69–70; and 'united front', 64 ff.; and 'primitive socialist accumulation', 99–103, 129–31; and the party's social composition, 128–9; and Third International, 146–51; and Jacobinism, 94–95, 129 n. 1, 311–14, 342–7, 436–40 and passim; history of, as interpreted by Trotsky and the triumvirs, 152–8; and arts and literature, 183–8 and passim; and American capitalism, 215–17; in middle 1920's, 228–73; and 'socialism in one country', 284–90, 300–2, 422 ff.; and Stalinism, 462–5 and passim. See also: Communist Party of the Soviet Union, Old Guard.
Bonaparte, and Bonapartism, 2, 94,

463 n. 1; Trotsky's alleged Bonapartist ambitions, 94–95, 161, 391–2; and Stalinism, 458–64.
Borodin, M., 319, 335, 338.
Bosch, E., 383.
Brailsford, H. N., 221–2.
Brandler, H., 78 n. 1, 109, 205; asks for Trotsky to come to Germany, 111, 141; his policy in 1923, 141–4; denounced by Comintern, 145–6.
Brest Litovsk Peace, 23, 79, 82, 156, 231, 284.
Briand, A., 362.
Bronstein (Sokolovskaya), A., 203, 372.
Bronstein, N., Trotsky's elder daughter, 372, 428–9.
Bronstein, Z., Trotsky's younger daughter, 372, 428–9, 453.
Bubnov, A., 31, 114, 321, 326.
Budienny, S., 458, 460.
Bukharin, N., 28, 30, 63, 155, 204, 206; his character, 82–83; and the triumvirs, 81, 83; defends non-Russian nationalities, 97–98; and the 1923 crisis in Germany, 142; and 'proletarian culture', 168, 181, 198–9; and Anglo-Soviet Trade Union Council, 221; his political evolution in middle 1920's, 230–2, 251 ff.; on Preobrazhensky's theory of accumulation, 238–40, 441–2; and 'neo-N.E.P.', 242–4; becomes Stalin's ally, 242, 246; his correspondence with Trotsky, 256–8; and 'proletarian democracy', 262; and Joint Opposition, 267–70, 273, 278 ff., 304–6; on socialism in one country, 290; and 'Soviet Thermidor', 304–6, 312–16; his attitude towards Kuomintang, 319 ff., 325–7, 332, 334, 337; his early dissensions with Stalin, 348; at XV Congress, 388; and 1928 grain crisis, 404 ff., 417 ff., 426–7; his secret meeting with Kamenev and revelations about Stalin, 440–2; at VI Comintern Congress, 444; sequel of his appeal to Trotsky, 445–51; as 'Thermidorian manqué', 457–8; objects to Trotsky's banishment, 469.
Bulgaria, 144.
Bureaucracy, 5, 110, 386–7; Trotsky and Workers' Opposition on, 52–54; and the Lenin–Trotsky 'bloc', 68–72; and the Forty Six, 113–14; attacked by Trotsky, 120–3 and passim; and

Fischer, Ruth, 64, 143, 145, 146, 255, 267 n. 2, 294 n. 1.

'Forty Six' (Opposition group), 113–17, 124–5, 132 n. 2, 139, 146.

Forster, Professor, 67.

France, 8, 11, 212, 213; and Rapallo Treaty, 58; occupies Ruhr, 111, 141; and Russia in 1926–7, 361–2; socialist Utopians of, 197; Paris Commune, 407, 467; 'unlearns' and 'relearns' freedom, 436–7. *See also* Revolution, French.

Communist party of, 28, 149–50; defends Trotsky, 140–1, 146; turns against him, 147 ff.; purged, 311.

Freud, S., 165, 178–80, 259.

Frossard, L.-O., 63 n. 1, 149.

Frumkin, M., 427.

Frunze, M., 55, 135, 162.

Futurism, 184–7.

Gamarnik, I., 303.

Genghiz Khan, 304, 441.

Georgia, and Russia, 49–51, 69 ff., 227; Lenin on, 71–72; Trotsky on, 90–91; debated at XII Congress, 97–98.

Germany, and Russia; in Rapallo Treaty, 56–58, 204; during French occupation of Ruhr, 111, 141–2; as model for U.S.A., 213–14; and Britain, 218; visited by Trotsky, 266; 'People's Democracy' of, 462 n. 1;

Communist party of, 3, 63–65; and occupation of Ruhr, 111, 141–2, 154; and abortive revolution, 142–5; leadership of, changed, 145 ff.; Radek's role in, 204–5; in aftermath of 1923 defeat, 212.

Gippius, Z., 182.

Glazman, 383.

Goethe, W., 191, 197, 403.

Gogol, N., 396.

Gosplan (State Planning Commission), 40–42, 46–47, 67–68, 111.

G.P.U. (Cheka), 84, 107; acts against inner party dissent, 108–10, 357–9; and Yoffe's suicide, 383; deports Trotsky, 391–4, 396–8, 468–70; quells peasant risings, 441; and right opposition, 443.

Gramsci, A., 185 n. 1.

Great Britain, 43, 56, 209, 212; and U.S.A., 213–14, 218; general strike in, 266–7; and Soviet Union, 333, 335–6, 361–2. *See also* Anglo-Soviet Trade Union Council.

Communist party of, 147, 149, 217, 220–1; against Trotsky, 223, 269.

Guérin, D., 463 n. 1.

Hazlitt, W., 27.

Heine, H., 259.

Herriot, E., 58, 212.

Hitler, A., 216, 217.

Hungary, 462 n. 1.

Imagists, 183.

International, Third, 4, 28, 31; participates in conference of three Internationals, 30; and Zinoviev, 77 ff., 267, 335; and 'united front', 59–65, 267; and Bolshevik inner party struggles, 140 ff., 267; and abortive German revolution, 142–5; 'Bolshevization' of, 146–51; its tactics in 1923–4, 212; and 'United States of Socialist Europe', 216; its tactics in 1925–6, 217, 267; and Joint Opposition, 274, 303, 310; and Chinese communism, 316–37; expels Trotsky, 359–61; its *Critique* by Trotsky, 425–6, 444.

Congresses of:
second (1920), 316–17.
third (1921), 63–64.
fourth (1922), 65.
fifth (1924), 146–7.
sixth (1928), 425–6, 444.

Italy, 60, 185:
Communist party of, 149, 444.

Jacobinism, and Jacobins, 11–12, 58; compared with Bolshevism, 94–95; 129 n. 2, 244, 311–14, 342–7, 436–7, 448–9, 462–3.

Japan, 213, 323.

Jaurès, J., 151.

Jefferson, T., 197.

Jews, 258–9.

Joint Opposition, its formation, strength, and views, 262 ff., 271–7; appeals to rank and file, 281–2 ff.; makes 'truce', 293 ff.; at XV party conference, 298–308; and after, 308–16; on 'Soviet Thermidor', 311–16; and Chinese Revolution, 321–37; early reprisals against, 338 ff.; its *Platform*, 356–9; its defeat, 367–70, 372–9; its disintegration, 384–9.

Kaganovich, L., 246, 306–7.

Kalinin, M., 66, 153, 331, 468; and Trotsky, 84, 254, 396, 397; at XV Congress, 388; against Bukharin, 441, 443.

Kamenev, L., 36, 56, 66; in triumvirate, 76, 90–92, 94; his character, 79–81; at XII Congress, 95 ff.; and Lenin's will, 137; attacked by Trotsky, 153 ff.; urges reprisals against

Freud and Pavlov, 177–80; as critic of art and literature, 180–97; his vision of socialist culture, 195–7; drive against Trotskyism in culture, 197–200.

during 1925–6 interval: disbands Opposition, 201; repudiates Eastman's disclosure of Lenin's will, 201–2; and Radek, Preobrazhensky, and other Trotskyists, 203–8, 236–8; advocates industrialization, 209–11; opposes economic isolationism, 211–12; on U.S.A. and communism, 212–17; *Where is Britain Going?*, 217–21; controversy with British critics, 221–2; on socialism in one country, 242, 285–91 and *passim*; on 'Soviet Thermidor', 244–5; and Leningrad Opposition, 248–52, 261–2; his silence at XIV Congress, 253–6; his correspondence with Bukharin, 256–8; his 'Jewishness', 258–9; is re-elected to Politbureau, 260 n. 1; forms Joint Opposition, 262 ff.; his reconciliation with Zinoviev and Kamenev, 263–5; on a trip to Berlin, 266–7.

at the head of the Joint Opposition: a general view, 271–3; the merger of two Oppositions, 267–8; speaks for Joint Opposition, 275–7; attacks Anglo-Soviet Council, 267–9, 278; appeals to rank and file, 281–2 ff.; offends nationalist feeling, 288 ff.; on 'Soviet Thermidor', 292, 311–14, 342–7; makes truce with Stalin, 293–4; and Eastman's publication in full of Lenin's will, 294–6; describes Stalin as 'gravedigger of revolution', 296–7; speaks at XV Conference, 299 ff.; 'to die like Lenin or like Liebknecht?', 309–10; his inner party tactics, 314–16; presides over Politbureau's Chinese Commission, 322–3; his attitude towards Chinese Revolution, 320–1, 324, 327–9, 332–3; and the 'Eighty Four', 334; appeals to Comintern, 334–7; at the Yaroslavl Station, 339; calls for pre-Congress debate, 339–40; is charged with indiscipline, 340 ff., 351–3; makes 'Clemenceau statement', 349–51; his self-defence, 352–5; expelled from Comintern Executive, 359–61; on relations with Britain and France, 361–3; and Rakovsky, 362; criticizes decree on seven hours' day, 364;

and preliminaries to November 1927 demonstrations, 365–6; expelled from Central Committee, 366–7; his behaviour and mood, 368–71; during November demonstrations, 376–8; expelled from party, 379; and Yoffe's suicide, 380–4; and dissolution of Joint Opposition, 385 ff.; persists in Opposition, 390; on eve of deportation, 391–3; deported, 394.

in exile at Alma Ata: *en route*, 395–6; early months of exile, 397–401; and the Opposition in exile, 401–3; and 1928 grain crisis, 405 ff.; on Stalin's 'left course', 407–9, 417, 426–8, 437, 446–7; compares Opposition with Paris Commune, 407; and Opposition's morale, 411–12; and dissensions among his followers, 414 ff.; his differences with Preobrazhensky, 418–19, 422–5; writes *Critique* of Comintern programme, 425–6; death of his daughter, 428–9; further dissensions among his followers, 429–34; his friendship with Rakovsky, 434–5; and Rakovsky's 'Letter to Valentinov', 436 ff.; thinker or man of action?, 439; receives report on Bukharin–Kamenev meeting, 440 ff.; rejects Kamenev's appeal and Stalin's vague blandishments, 443, 447–50; and VI Comintern Congress, 444; on agreement with Bukharin, 445–7; his further differences with Radek and conciliators, 450–2; his illness, 452; is placed in stricter isolation, 453–4; discourages hunger strike by followers, 453; his 'clothes stolen' by Stalin, 454 ff.; his influence in exile, 456 ff.; on two prospects of Russian Revolution, 457–60; rights and wrongs of his view, 460–8; his deportation to Constantinople, 468–71.

his *Archives*, their reliability, 67 n. 3.
Trotskyism, in 1923, 119 ff.; and literature and the arts, 197–200; its leading team, 203–8; and Bukharin's 'neo-Populism', 230 ff.; and Preobrazhensky's theory of accumulation, 234–8; and Zinovievism, 249–56, 268–9; and 'socialism in one country', 284–90, 299–302; and the Chinese Revolution, 321–7; as an opposition in exile, 389–90, 401–3 ff., 456; cross-currents in, 405–10, 413–26, 427–34, 438–9; 448–52; a general view, 460–8.